Origins of modern algebra

SCIENTIFIC EDITOR

L. Procházka

Professor at the Charles University, Prague

REVIEWER

J. Folta

Institute of Czechoslovak and general history

Origins of modern algebra

Luboš Nový

Institute of Czechoslovak and general history

Noordhoff International Publishing
Leyden, The Netherlands

ACADEMIA
Publishing House
of the Czechoslovak
Academy of Sciences
Prague

Noordhoff International Publishing, Leyden, The Netherlands
© 1973 Luboš Nový, Ph.D., Prague
Translation © 1973 Jaroslav Tauer, Ph.D., Prague

ISBN: 90 01 64560 7

Library of Congress Catalog Card Number: 76-51041

Printed in Czechoslovakia

CONTENTS

VI

PREFACE

This book is concerned with the evolution of algebra roughly between 1770 and 1870. It is the author's opinion that this hundred year period represents one of the important stages in the development of modern algebra. In the course of this period structural thinking in algebra, which was earlier included under differently formulated tasks, became conscious and the fundamentals of its terminology were established. The establishment of the principal algebraic structures and concepts, which is one of the fundamental results of this stage, expresses a change which took place in the contents of algebra in the hundred years after 1770 and which became the basis for the evolution of modern algebra.

However, it is not intended that this book should be an exhaustive history of algebra in the period considered, nor is it intended to present all the important algebraic results. The historical data are only analyzed in so far as is necessary, to discover, illustrate, or discuss trends in the evolution of algebra. An attempt is made to prove, or at least convince the reader, that the author's view of the evolution and the stages of development of modern algebra is justified.

Some of the partial problems included in this book were already presented for discussion at various other occasions; the opinions and comments regarding them helped me in writing this book. In this respect the discussions held in the Department of History of the Natural Sciences and Technology, Czechosl. Acad. Sci., were particularly valuable, and especially Dr. Jaroslav Folta helped in preparing the whole book, the creation of which he followed from the first outlines to the final text, which he influenced many times by giving advice. The comments of the scientific editor of the book Dr. L. Procház- ka were also stimulating.

The book could not have been written without the working environment provided by the Department mentioned above and without the possibility of systematic work, with which it provided me for a number of years. In some cases studies abroad were stimulating and necessary for filling in the data, particularly in Moscow, where I was able to make use of the abundant riches of the Lenin State Library, and in Paris where, thanks to Section VI of the Ecole pratique des hautes études, I was able to draw on the library fund of the Bibliothèque Nationale in particular.

Finally, I would like to thank all who took care of the technical side of publishing the book, whether it be the Publishing House of the Czechoslovak Academy of Sciences,

Academia, or those who typed, arranged and translated the manuscript. My particular thanks are due to Z. Knichalová, head of the mathematical department of editors, who encouraged with understanding and the necessary consistency in finishing the manuscript.

Praha, September 1970 *Luboš Nový*

CHAPTER 1

INTRODUCTION

1.1 Purpose and method

The nature of mathematics, more than of any other discipline, tempts one to interpret the history of mathematics only as a sequence of logically linked discoveries which culminates in the present state of science, this history being variagated and supplemented by data on the personalities of the individual scientists.

This concept of history has considerable advantages. First of all, there is the simplicity of the method, in which the easily defined sequence is supplemented by pointing out the mathematical connections of the discoveries. Soon one finds, and this is another advantage, that heuristics in this concept may be restricted merely to selected principal discoveries, which are more or less known for all mathematical disciplines, and that reading other works will have no effect on the general picture.[1]

Previous research into the history of algebra has used this procedure, indicated schematically. The fundamental facts which characterize the evolution of algebra from Newton to the propagation of Galois' theory are known. The work of the three mathematicians who published their papers around 1770 (Waring, Vandermonde, Lagrange), the subsequent work of Ruffini and Gauss (about 1800) and, finally, that of Abel and Galois (about 1830), form milestones which apparently characterize the evolution of the theory of algebraic equations up until the significance of their work for modern algebra was really understood and began to be elaborated. Any research into other algebraic papers of this period will soon show that they only contain results of second- or third-order importance and, therefore, are only capable of filling in the existing image. Under detailed scrutiny, however, a number of problems will appear which cannot be explained in this way because of the simple method and restricted selection of data.

---

[1] During recent years, several more extensive works on the evolution of mathematics in the 19th century have appeared. They mostly treat the data in detail, and usually only deal with the problem of the historical significance and contributions of the individual mathematical works, which, as pointed out earlier, cannot contribute to filling in a pattern by studying second or third-rate papers. This can be seen, e.g., in Paplauskas' work [326] on trigonometric series, where the extensive bibliography is no longer functional and where some of the papers are not reflected in the text itself. Some authors try to find the principal period milestones. They encountered difficulties in their endeavour and therefore sought methods more appropriate to these objectives. This includes, e.g., the works of Kramar [241, 242], Wussing [410], Crowe [108].

The main problem is that this method substitutes the actual time sequence of the evolution with a mere sequence of selected facts. The result is that it is difficult to explain why this evolution took place over such a relatively long interval of one hundred years, especially since one finds close intellectual connections in the fundamental ideas of the mathematicians mentioned above. Neither does this method explain why further comprehension and elaboration of each man's ideas was always left to the following generation in which several mathematicians, again independently, present new but similar ideas. Just as unexplainable is the fact that even after Abel's well-known paper on the insolvability of 5th-degree equations, further investigations were made into the problem of its algebraic solvability. Moreover, these initiatives came from first-rate mathematicians of the time, men who, like Bolzano, had an understanding for new ideas which deviate from tradition.

The situation is only slightly different in research concerning the evolution of most branches of mathematics, in each the features forming the characteristics of modern mathematics appeared in roughly the same period. This similarity becomes outstanding in the origin and spreading of non-Euclidean geometry on the one hand, and in revising the foundations of mathematical analysis on the other. It should be mentioned that also in these branches new basic results tend to appear in the work of two or more authors at the same time, quite independently of each other, after long periods during which no substantial advances are made. As in non-Euclidean geometry where Gauss, Lobachevskii and Bolyai progress independently of each other, Cayley and Grassmann publish in 1844 independently of each other their concepts of n-dimensional space. Likewise Galois made progress with his theory without knowing of Abel's work in pursuit of the same ends. The fact that these discoveries took place at the same time cannot be explained by the vague cliché that the time was ripe, because one then has difficulty explaining the lack of understanding for these results shown by other mathematicians of the time. At the same time it is conspicuous that a large number of significant discoveries of the first half of the 19th century had to wait quite a long time to be acknowledged. Disregarding Bolzano's work, in which special conditions played a decisive role, one could point out the function presented by Cauchy, the Taylor development which does not converge in every point to its values (a serious objection to Lagrange's explanation of the analysis, which was, nonetheless, ignored by contemporary mathematicians), or the concept of uniform convergence, etc. One must admit, however, that even in the first half of the last century extensive new regions were discovered which nearly immediately were brought into the limelight of interest and which were treated with considerable intensity. As examples one could name the theory of elliptical functions, or, with slight changes, also the study of the functions of the complex variable as a whole. Another example is the restoration and advance of projective geometry by several mathematicians of the second decade of the 19th century, work which led up to results in France and especially in Germany in the two subsequent decades, far exceeding the fundamental niveau, represented, e.g., by the work of Poncelet.

These considerations determined the limitations of the present work. The intention was to investigate the development of algebra approximately between 1770 and 1870. However, it was in no way intended to present a chronologically arranged and evaluated set of all algebraic results reached in the course of that hundred years. Its objective is to understand the essence of changes in algebraic thinking, the time, the intensity and the influences under which these changes took place.

The study of the topic, roughly outlined in this way, is based on several more general concepts and hypotheses. It is clear and has generally been acknowledged in literature that the period in question was a time of considerable change in the understanding of the objective of mathematics itself, as well as of its methods. In spite of its vagueness this is also substantiated by a study of the periodization of the history of mathematics. However, the stages in which these changes took place, and what their substance was, are not clearly defined. If one disregards the external influences (i.e. those entering mathematics from other sciences or the whole society), which basically had a negligible effect on the content of the changes in mathematics, the decisive role in the gradual execution of change can be attributed to certain concrete, sometimes even detailed, problems. A general outline of the development may be presented as follows: With a view to the overall state of mathematics (inclusive of the heterogeneous ideas, terms and methods used) at a certain moment, there exist unsolvable problems and some results which are in contradiction with the prevailing concepts. These results not only accumulate, but at certain times they are generalized. This gives rise to contradictions with the prevailing general concepts as well as allowing for new formulations of problems and methods. These results, originally detailed but contradictory to the existing order, gradually form a broader and more general conception which in turn changes the general concepts and ideas. New abstract and general concepts, therefore, are not created suddenly by a single creative act. The progress from the first ideas, which are only vaguely formulated and understood in a relatively narrow sense in close connection with certain unsolved problems, through more general ideas of different degrees of clarity, to accurately defined concepts and to their further generalization is historically very complicated and tedious.[2] Nevertheless, the under-

---

[2] It cannot be said that this problem is unknown in the historiography of mathematics. Bell ([26], 246) not only mentions it, but he differentiates between five stages of evolution on the process of creating a new abstract concept. However, with Bell the knowledge of the complexity of this process did not become a methodical means. On the other hand, in works on the history of mathematics one frequently finds that the first stages of evolution are arbitrarily considered as the period when a certain concept had already appeared, although it would be more accurate to speak of anticipating this concept. The author is of the opinion that this manifests an identification of mathematical considerations, expressed at the time but considered with a certain amount of uncertainty, with the mathematical substance of the problems or theories as understood later on; or, considered in reverse order, as attributing contents, by which a mathematician knowing present results, relates an older text to the ideas of authors of the said historical texts.

This leads one to try and understand authors in relation to their own words and opinions, and to appraise them with a view to the period they lived in. Therefore, quotations and paraphrases of the

standing of this process is the methodical key to a more precise understanding of the real development. It also seems that a new concept in process of being established, no matter how much it points to new results and conceptions, may in many respects find itself at a disadvantage with respect to older concepts, which may be said to be traditional. The latter are a part of a system which has been recited many times and repeatedly deliberated (a part of educational training), whereas the new, in many instances obscure ideas, are created separately, without apparent mutual interconnection and without support from the existing system.

It seems, therefore, that the whole system (or at least a relatively independent part of it) will have to change. The hypothesis that the change of these general ideas and concepts takes place in two stages seems to be plausible. In the first stage the new conceptions are first constituted, the mathematicians gradually progressing from vague feelings to a clear understanding of the conception and to its definition. During this stage the old terminology, which is relatively justifiable, still exists and prevails for a long time. The second stage is characterized by the new terms prevailing and becoming the foundation for developing the branch of science and for further creative work.

The conviction that the years 1770−1870 represent the first stage of the development of modern algebra, when structural algebraic thinking was established and when the whole concept of algebra, its objectives and methods, changed, is based on this general point of view.

However, one can hardly speak of the continuity or discontinuity of this process. There are several reasons for this. First of all, in the time interval chosen, there is not a sufficiently large number of short treatises in the individual branches of mathematics to form a more or less continuous flow, as is the case today. However, more important is the fact that attempts at generalizing (which may also be considered as attempts at solving more general problems, or attempts at formulating these problems) became more effective only in a quite small number of papers.[3] Moreover, in mathematical literature, as opposed to other sciences, general concepts which are to be fitted by the investigated terms are rarely formulated; it is more a case of creating a framework against a silently accepted background. To what extent a special treatise agrees with prevailing ideas can only be determined with difficulty and, therefore, represents an important methodological problem for mathematical historians. One may also express the hypothesis that disengaging one's self from inherited general ideas and concepts may help to open up

---

original text will frequently be used without interpreting the text in terms of present concepts. One cannot avoid obscurities included in the original text in this way, but this is not our intention. The terminological difficulties are similar. The unique use of present terms would mean the distortion and modernization of words used earlier; therefore, we frequently revert to less accurate, or more unusual terminology of the period for the purpose of expressing the contemporary state of knowledge.

[3] One may consider the studies of the authors from Vandermonde to Gauss, mentioned at the beginning of the section, to be works of this kind. This also yields a certain degree of justification for the fact that the historiography devoted its attention nearly exclusively to them.

a door to new understanding. The disengagement may come about in several ways: The decisive impulse may come from imagination and from the generalizing capabilities of a mathematical genius, but a certain separation, ignorance, or partial knowledge of the tradition may also help. However, most papers are based on tradition, and new ideas are created only slowly. These considerations indicate some of the consequences flowing from the choice of the methods. First of all, one could concentrate on the detailed reconstruction of the way new ideas were created and eventually led to profound changes, affecting the whole of mathematics. This side of the development, which may be called "positive", certainly deserves attention; nevertheless, a historian's approach cannot be restricted solely to this. As will be seen from the review of literature, most of the material, in spite of all the insufficiency, has been prepared and studied in detail in a number of treatises with this development in view. It can also be shown that the description of this positive side is insufficient to answer a number of questions already mentioned.[4]

Considerable attention must be devoted, on the other hand, to the "negative" side of the development, i.e. to the investigation of the ideas which were gradually discredited and why this was so.[5] Methodical considerations require a further conception which, like a number of the previous, can only be fully elucidated in the course of an actual analysis. Mathematics in the period considered, generally cannot be taken as a realm of consistent and accurate facts. Especially if one speaks of general ideas and conceptions, it is important to use terms of "traditional understanding" or of "prevailing understanding". It is known that in mathematics sometimes two (or even more) different concepts and explanations existed, among which it was very hard to decide, in spite of their incompatibility. At the beginning of the last century this situation occurred in the foundations of infinitesimal calculus when a kind of equilibrium existed between Lagrange's explanation and the upholders of the concept of limit.[6] During this period the arguments to prove or disprove one of the points of view were insufficient. Similarly in algebra one can observe that certain opinions about the algebraic solvability of equations of the fifth and higher degrees prevailed while simultaneously the first doubts were being voiced. An important role, in this connection, is played by the understanding of the term "solvability of an equation".

---

[4] This includes, e.g., questions as to why there were such long intervals between the individual advances and why contemporary methematicians had so little understanding for some of the important discoveries.

[5] In the actual analysis the difference is not so substantial between the "positive" and "negative" aspect mentioned. This is more a case of stressing a certain polarity, and in particular of expressing the methodical problem, that in interpretation one can describe the "positive" aspect of evolution and omit the earlier concepts and ideas which had to be discarded, or at least adapted to the new type of conclusions or considerations. This can happen because many ideas, in spite of their large influence, were not specifically mentioned in special mathematical treatises but were silently accepted, although they had not been defined with sufficient accuracy even then and their meaning and content wasn't quite clear. That is why the negative aspect is being stressed here.

[6] To these opinions one must also add the "naive" concept of "an infinitely small quantity".

Various opinions were also offered in the approach to numerical fields, the different meaning of which had to be characterized somehow. In an actual historical analysis however, one must take care that the use of expressions like "prevailing", should be specified by evaluating a more extensive set of data. These aspects, therefore, must be taken into account in selecting the material. General statements are most frequently made in textbook literature. In the period considered, a number of works were published which attempted a unified, logically assembled interpretation of mathematics and were designed more or less for educational purposes. These compendia only rarely publish individual discoveries, they are not, therefore, substantially affected by didactic objectives.[7] At the same time the number of books published which dealt with the special mathematical discipline increased. These are on the boundary between monographs, recording the contemporary state of mathematics (supplemented by isolated original results), and university level textbooks.[8]

These ideas on the development of mathematics, and certain previously mentioned methodical considerations, have understandably influenced not only the outline of this work, but also the selection and classification of source material. In its way this work is a methodological attempt to verify the scope of certain historical mathematical hypotheses and to show whether they can be established and how.

As regards the classification of the material, it seems necessary to combine chronological and subject divisions. The predominance of either of these principles breeds difficulties. The subject division would disrupt mutual relations and prevent one from observing the state of the science at a given moment; chronology, on the other hand, makes it difficult to grasp the principle trends. Therefore, a certain compromise between these possibilities seems to be most suitable.

1.2   Literature

Literature concerning the development of algebra in the period investigated is relatively plentiful and varied as regards standard and purpose.

The overall-picture can be found in surveys of the history of mathematics. These works differ, of course, in the significance they attribute to the establishment, e.g., of the Galois theory and to modern algebraic thinking. Although he gives the fundamental

---

[7] A whole series of German works had no didactic objective (e.g., Ohm [323]), or if the extent of the work represented a kind of compendium, these objectives dropped into the background (e.g., Kästner [221] and Karsten [223]). French works, in which the attemptat for monumentality and comprehensiveness of the system is not so outstanding, were affected by educational requirements to a larger extent. However, even in this case one finds differences. A larger influence of didactic objectives can be found in de la Caille's book [110], and to a lesser extent, e.g., with Bézout [35].

[8] For example Wood's algebra [405] was intended for English universities, whereas Lacroix's supplements [263] could have hardly played the role of university textbooks. Euler's algebra [141] is interpreted in its algebraic part with considerable didactic feeling, which improved considerably its illustrative qualities in the gradual introduction of different concepts.

facts, Klein [234], devotes little space to algebra in comparison with other branches (e.g., geometry); in the evaluation of algebraic results, Klein's point of view corresponds to the period of his own creative upsurge in the seventies of the last century. Whereas he places the study of algebraic structures in the background, Klein points out very interesting connections between the development of algebraic ideas and transformations in other mathematical disciplines (in particular special functions and geometry). A substantially different and more modern evaluation of the development of algebra is upheld by Bell [26] in his history of mathematics and by Bourbaki [46] in his historical notes. With both, the stress on the modern point of view leads to a unilateral selection of isolated facts which anticipate todays's algebraic ideas. However, one must admit that Bell was successful to a large extent in understanding the way the development progressed and he was thus able to avoid a modernizing interpretation of the words of the algebraists of the first half of the 19th century. Both works are valuable (in spite of their problematic procedures) in that they strive to indicate the trends in mathematics, the result of which is the present state of mathematics. The argumentation and documentation in these works is understandably very incomplete, as one might expect with surveys.

Apart from these works, data have also been accumulated in other historical reviews. Cantor's history [55] is still useful; for its fourth part (1759–1799) Cajori wrote a treatise on the development of algebra which widely refers to contemporary papers and gives a short indication of the results. Wieleitner's work [401, 402] is just as descriptive but much briefer; it goes up to the year 1850, but his description of the 1st half of the 19th century is limited to listing the principal results and works, supplemented by a selection of second-rate authors. Not even this work presents a reliable picture of the trends of development in algebra up to 1850.

Fundamental facts about the development of algebra can also be found in other reviews concerning the history of mathematics [361], as well as in the history of science (e.g., [203, 204]). The remainder of the literature, with a few exceptions which will be mentioned in the conclusion, can be divided into two groups: literature concerning individuals and literature on the development of a certain problem or complex of problems. The former is much more extensive, its interests being distributed quite non-uniformly. The literature on Gauss and Galois is particularly extensive, comparable only with the works on Abel. This particular body of work, dealing with the primary sources, accumulates with great care the data on the scientific work and life of these significant mathematicians. Therefore, the analysis and evaluation of the various aspects of their work has, on the whole, been carried out reliably and even recently mathematicians and historians have been drawing on their work, especially at times of their life anniversaries. These data, the repeated statements and serious analyses, also contain a lot of valuable material for the present book. In this connection one must point out Bell's materially well founded work, convincingly systematic in presentation, on the algebraic numbers of Gauss [25], or the papers [238, 116] emphasizing the relation between Galois' algebraic studies and Gauss', which, among other things, exactly define the elements (and their form) of Galois' theory which appeared in Gauss' work, especially in the art of Gauss' *Disquisitiones arithmeticae* on binomial equations. However, these papers also have their weak points; they were written for Gauss' anniversary and, therefore, they do not discuss the other effects on Galois (e.g., Lagrange) and they only view Gauss' words from the point of view of Galois' theory. Also Bashmakova's study [20] is very interesting; it shows the conceptual novelty of Gauss' second proof of the fundamental theorem of algebra and its connection with Kronecker's later treatment of the construction of algebraic fields. Although one could name a number of good treatises, dealing with Gauss' contribution to algebra, nevertheless, one cannot consider the analysis of Gauss' algebraic work as finished. For example, even the extent to which Gauss was aware of the general properties of the Abelian groups which he encountered on various occasions, is mostly unproved. The situation is similar with Gauss' opinions on the nature of mathematics, or special algebra, and on the method of mathematics, the material published hitherto being insufficient to answer this question.

On the other hand, Galois' literature, apart from the understandable interest in the life of the author, stresses the importance of Galois' results for the development of mathematics, beginning with the last few decades of the 19th century. In this context Galois' name is being connected more closely with the further advancement of his theory and with the theory of groups than corresponds to the emphasis on these subjects in the work itself. In the commentary to the editions of Galois' work, besides perhaps all the actual data, the most remarkable is the note made by Dieudonné ([154] V.) which appraises the marked difference between the work of Galois and that of his contemporaries and which goes further than the typical statements about unusual "moderness". The literature, published hitherto, has concentrated much less attention on the analysis of Abel's work. The recent monograph by Ore [324] indicates that attention is drawn, as in Galois' case, to the tragedy of his fate. Nevertheless, one can find a few articles which, under various circumstances, interpret the contents of Abel's work and its significance (e.g. [38a], [330], [364a]). A more detailed historical analysis, which is lacking for Abel's algebraic work, was carried out by Burkhardt [51] for Ruffini's algebraic work and partly also for Abbati's work at the end of the last century. The thoroughness of the analysis is substantiated, among other things, by the fact that nothing more substantial was added to the results.

It is surprising how little historical literature one finds on the algebraists of the second half of the 18th century. Lagrange can serve as an example in this case, because a reliable monograph on his life and work is lacking, as well as more partial analyses of his mathematical legacy. Therefore, there are several interpretations of his fundamental algebraic work [267], besides the data in the historical survey. Moreover, newer specialized literature on Vandermonde is lacking and only a few small articles on Waring's algebraic papers are available. Not even the more recent Swiss thesis [283a] contains a more thorough analysis of Waring's algebra.

The closer to the present, the more frequently the only mention of these individuals is in obituaries. Some, such as Valson's two-volume monograph [382] on Cauchy, are considerable in extent. It gives many data on his life, as well as an interpretation of his work. However, it stresses Cauchy's merits more than thorough historical analysis. The obituaries of other mathematicians, in spite of their valuable data, suffer the disadvantages of the occasion and time at which they were compiled. Some interesting distortions occur now and then; not only is work appraised with a view to the state of science to which it led, but it is also considered as a compact whole, while chronology is completely disregarded. This is the case in Weber's excellent treatise [395] on Kronecker's work which distorts the situation in algebra around the year 1850. One must regard several works similarly, of which some merely stress the famous results instead of evaluating the whole work. Posthumous studies have not, however, been devoted to many individuals. Sometimes treatises only consider their subject's fate.

These statements, which indicate the overall insufficiency of the material on individuals, 19th century mathematicians, should be supplemented by pointing out other literature regarding these men. First of all, there are short articles on significant individual which summarize (and also establish) data on the life and work of individual men (e.g. [27, 311, 334]). These items frequently form the basis and example for other reminiscences. These works also often determine the extent and nature of the data which are subsequently most frequently repeated. An exceptional position in literature on significant individuals is held by Graves' work on Hamilton [175]. The extensive three-volume publication is more than a scientific biography. It not only deals with Hamilton's life in connection with the origin of his scientific discoveries, but also supports everything with a selection from personal or scientific correspondence, and includes an extensive edition of correspondence between Hamilton and de Morgan ([175] Vol. 3). In this way Hamilton is one of the few algebraists of the period in question (apart from Galois, Abel and Gauss) whose scientific manuscripts, as well as his other manuscripts, have been published. However, even in this case modern editorial principles were not adhered to. In Graves' work one finds only excerpts of the correspondence and the author does not even mention in the notes what parts of the correspondence have been omitted and why. It seems, in fact, that special mathematical parts have been omitted (cf. [175], preface), so that one cannot rely on the deductions made from these letters about Hamilton's examples and motives. However, this may also be said of Gauss' letters, hitherto published.

In this connection one cannot ignore editions of collected works. Their objective mostly is to make it easier for the reader to acquaint himself with a body of scientific work without having to search in journals, which are often difficult to come by. Here again, with the few exceptions mentioned, there are no published inheritances of manuscripts.[1] In this respect even the extensive editions of Lagrange, Cauchy,

---

[1] Considering the very small size of the inheritances of Galois and Abel, it was not difficult to publish them practically completely. However, these complete editions are an exception. The large manuscript inheritances of other authors results, in the most favourable case, in the mathematical discoveries or

Jacobi, Kronecker, et al., are incomplete. Excerpts of correspondence included in collected works are not always subject to clear editorial rules and the character of omitted material is usually not indicated, not to mention evidence of its deposition.

As already mentioned, the second group of works on algebra in the period considered, includes studies dealing with the development of a certain problem or complex of problems in the advancement of algebra. Although this group is less numerous, it is very important. It is mostly based on particular and definitely incomplete data, but it indicates comparatively accurately the main trends of development and the important moments in the trends. Dubreil [132] captured this development in the sense indicated very nicely, having chosen the interval 1770–1910 (i.e. from Lagrange's work to Steinitz' interpretation of the theory of fields). He devotes most of his attention to Abel and Galois. The history of the solution of equations of the 5th degree was considered by Pierpont [332] at the end of the last century. He analyzed appropriate parts of the work of several more important mathematicians and compared their results. He also touches on the problem of non-algebraic solutions and his work in many places understandably exceeds the scope of the problems of 5th degree equations. His comments on the period and the reaction of contemporaries, e.g., with regard to the publication of Abel's proof, are also valuable. Pierpont's work has retained its value in many respects. In the same way, in spite of its popular nature, Hancock's study [195] on the development of the theory of algebraic numbers is interesting, whereas Funkhauser's work [151] only includes fundamental data on the development of the theory of symmetric functions which were known from elsewhere.

Attention has been devoted by historical literature to two problems: first, the history of the theory of groups; and secondly, the beginnings of vectorial calculus and the geometric representations of complex numbers connected with it, and the origin of the theory of quaternions.

The study of the development of the theory of groups, one of the fundamental theories of modern algebra, has a longer tradition. Many data can be found (apart from Burkhardt's study [51] mentioned earlier) in the *Encyklopädie der mathematischen Wissenschaften* [52]. The fundamental study was written by the American mathematician Miller ([286–294]), who indicated the main trends of work in the 19th century as well as the development of the concept of the group itself. He is ironic about the nationalistically tinged argument about the priority of the definition of a group (Cayley or Kronecker). He rewrote some of his historical papers for his collected works [285]. The substance of his historical work is in an accurate description of even the small discoveries which formed the development of the theory of groups in the 19th century. A thorough analysis of the possibilities provided by the development of the theory of groups in the second half of the 19th century, together with

---

documents pertaining to some interesting turns in life being selected. But correspondence could throw light on the way ideas were propagated and how they crystallized, none of which is ever mentioned in a concluded scientific paper.

mention of indications of discoveries and methods, is contained in the commentary to the edition of Jordan's works by Dieudonné [119]. This, of course, is rather more concerned with the development beginning with Jordan's *Traité des Substitutions* (1870 [219]).

The picture provided by these and subsequent historical studies is reliable as regards their data and only requires the inclusion of certain details, in particular those leading up to a more accurate definition of individual period milestones. With this in mind [314 – 7] are also included. H. Wussing considered the interpretation of the theory of groups on a much broader front; his monograph [410] was preceded by several studies [407 – 9]. As he also aimed at uncovering the decisive moments in the development of modern algebra, which is essential for a compound work on the development of the theory of groups, it should prove advantageous to look at his book [410] more closely.

Wussing compiled a large body of data on the development of the theory of groups from various historical works. He interprets with understanding and clarity the individual discoveries and results, and tries to find connections between them. As regards Miller (to some of whose historical works he refers) he differs in that he does not only consider discoveries pertaining to the theory of groups in the narrower sense, but also how the theory of groups and its concepts or theorems were constituted in various connections and as part of various requirements in different branches of mathematics. His principal thesis could perhaps be briefly expressed as follows: The theory of groups was constituted in algebra (in particular the theory of equations), in the theory of numbers (e.g., composition of classes of quadratic forms) and in geometry (geometric transformations), and by connecting these together the abstract theory of groups was created.

As regards the data, as well as the interpretation of the principal trends, one can hardly add anything to Wussing's work. However, in method the monograph is very traditional. Miller's studies are in some respects better as regards completeness of data. However, the author did not quite decide whether he was investigating the abstract concept of a group (as indicated by the title), or the creation of an abstract theory (as stated on the book cover and in many places within the book). The principal thesis of the study indicates the latter. However, the author has not discussed this problem in detail. To the connection of the three trends mentioned he adds a connection with Klein's Erlangen Programme and with Kronecker's work of the same period. He did not try to determine the role the three trends played in the creation of structural thinking in algebra and he did not consider the possibility of a specific, even though partial, achievement of the objective by the internal development of algebra. This is perhaps the reason for his underestimating Cayley's works, including his definition of a finite abstract group, or even the influence of Jordan's work on the further development of algebra. Wussing was more interested in how the idea of structures, equivalent with groups, was applied in various branches of mathematics and what role it played in them, than how and why the concept of a group was formed in algebra. The creation and application of the concept of a group in algebra is closely connected with the whole change in algebraic thinking and with the creation of other abstract and structural ideas. In this respect, our understanding of the development of algebra differs in some points with that of Wussing. Especially in paragraph 2 of chapter 7 the reader will find different conclusions from those in Wussing.

Another problem frequently treated in literature is the beginnings of the vector analysis. Crowe's recently published monograph [108] presents an excellent summary of abundant historical literature[2] (enriched by the study of sources). It is a pity that the analysis of the development from quaternions to linear algebra has been pushed into the background, although the fundamental data on B. Peirce's work of 1870 are mentioned here along with his relation to quaternions. This gap, however, is not unintentional and it is connected with the objective of the monograph, which rather tends towards the evolutionary contingency of the theory of vectors and its extra-mathematical exploitation, than towards the role of the theory of vectors in the development of mathematics in the 19th century. Crowe's description of the evolution prior to 1844 is suitably supplemented by Kramar's studies [241, 242]. Kramar is bolder than Crowe in stating general hypotheses on the development of algebra. Kramar's point of view, of course, is closer to the study of problems of algebra. Bachelard's recent lecture [19] on the geometric representation of complex numbers, some of the conclusions of which would call for further verification, was very stimulating.

As regards other literature on the evolution of algebraic problems, it is necessary to mention Muir's multi-volume work [307, 308]. This, however, is only a bibliographical survey with annotations pertaining to the contents of papers on the theory of the determinant. Similar data were compiled by Dickson [118] on the theory of numbers, but in some of the chapters he touches on problems investigated in the present book. The study of the development of the theory of invariants [284], unfortunately is very brief in its introductory part on the first half of the 19th century, but has recently been supplemented by Fisher's study [142] and by Weyl's historical notes [400]. Molodshii (in particular [295]), whose conclusions can be relied on, especially up to the end of the 18th century, dealt with the questions of the development of arithmetic, including the development of the concept of algebra and of the methods of the whole arithmetic-algebraic region. The subsequent part of his work, however, is not so reliable.

Vuillemin's study, especially his extensive book [387] of 1962, some of the conclusions of which are repeated in [388], stands apart from all this literature as regards its objective. The author tried to deal with more general problems, resulting from the connection between the origin of modern algebraic thinking and philosophy. However, as regards the evolution of algebra, he did not present any substantially new data. His excessively detailed interpretation of algebraic problems distorts Lagrange's concepts of

---

[2] On the understanding that Crowe is attempting to narrow down as much as possible the extent of the problems investigated and since, as far as possible, he does not discuss broader mathematical aspects of the data, his data on literature, including its evaluation, are very reliable. His way of giving references, however, due to its briefeness, is not sufficiently well arranged.

"apriori" and "metaphysics", which he interprets in their original philosophical meaning.

To conclude this section, which does not give an exhaustive review of the literature, brief comment should be made about the nature of the original sources which were used. Only printed sources are considered, Bolzano's manuscripts being the only exception. Clearly, the sources mentioned are not always fully sufficient to provide an understanding of the way the authors reached their published results, under the influence of which works and ideas they were, and, on the other hand, in what way their ideas and results were propagated. In this respect, it seems that there is much left undone and great help can be provided by special editions, thorough mathematical studies devoted to individuals, and to important, precisely defined problems.

CHAPTER 2

PRINCIPAL FEATURES OF ALGEBRA
OF THE 18TH CENTURY

2.1 Status and methods

Mathematics in the 18th century[1] was a science of quantities, a quantity being understood to be all which could be increased or decreased. Therefore, one could speak of mathematical sciences, which included all those branches in the 17th century in which one calculated or measured, although the quantities investigated by them were very "concrete".[2] In this wide mathematical region which corresponded to a considerable extent to the range in which leading mathematicians were capable of creative activity, mathematical problems themselves were being divided in quite diverse ways. However, the principal division into geometry and the remainder of mathematics was not usually violated. This division reflects the traditional difference between a geometric quantity (represented, e.g., by a line segment) and an arithmetic quantity, represented by a number.

The relation between these quantities began to lose the simplicity it had had when the geometric quantity was considered superior to the arithmetical (when in fact the latter represented the former's special case). Analytical geometry indicated the greater

---

[1] If in this chapter one speaks of mathematics of the 18th century, one should not consider the interval as strictly defined. By this one understands mathematics prior to the period which is discussed in the following chapters, i.e. more or less prior to 1770. Neither are the dynamics of mathematics in this period of interest. The purpose is only to remind the reader of the status of mathematics as a whole and of algebra as its part, of the opinions that predominated and of the elements of new trends which mathematicians to come would find in it. This, then, is a reminder of the source from which the algebra of the period to be discussed in the following chapters, developed. Because of this nature of the chapter it is restricted to the principal features, which may be found in current literature anyway (e.g., Cantor, Matthiessen, Molodshii, et al.) in much greater detail and in more concrete form.

[2] Apart from "pure" mathematics, the mathematical sciences also included mechanics, optics and astronomy, as well as geodesy, architecture, theoretical and practical ballistics, etc. These "mathematical" disciplines were included, therefore, in large mathematical compendia and dictionaries, which were published from the second half of the 17th century (Hérigone, Schott, Dechales, et al.), including the well known works of Kästner, Klügel, Bézout, et al., in the second half of the 19th century. Only then could one observe the narrowing of the scope of "mathematical" disciplines which created the tradition for the whole of the 19th century. Mathematics was divided into pure and applied at the end of the 18th century, as substantiated by the Encyclopedia [136] published at the turn of the 19th and 20th centuries, and by the contents of the journal *Jahrbuch über die Fortschritte der Mathematik*.

usefulness and larger scope of a non-geometric solution to geometric problems relative to traditional geometric methods. The apparent difference between the "continuity" of geometric quantities and the "discontinuity" of arithmetic quantities was eliminated by the "arithmetic" of infinitely small quantities, which (like the concept of a function earlier) in its way proved the continuity of arithmetic quantities at a time when concept of limit was not widely acknowledged.

These new elements in the status of arithmetic quantities did not solve everything. Much remained obscure as regards their definitions and expressions, as well as their calculus. However, at least one thing was beyond: Antique geometry could not be considered the only foundation of mathematics and, in spite of all the obscurities and difficulties, arithmetic was granted a similar status. However, geometry had one substantial advantage compared to the other mathematical disciplines; it had its axiomatics, admittedly imperfect from today's point of view, with its definitions, axioms and postulates and a technique for deriving theorems. Together these factors moulded the concept of algebra in the 18th century.

Algebra originally was a science dealing with the solution of problems by means of equations. The solution of equations only formed a part of it and could be considered a method of algebra.[3] In the 18th century the study of the properties of equations and of their solutions was gradually separated from the compilation of these equations, which remained the domain of didactic books. However, algebra, wanted to present a general theory of solving equations. The generality of the equation was guaranteed by the general significance of the coefficients. For the sake of generality (and for the sake of lucidity and simplicity) letters were introduced at the time of Viète and Descartes. The letter represented a number of a certain realm, a number in general. This, of course, brought into the forefront the other approach to understanding algebra which Newton expressed very clearly with his concept of algebra as "arithmetica universalis". Newton referred to an older tradition which differentiated between numerical or ordinary arithmetic (arithmetica vulgaris) and the arithmetic of letters (arithmetica litteralis). However, his concept of general arithmetic and that of his followers was acquiring a new and more general nature. The letter no longer represented a number of some, perhaps inaccurately defined, realm, but rather a quantity in general, *i.e.* arithmetical or geometrical.[4] This concept became current and in the

---

[3] For example, Ozanam, in his work *Nouveaux éléments d'algèbre* [325] which bears the subtitle of "principes généraux pour résoudre toutes sortes de problèmes de mathématique", understands it as a general method and also calls it "analyse" (analysis). As regards the relation between geometry and "analysis", he intentionally adopted a compromising point of view; analysis with the help of "analytical expressions" and "letters" facilitates the discovery of truth, whereas geometry gives proof. These opinions were probably strongly influenced by the Descartes tradition.

[4] The development leading to this concept, as well as its consequences, has not yet been historically analyzed. Even Molodshii's otherwise very good historical work [295] does not give this problem any substantial attention, although it does give arguments for the status of arithmetic in the contemporary system of mathematical sciences.

middle of the 18th century it appeared in very widely used textbooks in various forms.[5]

Therefore, in the 18th century there existed two substantially different, but mutually supplementary, concepts of algebra.[6] One of these considered algebra to be a science of equations and of their solutions, the other a science of quantities in general.

The latter concept, algebra as "calculation with letters", became the foundation of all mathematics. However, it should be pointed out that this foundation was very labile and problematic. This "universal arithmetic" had only just begun to grasp its logical structure and elementary geometry was its example. As early as the 17th century attempts were made at discovering the logical structure of contemporary arithmetic. This endeavour accelerated in Germany at the beginning of the 18th century in the works of Sturm, Wolff and their followers. The objective was an axiomatic system. During the 18th century, however, only the "general" axioms of Euclid's work were extracted and partly multiplied. Even statements such as "A whole is larger than any of its parts", or "A whole is equal to the sum of all its parts", appeared as axioms at the time. Of course, one can also find[7] the following axioms:

---

[5] De la Caille's textbook [110], published several times and also translated, states: L'algèbre est une Arithmétique universelle, c'est la science de la grandeur en général; comme l'Arithmétique est la science des nombres (p. 38). On the other hand, the widely read and highly influential compendium of Kästner [221] gives a somewhat different opinion. At the beginning of the chapter called *Die Buchstabenrechnung*, one reads ([221] Vol. 1, p. 81): Will. Satz. Eine Zahl überhaupt anzudeuten, bediene man sich eines Buchstaben. Kästner considers letters either as integers, fractions, irrational numbers, etc., and the operations and their signs remain the same.

However, the opinion expressed in the text only characterizes the main trend along general lines. Many other opinions existed as well. This is only a question of the fundamental general orientation as regards the trends which he drew on and which he encounteres in the further development, which is the case in the whole of this chapter.

[6] This only characterizes two substantial types of understanding of algebra. A large number of subtle variations and other opinions also existed. One should also mention a further modification of Newton's concept which Euler expressed in his algebra ([141], Chap. 16—17); he briefly writes that each quantity can be expressed by a number and, therefore, the whole of mathematics is based on the science of numbers, which he calls "Analytik" or "Algebra". Whereas arithmetic considers numbers and operations as encountered in everyday life, "begreift die Analytik auf allgemeine Art alles dasjenige in sich, was bei den Zahlen und der Berechnung derselben vorfallen mag".

[7] Cf. Molodshii ([295] p. 28), or Nový ([313] p. 41). For the sake of comparison, Euclid's axioms after Heiberg's edition are given:

1. If $a = b$ and $c = b \Rightarrow a = c$
2. If $a = b$ and $c = d \Rightarrow a + c = b + d$
3. If $a = b$ and $c = d \Rightarrow a - c = b - d$
4. If $a \neq b$ and $c = d \Rightarrow a + c \neq b + d$
5. If $a = b \qquad \Rightarrow 2a = 2b$
6. If $a = b \qquad \Rightarrow \dfrac{1}{2} a = \dfrac{1}{2} b$
7. Things which coincide with one another are equal to one another;
8. The whole is greater than a part;
9. Two straight lines do not limit an area.

1. $a = a$
2. $a = b \Rightarrow b = a$
3. $a = b, a = c \Rightarrow b = c$
4. $a > b, b > c \Rightarrow a > c$
5. $a = b \qquad \Rightarrow a \pm c = b \pm c, \qquad a > b \Rightarrow a \pm c > b \pm c$
6. $a = b \qquad \Rightarrow a \cdot c = b \cdot c, \qquad \dfrac{a}{c} = \dfrac{b}{c} \; (c \neq 0)$
7. $a > b \qquad \Rightarrow a \cdot c > b \cdot c, \qquad \dfrac{a}{c} > \dfrac{b}{c} \; (c > 0)$

Axioms formulated in this way assume knowledge of the operations which occur in them. Their definitions, however, present two difficulties. First of all, the basis of these operations are elementary arithmetical operations. These are intuitively known as regards calculations with real numbers. Attempts at a logical elaboration of arithmetic operations[8] lead to attempts at their explicit verbal definition in a way similar to the definitions in Euclid's Elements in which they had, as was currently believed, been introduced. In spite of all the doubts concerning these definitions, mathematicians of the time were aware that definitions, plausible for operations with integers, lose their meanig when applied to rational numbers. A well known example is multiplication, which requires that the multiplicand be added as many times as there are units in the multiplier. This definition of multiplication satisfies the definition of a number as a sum of unities found in Euclid. The difficulties connected with it lead to changes in the definition of a number, which Newton, e.g., defines as a ratio of two quantities.

The purpose is not to describe the difficulties in the logical structure of arithmetic in the 18th century, but merely to state their existence. Calculations with letters were derived from calculations with numbers. So long as calculations with numbers or within numerical realms were not defined accurately, calculations with their general symbols, representing numbers, also remain undefined and unexplained. This was the first difficulty. The other was the generalization of the validity of the symbolic rules obtained. Even the axioms mentioned were generally valid for quantities in general. This included arithmetical as well as geometrical quantities. If one disregards the meaning the operations had for geometric quantities,[9] the unlimited range of validity for arithmetic quantities becomes an idea which, while fertile at the beginning, later leads one astray and hinders the differentiation of some numerical realms. As early as the 18th century one can see that many writers, according to the rules for calculating with

---

[8] This topic has recently been treated very well by V. N. Molodshii [295]. Especially his analysis of development in the 18th century is quite convincing and based on a wide range of data. He also discusses in detail attempts at defining the concept of a number. The present book in many cases draws on his results.

[9] It is possible that in between lines of contemporary considerations an obscure possibility of a geometric calculus was constantly being preserved, or rather the endeavour to create it cropped up from time to time.

quantitities "in general", were calculating with infinite expressions, or with "impossible" or complex expressions. It should not be necessary to point out the paradoxical statements they were bound to arrive at.[10]

Such difficulties do not represent anything exceptional in the evolution of mathematics. Also in this case, the generalization connected with or controlled by the symbolic (formal) expressions becomes a means for further generalization and, simultaneously, a search for the limits of the justification for the generalization. However, in the 18th century "arithmetica litteralis" had yet another role as a method of proof. If the expressions or equalities were proven as valid for letters, they were "proven" or considered as valid for any arbitrary quantity without further differentiation. It became unnecessary to "prove" them for individual types of numbers or for numerical realms.

As to the elaboration of the logical structure of arithmetic, this methodical element did not produce outstanding results. However, it represented a methodical guide for the theory of equations. The formal, symbolical approach made it possible to study equations generally while simultaneously obscuring some of the more important algebraic elements. That is why algebraic considerations adopted facts of numerical elements in the form in which they had earlier been prepared by arithmetic. Arithmetic differentiated among natural numbers, integers, fractions (identical with the concept of rational numbers), and irrational numbers. The problem of whether there exist other real numbers besides irrational numbers acquired a concrete form only in the work of Euler and later in the work of Lambert. The problem of complex numbers was different; from the point of view of contemporary arithmetic these numbers were hard to understand and "imaginary", but also accepted as a necessarily evil result of expanding the numerical realm. They were created just as naturally by taking the root of a number, as negative numbers were created by subtracting. By accepting these numbers, statements concerning the properties of equations and their solutions became general and elegant. This also displays the significance of relations; expressed by algebraic symbols they are valid for an arbitrary realm, thus elements of an arbitrary realm can be substituted for the letters. Provided one upheld the general "algebraic" expression, there was no differentiation between the elements of the various realms because this would diminish generality.

However, let us go back to the consequences of algebraic symbolism for the development of the contemporary algebraic method. Authors had different views

---

[10] Euler also ([141] p. 161 nn) arrived at a solution of the equation

$$x^\infty = x^{\infty - 1} + x^{\infty - 2} + \ldots.$$

Stepling (cf. [313], p. 75 nn) calculates with powers of "infinity" according to current rules, i.e. ∞^n, for n rational and explains infinitesimal calculus "algebraically" in this way. Another example of "unusual" statements is obtained if in the 2nd part of axiom 5 one puts $c = \sqrt{-1}$, so that for any arbitrary real a and b one arrives at

$$a > b \Rightarrow a + i > b + i.$$

on the logical structure of contemporary algebra, nevertheless, the relation between the letters (i.e. the formula) had a dominating significance. Newton characterizes this in the very first pages of his "Arithmetica universalis" very precisely: Calculations in algebra are carried out by means of numbers, and therefore, nearly all statements and, in particular, conclusions which can be found in algebra, can be called "theorems". Theorems are established, therefore, by calculating with letters. This is the way it was understood in the Newtonian tradition in algebra. Basically the same result is reached in a somewhat different way by Clairaut in his *Eléments d'algèbre* [102] which was first published in 1746 and then several more times in the subsequent half century.[11] He understood algebra as a kind of language for solving problems (§ 1). He avoids deriving theorems, but he proceeds from a problem which he expresses "generally" (i.e. in terms of letters) and arrives at a general rule[12] which is "generally" valid.

The methodical procedures indicated were originally connected with the endeavour to elaborate on the logical structure of the arithmetico-algebraic region, but their effect in this respect was minimal. Because of the relative disinterest in fundamental problems during the 18th century, not even the quite clear logical deficiencies (some of which we pointed out) were uncovered. The operations indicated, however, mainly because they provided for a certain generality and elegance of form, controlled algebraic thinking to a large extent. In particular this meant that algebra, provided its theorems had not been derived from theorems of other mathematical disciplines, was satisfied with finding plausible algebraic formulae, or relations between formulae, and accepted their validity without further proof. Theorems were "illustrated", or in other words, proved by numerical examples. Complete mathematical induction was a little used tool.

There were not many who stood apart from the overall trend and who were aware of at least some of the deficiencies in contemporary algebraic interpretations. One of these was A. G. Kästner[13] who described some of the methodical errors of the others, accurately on the whole, at the beginning of the second half of the 18th century. However, in his own papers he was not able to completely avoid logical inconsistencies

---

[11] They represented one of the most widely spread algebras of the second half of the 18th century; their 5th impression was published in Paris in 1798 in two volumes with a large number of appendices and in 1801 a further impression was even expanded.

[12] Not only Clairaut proceeds in this way, but also others. A comparison with the method of contemporary elementary geometry and its choice of a general example offers itself.

[13] The problem of proofs of algebraic theorems is discussed by Kästner in the preface ([221] III/I, 1759). Among other things he wrote: "Ich erfuhr aber wirklich, dass nicht alles in der Algebra wenigstens völlig allgemein und mit allen Umständen demonstrirt wird, was man dafür ausgiebt. Verschiedene dergleichen Errinnerungen fand ich bei dem gewöhnlichen Vortrage der Lehre von den Gleichungen zu machen. Wenn man ihren Ursprung aus der Multiplication erklärt, so giebt man meines Erachtens den Grund nicht allemahl deutlich genug an, wesswegen man die Wurzelgleichungen auf O bringen muss; und wie einerley Buchstabe zugleich verschiedene Grössen bedeuten kann...". Further on he wrote: „Man begnügt sich in der Algebra sehr oft allgemeine Gesetze anzunehmen, wenn man sich von der Richtigkeit derselben bey einigen besondern Fällen versichert hat".

and jumps, which was then apparently impossible. He was able to avoid many difficulties by always trying to determine the realm of an element represented by a letter; because of this his results did not always proceed toward generality and he was apt to connect algebra with arithmetic more than usual.

2.2 Main trends

The principal purpose of 18th century work in algebra was, without doubt, the solving of equations. Of the more general problems attention was devoted only to the properties of complex numbers[1] and to calculation with irrational numbers.[2] The fundamental theorem of algebra was only just becoming the object of investigation and of the first unsatisfactory proofs.[3] It was not quite clear whether it was necessary to prove the existence of the root of an algebraic equation. Experience indicated that linear, quadratic, cubic and biquadratic equations had roots; moreover, it was possible to find expressions which, when substituted into an equation, zeroed it out. The equivalence

$$(x - x_1)(x - x_2)(x - x_3) \ldots = $$
$$= x^n + a_1 x^{n-1} + a_2 x^{n-2} + \ldots = 0 \tag{2.2.1}$$

(for roots x_i) seemed quite clear. It lead to the statement that an equation of the n-th degree has n roots and that the well known relation between roots and coefficients holds. This was all explained by a kind of incomplete induction and accepted as early as the 17th century. When the neccessity arose for these truths to be proved, it led to advancement toward the ideas which were finally realized in the 19th century.

Apparently, the most important objective of algebra was to present an algebraic solution of a degree higher than the fourth.[4] This task, which had been a logical requirement tied up with the results of the Italian school since the 16th century, remained more or less under the surface. There were very few direct experiments.

---

[1] The knowledge and proof of the fact that each algebraic expression of the type $f(x + iy)$ can be expressed by a sum $u + iv$, where u and v are real, was important. At this point it is necessary, in particular, to stress d'Alembert's contribution. However, this did not terminate the discussion, as substantiated by what Gauss wrote in his thesis [160].

[2] Calculating with irrationalities was not only a regular part of all interpretations of algebra, but also the topic of specialized treatises. The discovery of the n-fold multiplicity of the n-th root was already a comprehensive result of the theory of algebraic equations, as well as of the fact that

$$a^2 = (+a)(+a) = (-a)(-a)$$

[3] Cf. Bashmakova [20].

[4] Even during the first half of the 19th century the differentiation between a numerical and algebraic solution of an algebraic equation was often rather intuitive, without an accurate definition. Therefore, vague formulations are used which can be understood sufficiently but which suggest that contemporary mathematicians had not thought of an accurate definition. The situation was then also complicated by the way "generality" of the equation solved was understood. It was then sufficient for the coefficients to be represented by "letters".

This could have been, to a certain extent, a manifestation of the limited interest in algebra.[5] In all events, the first more serious attempt after Tschirnhaus, apart from interpretations of other authors, was Euler's publication of 1732 [140] which is justified in claiming some of the new ideas. Euler himself returned to this problem in a treatise published in the sixties [137], when the works of Bézout [37] and the first algebraic studies of Waring also appeared.

Tschirnhaus' method is sufficiently well known. Its author wanted to determine a suitable substitution of the form

$$y = b_1 x^{n-2} + b_2 x^{n-3} + \ldots + b_{n-1} \qquad (2.2.2)$$

which would transform equation

$$x^n + a_1 x^{n-1} + a_2 x^{n-2} + \ldots + a_n = 0 \qquad (2.2.3)$$

to

$$y^n + c_n = 0. \qquad (2.2.4)$$

Tschirnhaus found a new solution of the cubic equation in this way. However, for $n \geqq 5$ the calculations used to determine the coefficients b_i lead to equations of higher degrees.[6] In 1732 Euler proposed the following form for the root of Eq. (2.2.3):

$$x = \sqrt[n]{A_1} + \sqrt[n]{A_2} + \ldots + \sqrt[n]{A_{n-1}}, \qquad (2.2.5)$$

A_i being the roots of the resolvent (aequatio resolvens, first named by Euler) of the $(n$-1)st degree. In his propositions Euler considered that one can reach a solution for $n = 2, 3, 4,$ but for $n = 5$ he only believed it possible to find suitable means and reach the objective. He did not, however, treat this case himself.

Without letting it be known in the following years that he had reached a different opinion (although it is quite clear from his manuscript inheritance that he was still interested in these problems [151a]) he suggested in 1762 a new form of the root:

$$x = w + A \sqrt[n]{v} + B \sqrt[n]{v^2} + C \sqrt[n]{v^3} + \ldots + Q \sqrt[n]{v^{n-1}}, \qquad (2.2.6)$$

where w is a rational number, $A, B, C, \ldots, Q$ are either rational numbers or numbers "which contain no root of the n-th degree", and v is the root of the resolvent of the

---

[5] It seems that in the later decades of the 17th century and in the whole first half of the 18th century the interest of mathematicians was mostly absorbed by infinitesimal calculus and by the problems of mechanics and geometry connected with it. There were very few algebraic works of the time. Only with the quantitative increase of mathematical papers, which became pronounced in the middle of the 18th century, did the interest in solving equations also increase. In the sixties it had already become relatively widespread.

[6] Leibniz immediately became skeptical about the possibility of using the proposed method for solving a general equation of a degree $n \geqq 5$. Tschirnhaus' method already bears characteristic features of later attempts. It generalizes a partial case, i.e. the substitution, which eliminates the second term, and it deals with the proposed method for $n = 2, 3, 4$. However, Tschirnhaus did not go further in his work.

$(n-1)$st degree. He again showed how this form satisfied equations of degrees $n = 2, 3$ and 4, and he was convinced that there was the greatest probability that the root of the equation has a general form.

Incomplete induction and the failure to solve further cases also characterize similar work of Bézout [37] of the same period. On the whole, however, these attempts at a kind of "experimental" solution uncovered quite clearly the weak spots of this method, at the same time yielding some important facts (e.g., the significance of n-th roots of unity stressed by Euler, the existence of auxiliary equations and the necessity to study irrationalities). However, under the influence of the problems of solving equations, progress was made in these regions only in the study of n-th roots of unity. The principal credit is again due to Euler who, at least in some of his results (if not in method and accuracy), anticipated Gauss' results regarding the properties of these roots, without reaching a solution of the algebraic equation

$$x^n - 1 = 0 \qquad\qquad (2.2.7)$$

(for any arbitrary natural n), which only Vandermonde accomplished in the 18th century.

As regards the other problems of the theory of equations, attention was devoted in particular to the relation between the coefficients and the roots of equations, i.e. to symmetric functions. Newton's work was largely responsible for this; among other things it shows how it is possible to express the sum of n-th powers of the roots by means of the coefficients of an equation. This led to the establishment of the so-called Newton formulae,[7] which were not expressed generally enough and were without proof. In the 18th century facts about symmetric functions were frequently included in textbooks on algebra.[8] The interesting fact that any arbitrary symmetric function could be expressed as a function of elementary symmetric functions (i.e. of the coefficients of the equation) was also established, but these more or less arithmetical results were not exceeded.

Apart from the problems connected with contemporary theories of equations, mathematicians also searched for other methods of solving equations. Some investigated Descartes' procedure. He resolved a biquadratic equation into two quadratic equations and tried to find out whether one could decrease the degree of an equation in this manner generally. These attempts stressed the problems of elimination from a completely different aspect. The climax of the study of elimination, also considered by the foremost mathematicians, is represented by the work of Bézout and by some parts of Waring's algebraic work. The main problem in this case was to determine the minimum degree of the auxiliary equation during the process of elimination from

---

[7] H. Funkhouser [151] pointed out that the results connected with Newton's name could already be found in Girard's *Invention Nouvelle en l'Algebre* (1629) and that in the 18th centuries some mathematicians tried to prove these relations; in this connection he drew attention in particular to Euler's 2nd proof.

[8] Cf. [151] p. 363.

non-linear equations. But, attention was also devoted to a system of n linear equations of n unknowns, the study of which, especially by Cramer, lead to the elements of the theory of determinants. However, determinants, symmetric functions and the theory of elimination have one thing in common which hardly ever intentionally reaches the limelight during the 18th century. With all three the necessity for, or the exploitation of, combinations of various elements could be observed, thus providing one of the impulses for advancement in combination calculus.[9]

Other trends in algebra in the 18th century rather tended toward numerical solutions. For example, the study of the problem which is created by Descartes' rule was important here, i.e. the study of the number of complex roots. Further attention was devoted to "arranging" equations, i.e. to substitutions, which was to provide the roots of the equation, created by substitution from a given equation with the required properties, e.g., all roots positive (or negative), etc. Indivisible from all the problems hitherto mentioned were methods of approximate solutions, to which most of the scientific work of the time was devoted.

2.3 Historical significance

From the propagation of Descartes' algebraic knowledge up to the publication of the important works of Lagrange, Vandermonde and Waring in the years $1770-1$, the evolution of algebra was, at first glance, hardly dramatic and one would seek in vain for great and significant works of science and substantial changes. Although algebraic facts were frequently used in the other mathematical disciplines and, therefore, one might even find a certain number of systematic textbook interpretations of algebra, nevertheless, at the time relatively little attention was paid to research into algebra and only toward the end of this period could one have observed an increase in the number of papers attributable to algebra. However, this opinion is not quite justified. If one considers the matter in greater detail, one will find that algebra, in indivisible connection with the whole of arithmetic, was achieving, or had already achieved, a new status. It gradually became the fundamental element within the whole of mathematics. Algebra, as a general "theory of quantities", also became the language of mathematics. This meaning of algebra, which was then established, cannot be sufficiently appraised.

The algebra of the time, however, must be understood also in the narrower sense of the word, not only as universal arithmetic, but also as general arithmetic which forms a compact arithmetico-algebraic region together with the arithmetic of numbers. At this point, under an apparently calm and insignificant surface, a battle was

---

[9] It is undoubtable that the coefficients in the binomial development, or some of the questions connected with the theory of probability, not only were more significant, but also had a more direct effect.

waged over understanding numbers, various numerical realms, and operations with the elements of these realms; and not only over understanding, but also over definitions. In this way a traditional set of problems, definitions and difficulties was created which the contemporary arithmetico-algebraic region could not escape of its own accord, but which it could at least begin to realize.

Many paradox situations were created in this way. The transposition of the Euclidean axiomatic method into the arithmetico-algebraic region with verbal definitions of the elements of the numerical realms and of their operations, together with some of the axioms, did not even satisfy the requirements of the time as regards accuracy. On the other hand, the formal algebraic language was a very doubtful "proof" open to attack. A numerical calculation was correct and to prove it exceeded contemporary mathematical possibilities. Nevertheless, general statements, expressed by algebraic language, were proved numerically or by means of a special case. The endeavour to obtain proof only leads to partial results, to a logical connection of only some of the statements.

Under these conditions some mathematicians tried to treat the main problem of the science of equations which was then quite rightly called algebra. Their attempts, as regards the standard, were fully compatible with the overall methods of the whole arithmetico-algebraic region. Moreover, they also corresponded to those concepts which were created in arithmetic, i.e. in calculating with numbers. However, not only did these attempts fail to reach their objective (which was no wonder) but it was also easy to point out errors and gaps in the experiments. In this way an abundance of partial results was accumulated concerning equations and their solutions, based rather more on calculations than on proof. The obscure formulations of the procedures for calculations used, which in many cases were not reliable, did not hinder the abundant exploitation of algebraic knowledge in solving problems in other fields.

CHAPTER 3

SOLVABILITY OF EQUATIONS

3.1 Introduction

It is a known fact that the search for a general solution of algebraic equations provided the main thrust from which the whole transformation in algebra evolved. The study of algebraic structures and the establishment of the necessary new terminology and methods took place only in connection with the elaboration of the so-called Galois theory, while other influences also strongly affected the transformation of algebra. This process took place well inside the 19th century when the new concepts and methods had already begun to crystallize. However, this chapter deals with a considerably earlier interval. It discusses the creation of the background of the work of Galois (and of Abel), beginning with the seventies of the 18th century. Also discussed are the beginnings of certain changes and their subsequent acknowledgement in the work of mathematicians in the first decades of the 19th century.

The authors of the period founded their work on the state of algebra which was characterized in the foregoing chapter. However, they sought new means to treat old problems, among which the foremost was the problem of solving equations of the fifth or higher degrees. Their papers contained the ideas of the theory of algebraic equations in embryonic form, the theory being later established in the work of Galois and Abel. There is a question as to how the understanding of the outlines of the new ideas was born, or, vice versa, what obstacles existed in the thinking of mathematicians of the period, and which of these hindered their understanding.

This topic is divided into three sections. First of all, we shall consider the solvability of algebraic equations with Lagrange and his contemporaries, and in the next section (3.3) Gauss' and Ruffini's results, which might be considered a continuation of the group of problems selected by Lagrange. The fourth section is devoted to the way in which contemporary mathematicians and the mathematicians of the beginning of the 19th century understood the work of Lagrange, Gauss and others and what they considered as substantial and contributive. However, the latter section cannot cover the problem in question completely; it will only contain a picture of the various types of reaction to the works of Lagrange, Gauss and the others and will interpret the most outstanding cases.

Neither are the second and third sections intended as a detailed interpretation, analysis and evaluation of the work of Lagrange, Gauss and the others. The literature, hitherto available, is reliable and quite extensive as regards the evaluation of their algebraic work, and it is sufficient to refer to it. Therefore, only the contents of the individual works and their most significant contributions and evaluation will be mentioned. Simultaneously, an attempt will be made to point out the new features in the approach and method, i.e. the features which predestine further development.

3.2   Lagrange and his contemporaries

The problem of solving[1] an algebraic equation of the 5th or higher degree became outstanding for all mathematicians from the time of the Italian school of algebra of the 16th century. Nevertheless, the progress in solving it was negligible and work rather concerned individual details of the theory of equations. At this point, three mathematicians working in three different countries tried nearly simultanously to present a more detailed analysis of the methods of solving equations. Lagrange, then working at the Berlin Academy, published an extensive treatise, *Réflexions sur la résolution algébrique des équations* [267], in the local academic journal in the years 1770—1771. In 1770 Vandermonde also wrote papers which were published in the transactions of the Paris Academy under the title *Mémoire sur la résolution des équations* in the years 1774 [383].

The chronology does agree as well in the case of the third mathematician, E. Waring, who studied similar problems at Cambridge. Waring published his book *Miscelanea analytica de aequationibus algebraicis et curvarum proprietatibus*[2] in 1762, the first algebraic part of which (65 pages) he rewrote and expanded under the title of *Meditationes algebraicae* [390a] in 1770. He himself wrote in the preface to the third impression [390b] in 1782 that he had sent a copy of the second impression to Lagrange in May of 1770, but that he could not say whether Langrange had actually got it.[3] However, one may assume that Langrange's, Vandermonde's and the second impression of Waring's book originated independently of each other.[4]

---

[1] In this and subsequent sections, unless expressly said so, it is always a question of algebraic solutions of algebraic equations. The formulation of the question at the time, however, did not coincide in several details with the present one, but this shall only be pointed out where the difference is important.

[2] [391] of which the algebraic part consists of the first 65 pages or about two fifth of the text.

[3] ...et Mense Majo anno 1770 exemplaria secundae editionis hujusce operis ad summos mathematicos D'Alembert, Bezout et Monteclu; Euler, Le Grange, Frisi etc. mittebantur; utrum ea receperint, nescio; recepisse antem numquam literis ad me missis agnoverunt, celeberrimo Frisio excepto..., [390b] Praefatio, p. XXI.

[4] Vandermonde expressly mentions the independence of his study from the publications of Waring and Lagrange. His paper was presented at the Paris Academy in November 1770, but prior to its publication he had the opportunity to acquaint himself with the studies of his "competitors". He then pointed out the similarity of Waring's results which were contained in the 5th part of his own treatise, and the agreement with Lagrange's study. In order to indicate that Waring and Lagrange were not necessarily the only ones to reach the same results or who considered similar problems, he pointed out

The works of the three mathematicians mentioned differ in some features from previous investigations, and in this they resemble each other. They are all publications which are aware of the fact that previous attempts were unsuccessful and they understand in one way or another why the methods used hitherto had been unsuccessful. The authors, having been enlightened, attempted a more general approach which was expected to enable them to progress. At the same time, in contrast to the individual ideas and approaches which their predecessors employed in attempting to overcome the difficulties, they concentrate more on the fundamentals of the problems of the solvability of equations. They approach the nucleus of the problem and, therefore, their work also gives a more logically compact impression. What actual form their approach had is irrelevant because it always tended toward the same result. Waring's objective was to investigate the degree of the resolvent, i.e. the equation, the roots of which are functions of the roots of the given equation.[5] Vandermonde formulated his objective slightly differently in that he wanted to find the most simple, general expression which could in general satisfy an equation of a given degree.[6] The general approaches of Waring and Vandermonde, however, were superseded by Lagrange's ideas, which had the most influence and, therefore, will be considered first in order to be able to compare them with the results of Waring and Vandermonde.

Lagrange's fundamental idea, with which he approaches research into the solvability of equations and which he proclaims at several points in his work, is simple: Hitherto several methods have been known which lead up to the solution of equations of the 3rd and 4th degree, and which mathematicians, so far unsuccessfully, have tried to apply to 5th-degree equations and equations of higher degrees. They invariably got lost in complicated calculations and it is hard to determine the way in which to continue.[7]

---

that de Marguerie's study was being printed as part of the Proceedings of the Royal Naval Academy at the same time and that the 5th volume of the Proceedings of the Torino Academy would include the results of Condorcet concerning the same topic. However, Lagrange admitted ([267] 369—370) that he knew Waring's work. He wrote at the end of one paragraph in the fourth part of his treatise, which was included in the proceedings of the Paris Academy after 1771, about the *Meditationes algebraicae:* Ouvrage rempli d'excellentes recherches sur les équations. However, the material effect cannot be determined and Lagrange's comment is more or less marginal although he could have referred to Waring's work elsewhere as well.

[5] Waring fomulates this problem, which forms the topic of the first chapter of the *Meditationes algebraicae*, in these words: De methodo inveniendi aequationem, cujus radices sint quaecunquae algebraica radicum datae aequationis vel datarum aequationum functio. He formulated the task similarly in his study of 1762 ([391] Praefatio, p. II).

[6] On demande les valeurs les plus simples, qui puissent satisfaire concurrent à une Équation d'un degré déterminé. As regards the interpretation of this sentence, which is the first paragraph of Vandermonde's study, compare this with the introductory part to its German translation in which the translator points out certain historical connections [385].

[7] As regards the extensive calculations which were carried out for equations of the 5th degree, it might have frequently seemed only a question of a lucky idea or operation which would make further progress possible. It seems that Euler was of the same opinion. For details see Lagrange [267] pp. 305—7.

In contrast to this approach, which Lagrange calls an *a posteriori* approach, he adopted an *a priori* approach, i.e. prior to applying these methods to equations of higher degrees he tried to decide why they yield a solution for equations of the third and fourth degrees and whether they will yield a solution for higher degrees. Lagrange's success is not only in stating of the requirement for this new approach, but also that he was able to apply it extensively. His extensive treatise is composed of four parts. In the first he analyzes the various methods of solving a cubic equation and proves their common basis. He did the same in the second part for equations of the fourth degree, and in the third part he considered equations of the fifth degree. Drawing on the analysis in the foregoing parts, in the fourth part he presented general theorems on the solvability of equations of the n-th degree. The analysis of the known methods of solving equations of the 3rd and 4th degrees lead Lagrange to the following: Consider an equation of the n-th degree

$$x^n + a_1 x^{n-1} + \ldots + a_n = 0, \tag{3.2.1}$$

the roots of which are $x_1, \ldots, x_n$. All known methods of solution of this equation are based on finding suitable auxiliary equations, which he calls "réduite" and the roots of which are rational functions of roots of (3.2.1). It was found that in general the auxiliary equation, now called the resolvent, is of degree $n!$ He constructs the resolvent in the following manner: let $f(x_1, \ldots, x_n)$ represent a rational function of the roots of Eq. (3.2.1). Considering all permutations of $x_1, \ldots, x_n$, of which there are $n!$, this function can only yield a maximum of $n!$ different values. Denote them by $f_1, f_2, \ldots, f_{n!}$. Then consider the equation

$$(t - f_1)(t - f_2)(t - f_3) \ldots (t - f_{n!}) = 0 \tag{3.2.2}$$

This equation is called the Lagrange resolvent and clearly the roots of Eq. (3.2.1) are rational functions of the roots of Eq. (3.2.2). Thus, Lagrange drew the conclusion that all solutions of Eq. (3.2.1) depend on suitable properties of the function

$$f(x_1, \ldots, x_n).$$

The analysis of the cases for $n = 3$ and $n = 4$ indicated that, in the cases considered, the function may be chosen so that the resolvent is of degree six or twenty four, but that it may be reduced to an equation of a degree smaller than n. For example, for $n = 3$ the resolvent has the form

$$y^6 + b_1 y^3 + bc = 0. \tag{3.2.3}$$

This occurs in cases where the function f is of the type that it attains two different values for all permutations of the roots x_1, x_2, x_3. Thus, in the 3rd part of his work he asks how one may apply previously known methods to the solving of an equation of the 5th degree. He then goes on to investigate how one may, given a suitable function,

$$f(x_1, x_2, x_3, x_4, x_5), \tag{3.2.4}$$

decrease the degree of the resolvent, which in general is of degree $120 = 5!$. He drew the conclusion that its degree may be decreased to six, but he expressed doubt[8] whether the forms of function (3.2.4) used would lead to a further decrease of the degree of the resolvent below $n = 5$. In the fourth part of his work, in which the author summarizes the foregoing results, he reached an important general conclusion. If one wants to investigate the solvability of equations "*a priori*", to use Lagrange's terminology, one must devote one's attention to the study of functions $f(x_1, \ldots, x_n)$ and investigate their properties for the permutations of the roots $x_1, \ldots, x_n$, i.e. how many different values they attain. The discovery of this objective, to which Lagrange reduces all solutions of equations calculated hitherto, and which he denotes as the *a posteriori* procedure, may be called, in accordance with the author, the "metaphysics of the theory of equations" proper.

The main outlines of the trend in Lagrange's extensive treatise have thus been indicated, his procedure being segregated from a more modern interpretation. Of course, Lagrange's treatise is understadably much more extensive in content.[9] It has been said that his work forms one of the milestones in the study of the solvability of algebraic equations. As with most works of this historical importance, Lagrange's also summarizes the results of his predecessors on the one hand, and introduces elements of further evolution, on the other. It should be interesting to outline the relation of Lagrange's work with respect to the preceding development, however, this is outside the scope of the present objective.[10] Let us at least mention those elements of Lagrange's work which were the forerunners of the future evolution of algebra and in their way represent incentive for the work of other algebraists.

The most pronounced feature of Lagrange's work is his endeavour to proceed, as he himself said, *a priori*. Apart from the meaning of this method, of which mention has already been made, it also clearly shows the requirement for a proof of existence, i.e. whether the solution of the given problem in the sense of the methods at hand, exists at all. In the sense of this procedure the question of the solvability of algebraic equations is also presented. This does not mean that a new problem has been created. Even earlier a certain amount of scepticism appeared. However, Lagrange's attitude, without giving any special definition, explains what is understood by algebraic solution of an algebraic equation, the principal role being played by the resolvent (which may again be solved by other resolvents of lower degrees), and at the same time it indicates that this

---

[8] For example, on p. 342 he writes: "...mais nous n'entrerons point ici dans ce détail qui, outre qu'il exigerait des calculs très longs, ne saurait d'ailleurs jeter aucune lumière sur la résolution des équations du cinquième degré; car comme la réduite en z est du sixième degré, elle ne sera pas résoluble à moins qu'elle ne puisse s'abaisser à un degré inférieur au cinquième; or c'est ce qui ne me paraît guère possible d'après la forme des racines z', z", ... de cette équation".

[9] Several historical studies discuss Lagrange's treatise. To mention some, e.g., Cantor [55b], Matthiessen [282], Vuillemin [387, 388].

[10] It is necessary to say that this analysis was not carried out, nor can it be found in the papers referred to in the foregoing note. Pierpont's study [332] also considers Lagrange's work rather more from the point of view of its influence on further development, and although it also discusses the

endeavour may not be fruitful, i.e. it may not yield a solution.[11] Lagrange himself
did not make a unique statement as to the solvability of equations of the 5th and
higher degrees, he rather left himself a certain amount of scope in this respect. He showed
that the methods used so far could not be successful: he expresses himself, even in general
terms, quite sceptically, but, on the other hand he points out the possibilities which remain.[12] In actual fact he believed that the solution of a 5th degree equation depends on
a resolvent of the 6th degree, but as the resolvent of the 6th degree depends on the original equation of the 5th degree, its difficulties cannot be greater than those of the original
equation. Let us ignore the meaning of this statement, which is Lagrange's personal
opinion not discussed further.[13] Lagrange, therefore, did not have a unique opinion as to
the solvability of equations of the 5th degree and of higher degrees; he leaves both
possibilities open, although more recent literature is of the opinion that Lagrange was
sceptical.

The problem, which Lagrange apparently quite omitted, is the nature of the numerical
realms which he considered. Without saying so explicitly, he bases his considerations on
the then current differentiation of the fundamental realms, considerations concerning
complex numbers[14] he takes to be quite current and very frequently also the study of the
complex roots of the equation

$$x^n - 1 = 0 \qquad\qquad (3.2.5)$$

work of Tschirnhaus, Euler and Bézout, it does not mention in what way they anticipated Lagrange's
results. Burkhardt [51] also only interprets Lagrange. Nevertheless, the conjecture offers itself that
Euler and particularly Bézout displayed more general ideas, closer to Lagrange's, about the
problems of solving the equations of the 5th and higher degrees, although they never explained them
explicitly. The study of elimination and its conditions, binomial equations and the relation of its
roots of the general equation, as well as the study of symmetric functions, indicates this. However,
there exists one more argument of a historical nature in favour of this; three independent studies
could not have been created nearly simultaneously around the year 1770 without the evolution of
general ideas. However, the search for the actual stimuli and preparation for the turn from relying
on calculation techniques to investigating a priori, so typical of Lagrange, will probably be in vain.

[11] This by no means concluded the groping around the problem as to what one should understand
by algebraic solution to be discussed later on. However, an approach similar to Lagrange's in this
problem was mainly adopted by Vandermonde.

[12] Lagrange ([267] 305) even said: Le Problème de la résolution des équations des degrés supérieurs
au quatrième est un de ceux dont on n'a pas encore pu venir à bout, quoique d'ailleurs rien n'en
démontre l'impossibilité. Elsewhere (p. 357) he pointed out: "...si la résolution algébrique des
équations des degrés supérieurs au quatrième n'est pas impossible, elle doit dépendre de quelques
fonctions des racines, différentes de la précédente." At the end of the whole book he wrote (p. 403)
that the solution of an equation of the 5th and higher degrees would only follow from a certain kind of
"calcul des combinaisons", but "le succès est encore d'ailleurs fort douteux" ... "nous espérons cependant pouvoir y revenir dans un autre temps, et nous nous contenterons ici d'avoir posé les fondements
d'une théorie qui nous paraît nouvelle et générale".

[13] It is interesting that Lagrange criticizes Bézout, for example, elsewhere in his treatise, for
making similarly unfounded general statements.

[14] In the introduction to his study he considers already the theorems regarding complex roots,
inclusive of their form and number, to be significant successes of the algebra of the recent period.

It is certain that in between lines, as was realized by mathematicians much later, one could have found a differentiation of various algebraic fields. However, they were not differentiated explicitly by Lagrange and their differentiation was not exploited. The realm of the coefficients of a given equation was not defined more closely; they represent given, one may even say constant, quantities because the generality of the equation is understood in the sense that one does not assume any special relation between the roots and, therefore, neither between the coefficients, which would facilitate the solution of the equation. The rational (or entire rational) function of the roots, or of the coefficients, as well as the values which these functions attain, are stressed. Whether the initial elements are rational or algebraic numbers, Lagrange does not say; if one analyzes their nature today, it follows from the context that the elements created form an algebraic field, defined by the adjunction of the roots of the given equation or of the binomial equation to the field of rational numbers.[15] This opens up a large number of problems, the solutions of which Lagrange did not require, which, however, turned out to be fruitful in the course of further development. The focus on the study of the variety of realms by Lagrange is suppressed by the main problem he considers, i.e. the number of different values of the entire rational function of the roots of a given equation, considering all the permutations of these roots.

The fact is then decisive that Lagrange in fact had no interest in the realm to which the roots of the equation belonged, nor did he consider the relation between this realm and the numerical realm of the coefficients. Like Euler, Bézout and others, he considers the irrational expressions and the forms of the roots, but he considers the analysis of these expressions as included in the study of the resolvent.[16]

The coefficients of the resolvent are rational functions of the coefficients of a given equation. This conclusion, known earlier, is reached by studying the symmetric functions of the roots. However, as regards the properties of the roots Lagrange again states nothing more accurate. One is informed of their irrationality in analyzing the solutions of 3rd and 4th degree equations. Many more facts about the nature of the realm, among the elements of which the roots of the given equation belong, can be found from the study of the degree of the resolvent. Lagrange pointed out several times that "la résolution d'une équation du troisième degré est, à proprement parler, la résolution d'une équation du sixième degré" (p. 210), which leads to the solution of an equation of the 12th (p. 261), or even 24th degree (p. 288), in the case of a 4th degree equation. However, as regards Lagrange this does not mean anything else but that the solution of the given equation follows from the solution of the resolvent (the resolvent is defined in this way by himself), and he says nothing directly about the relation of the realms of the roots of both equations. The only indicator of the knowledge of this relation in his text is represented by numerical calculations from which it follows that in some cases the irrationalities are not of the same type.

---

[15] Cases even occur where the expressions created have been fractional rational functions.

[16] He, of course, differentiates between commensurable and incommensurable quantities (e.g., p. 405).

All this then yields the conviction that Lagrange prepared the ground for considerations of the realms of the coefficients or roots, but he does not mention them explicitly in his own work.

However Lagrange's considerable contribution is in the method which follows from his procedure a priori and which makes it possible to reduce the degree of the resolvent; it is in the study of the properties of entire rational functions of the roots, considering the permutations of these roots. The consequences of this contribution may be found in two areas, i.e. particular attention paid to special relations between the roots, and the beginnings of the study of permutations.

The first of these consequences Lagrange considers marginally, at the conclusion of the whole treatise (pp. 403–421). Of course, this is a question of the knowledge of the relations between the roots, as he says, a priori; these relations between the roots are either given by the form of the equation itself, or by the nature of the problem the equation solves. As is frequently the case in his papers, Lagrange studies this problem from the most general point of view. Therefore, in the introduction he asks what the consequences for the solution of an equation of the n-th degree,

$$x^n + a_1 x^{n-1} + a_2 x^{n-2} + ... + a_n = 0 \qquad (3.2.6)$$

with the roots x', x'', x''', ..., $x^{(n)}$,
can the relation between its roots (p. 404n)

$$x', x'', ..., x^{(\lambda)}, \quad \lambda < n$$

have, and he considers several possibilities. The result, however, is the determination of the degree of the auxiliary equation or equations, because the relation between the roots is always finally expressed by an algebraic equation. For example, let the relations between λ roots be given by an algebraic equation of degree λ. Its roots are also functions of λ roots of the original equation of degree n. Lagrange then asks about the possible relations of these λ roots with respect to the remaining roots of the original equation. For this purpose he uses permutations of the λ roots (or he substitutes them by other groups of λ roots of the original equation). Drawing on the properties of the functions of the λ roots, he derived the various conditions of solvability of the original equation and proved the commensurability (i. e. the rationality), or the incommensurability of the roots of the equations under the said conditions. For example, if only the roots x' and x'' are interchangeable, they depend on a second-degree equation:

$$x^2 - ax + b = 0, \qquad (3.2.7)$$

in which, as he says, a and b are commensurable. In this way he carries on considering the interchangeability of more than two roots of the roots considered. He then goes on to consider the interchangeability of the considered roots x', x'', ..., $x^{(\lambda)}$ with other roots $x^{(\lambda+1)}$, ..., $x^{(n)}$ and again arrives at auxiliary equations.

Inspite of the fact that Lagrange only outlined this procedure, it could have had a stimulating effect. He himself explained it in particular in the example of reciprocal equations, where he employed the quadratic equation

$$x^2 - a_i x + 1 = 0, \tag{3.2.8}$$

and studied the properties of the coefficient a_i.[17] Lagrange then goes on to give examples without further general study of the solution of equations under given relations among roots.

A further considerable contribution is the beginning of the research into the theory of permutations. One cannot, of course, say that one would not have found elements of the theory of permutations, or its exploitation in the theory of algebraic equations prior to Lagrange. However, Lagrange pointed out its fundamental significange for the further progress of algebra as a science of the solution of equations and began to study it in the sense that we encounter it in nearly the following hundred years. Let us repeat Lagrange's formulation of the problem. To each algebraic function

$$x^n + a_1 x^{n-1} + \ldots + a_n = 0, \tag{3.2.9}$$

with the roots

$$x_1, \ldots x_n,$$

it is possible to attribute an entire rational function of the roots

$$f(x_1, \ldots, x_n),$$

which, considering the permutations of the roots, in general has $n!$ values

$$f_1, f_2, \ldots, f_{n!},$$

These values are the roots of the resolvent

$$(t - f_1)(t - f_2) \ldots (t - f_{n!}) = 0. \tag{3.2.10}$$

The degree of the resolvent (3.2.10) can be decreased if a function f can be found which will display less different values when the roots are permutated. In this cnnection Lagrange established an important theorem which states that the number of values will always be a divisor of the number $n!$. Langrange[18] went on to investigate what these values could be, but he made little progress. Nevertheless, he introduced certain important concepts, which later played an important role in elaborating the theory of permutations. For example, Lagrange defined the concept of similar functions, i.e.

---

[17] If the reciprocal equation is of degree n, a_i are the roots of an equation of degree $\dfrac{n}{2}$ or $\dfrac{n+1}{2}$, depending on whether n is even or odd, respectively.

[18] The following interpretation of Lagrange's theorem in the theory of groups is usual: The order of the subgroup divides the order of the group. However, the terminology and context of Lagrange are quite different. Lagrange studied these problems in particular on pp. 373—4. Cf. below Chap. 7, (2).

those which change their values for the same permutations and also do not change their value together. He even puts this concept at the head of Section 4 of his work. Mention has been made above of Lagrange's study of the possible relations among roots; it is only a step from his deliberations to the concept of transitivity and resolution to irreducible systems. However, the centre of interest remains Lagrange's functions mentioned above; the concept of substitution (permutation) he considers to be intuitively sufficiently clear, so that he avoids interpreting it. He also leaves untouched the important concept of composition of permutations.

Like many other mathematicians of the 18th century Lagrange also devotes considerable attention to the n-th roots of unity. At this point it is necessary to differentiate among several problems.[19] In particular, he considers the solvability of the equation

$$x^n - 1 = 0 \qquad (n \text{ odd number}) \tag{3.2.11}$$

in the same way as other contemporary authors. Apart from the root $x = 1$, the other roots are represented by the roots of the equation

$$x^{n-1} + x^{n+2} + x^{n-3} + \ldots + 1 = 0, \tag{3.2.12}$$

which he solves as a reciprocal equation. He thus arrives at the solution of an equation[20] of degree $\dfrac{n-1}{2}$. He then points out the connection of the solution of this equation with circle graduation, referring to Moivre's papers.[21] Lagrange also considered the case wherein n was not a prime number and showed that Eq. (3.2.11) was solvable for

$$n = 2^\lambda \cdot 3^\mu \cdot 5^\nu \cdot 7^\mu, \tag{3.2 13}$$

(where λ, μ, ν, u are natural numbers). One of the parts of Lagrange's interpretation is the analysis of the relations between the roots of Eq. (3.2.11). Lagrange summarized the results which could then be found in the work of other mathematicians, especially Euler. He proved that all roots could be expressed as powers of a certain root $\alpha \neq 1$, i.e.

$$\alpha, \alpha^2, \alpha^3, \ldots, \alpha^n = 1. \tag{3.2.14}$$

As regards the equation for $n = 5$ one can adopt as the base (pp. 250−251) instead of α for which

$$\alpha, \alpha^2, \alpha^3, \alpha^4, \ \alpha^5, \tag{3.2.15}$$

---

[19] He considered Eq. (3.2.11) at several points. Its coherent interpretation can be found at the end of the first section (pp. 243—254) and in several places of the third section (e.g., pp. 309, 317).

[20] He maintains that the solution of the equation

$$x^n - 1 = 0$$

for $n = 11$ (p. 246) "demanderait la résolution d'une équation du cinquième degré".

[21] Although he mentioned that a solution was possible using trigonometric functions, he promised on p. 247 that he would present a priori reasons later, for which an extensive reduction of Eq. (3.2.12) was possible. In studying Eq. (3.2.12) he said directly (p. 246, 309) that it would lead to "la division de la circonférence du cercle".

e.g., the root α^2, so that the roots become

$$\alpha^2, \alpha^4, \alpha^6, \alpha^8, \alpha^{10}, \tag{3.2.16}$$

and, as Lagrange says, because

$$\alpha^5 = 1 \tag{3.2.17}$$

these roots represent

$$\alpha^2, \alpha^4, \alpha, \alpha^3, \alpha^5, \tag{3.2.18}$$

which, except for the order, is the same as (3.2.15). The same may be said for any of the roots (3.2.15) with the exception of α^5. Lagrange goes on to generalize the result for any arbitrary prime n. If m is natural, $m < n$,

$$\alpha^m, \alpha^{2m}, \alpha^{3m}, \ldots \tag{3.2.19}$$

will again yield, although in a different order, the powers

$$\alpha, \alpha^2, \alpha^3, \ldots, \alpha^n = 1. \tag{3.2.14}$$

In this connection he derived an interesting fact. Provided $(251-252)$ n is not a prime number, one may consider an arbitrary natural number m, such that $m < n$, m being relatively prime with respect to n. However, if it is not relatively prime, i.e. $(m, n) = l > 1$, one may also adopt the l-th powers of the root α,

$$\alpha^l, \alpha^{2l}, \alpha^{3l}, \ldots, \alpha^n, \tag{3.2.20}$$

each power in this order being considered exactly $\frac{n}{l}$ times. These roots represent the roots of the equation

$$x^f - 1 = 0 \tag{3.2.21}$$

under the assumption that

$$l \cdot f = n. \tag{3.2.22}$$

These considerations, which are discussed in detail in section three, were certainly suggested by Euler's results which the author refers to (p. 320).

As in other cases, it would be possible to give in even greater detail the various parts of Lagrange's work, which represented an incentive for further development.[22] However, these would be outside the main trend of the evolution of algebra. For example, the study of elimination, to which Tschirnhaus' transformation leads, was also the

---

[22] Lagrange's ideas were preserved, as far as possible, in their original state, using the original terminology and excluding modernization. Nevertheless, it was only possible to present fragments of his extensive study which, in spite of everything, means modernization. Lagrange's interpretation is based on the contemporary state of the problems and therefore the features mentioned in the text rather represent the exceptional, the principal interest being in detailed analyses of the methods of solutions known at the time, in which Lagrange proceeds with great detail and unnecessarily verbosely.

subject of many other contemporary papers as indicated by Bézout's work, but also by the fertile work of the algebraists of the 19th century. In spite of the new features and inventiveness of Lagrange's work, which was displayed in the overall approach, as well as in some of the partial problems, his interpretation rests with concepts current at the time which he has no particular need of explaining. As we have seen, he does not feel it his duty to investigate the numerical realms in greater detail. His slight scepticism as to the solvability of 5th order equations and of equations of higher degrees is in no way different from the contemporary, dominating attitude; Lagrange himself did not attempt to prove the insolvability.

Distrust of the capacity of the methods used hitherto for solving equations also provides the initial momentum in Vandermonde's treatise [383]. He did not reach this conclusion by way of an extensive analysis like Lagrange, but in the introduction to his paper he writes that the difficulties one encounters in applying these methods to equations of higher degrees are frequently in the "nature of the analytical methods used, themselves". For this reason he adopted a different approach; in spite of this, he arrived at results similar to Lagrange's in many cases. The differences between them may be attributed more to the difference in approach, method and formulation, than to the substance itself. Lagrange, in his endeavour to solve the given problems a priori, stressed the algebraic, formal and general approach, in which these three attributes of approach merge; they apparently represented the principal motives for pushing the arithmetic aspect and, therefore, also the study of the numerical realms, into the background. In contrast, the arithmetic aspect, indulging in nature and properties of the root of the equation, is outstanding in Vandermonde. His first paragraph has been quoted already in which Vandermonde establishes the prerequisite that "les valeurs les plus simples, qui puissent satisfaire concurrement à une équation d'un degré déterminé".[23] He then demonstrates, in particular in the analysis of the solution of the quadratic and cubic equations, that he is searching for a way to express all the roots of a given equation of the n-th degree with the help of its coefficients, which he considers to be constant, and of the n-th roots of unity. He does not mention the nature of this expression, whether it is rational or algebraic. As the coefficients are symmetric functions of the roots, he can express "a substantial condition of the general solution", i.e. a kind of "metaphysics of the solution", in three points:

1. Find the function of the roots, of which it can be said that it is equal to each of these roots according to the meaning which is given to the function.[24]

2. This function should be reduced to a form which would not change when the roots change.

3. Finally, express this function as a function of elementary symmetric functions of the roots, to use modern terminology, i.e. coefficients of the equation. By satisfying these

---

[23] Although Vandermonde did not explicitly say where this would lead him, one may assume that this was a further attempt at elaborating a method of solving equations of higher degrees.

[24] At this point he had in mind the different meaning of the n-th roots which may occur in the expression, thus yielding n different values.

three points, the latter considered to be the simplest, one arrives at the solution of the equation according to Vandermonde.

Vandermonde's prerequisites for determining a suitable form of the root, which reflect the approaches of Euler and Bézout to whom the author refers, are too general and too indeterminate without Vandermonde's own analysis of the various cases.[25] His first approach to satisfying the three points mentioned is no longer general but only illustrated by a kind of incomplete induction. As regards an equation of the n- th degree the roots of which are

$$x_1, x_2, \ldots, x_n, \tag{3.2.23}$$

assuming that the equation

$$x^n - 1 = 0 \tag{3.2.24}$$

has the roots

$$r_1, r_2, \ldots, r_n, \tag{3.2.25}$$

the first prerequisite is satisfied by the expression

$$\frac{1}{n}\left((x_1 + x_2 + x_3 + \ldots + x_n + \sqrt[n]{(x_1 + r_1 x_2 + r_2 x_3 + \ldots + r_{n-1} x_n)^n}\right)$$
$$+ \sqrt[n]{((x_1 + r_1^2 x_2 + \ldots + r_{n-1}^2 x_n)^n)} + \ldots + \sqrt[n]{((x_1 + r^{n-1} x_2 + r^{n-1} x_3 + \ldots +}$$
$$+ r_{n-1}^{n-1} x_n)^n) \tag{3.2.26}$$

which attains exactly n values. He then goes on to show that it is possible to find a satisfactory function for $n = 12$ which will only contain 10 square roots, etc. In this way he changes and reduces the problem of solvability to finding a function which has certain properties when its "elements" are permutated; it is in fact possible to reduce expression (3.2.26) in this way. In this manner (Section XXVIIInn) he also arrives at the auxiliary equations which he calls resolvents.[26] As regards 5th degree equations he again arrives at a resolvent of the 24th degree for which he only indicates the possibilities of fruther treatment; the latter (Section XXXIV) leads up to a resolvent of the 6th degree. As with equations of higher degrees he determined the degrees of the resolvent in some cases. An analysis of these cases permits him to draw the conclusion that he is incapable of finding a function of the roots of a 5th degree equation which would lead up to an equation of the 3rd or 4th degree.[27] He adds that he is convinced that no such function

---

[25] However, it would be excessive modernization if one were to see more than indications of the construction of the elements of a field in which all roots of a given equation exist, in Vandermonde's endeavour. Nevertheless, these indications, although in a more obscure form, may be observed with Vandermonde's contemporaries.

[26] He refers to equations ([383] Section XXVIII] which other authors call resolvents or reduite.

[27] Vandermonde's terminology is reversed; he found no equation of the 3rd or 4th degree the roots of which would be a function of the roots of the 5th degree, i.e., a function having the 5 gi ven values. This fact is being pointed out to indicate how specific Vandermonde's approach was· It would probably be possible to prove that the conceptual aspect was the source of Vandermonde's difficulties in many places, in particular as regards the expressing of new ideas using old and current concepts. This is manifest, e.g., in the transition from his deliberations to determining the degree of the resolvent.

exists, although he was capable of finding a function which would yield a quadratic equation. Nevertheless, a few pages further on, at the very end of his treatise, he is very optimistic about his results. He concludes that he was always capable of satisfying the first and third of the three conditions he presented for the general solution of equations in the introduction, whereas with the second he only reached a stage in the solution procedure which presented difficulties due to unavoidable length. Therefore, Vandermonde's conclusion nearly represents the same point of view as Euler and Bézout had earlier, although he had had a different procedure in mind. Vandermonde's words (cf. Section XX) also indicate that he was convinced of the existence of a general expression for all roots of any algebraic equation of any arbitrary degree n, of an expression which would include, at the most, roots of the n-th degree and n-th roots of one. Therefore, he assumed the existence of an algebraic solution in this sense.

However, special attention should be devoted to certain sections of Vandermonde's treatise. He was aware that one could avoid general formulae in solving equations when the relations between the individual roots of the equation were known. This applies to equations like

$$x^n - 1 = 0 \qquad\qquad (3.2.24)$$

He devoted special attention to these equations and maintained (Section XI) that there existed such relations between the roots of these equations in general that it was always possible to derive the values of all roots of one from the system of relations. This statement indicates that he anticipated Gauss' later result. But let us mention some of the principal features of Vandermonde's approach, which could have lead him to make this statement,[28] without discussing the problem of the relation of his study to Gauss'. It should be pointed out that Vandermonde's interpretation is not clear and systematic and that the treatment he used, to arrive at the general statements is only outlined briefly. As regards Eq. (3.2.24) he differentiates between the cases of odd prime numbers n, i.e.

$$n = 2m + 1$$

and cases when n is not a prime number. In the former cases he maintained that the relation

$$x^n - 1 = x^{2m+1} - 1 = (x - 1)(x^2 + x'x + 1)(x^2 + x''x + 1) \ldots \quad (3.2.27)$$

holds and that x' and x'' are roots of the equation

$$x^m - x^{m-1} + (m-1)x^{m-3} - \frac{(m-3)(m-4)}{1.2} x^{m-5} + \ldots = 0 \qquad (3.2.28)$$

---

[28] This question was studied particularly by A. Loewy [277] who drew the conclusion that Vandermonde achieved profound results, similar to Gauss', but less accurate and lucid. Moreover, he maintains that Vandermonde left some of the problems unsolved and some of the statements unproved. Loewy's analysis is reliable and it also contains an interpretation of Vandermonde's treatment of modern algebraic terminology together with an analysis and critique of the deficiencies.

which is easy to determine. If, on the other hand, n is not a prime number, it is even easier to simplify the procedure. Vandermonde does not however, prove the statements he has made but gives examples which indicate the possibility of complete induction.

In a similar way he indicates the relations among the n roots of equations (3.2.24) which are being considered. If $r^i, r^{ii}, \ldots$ are the roots of Eq. (3.2.24), Vandermonde gives the following relations valid for them (Section XI):

$$
\begin{aligned}
&1 + r^i + r^{ii} + \ldots = 0; \\
&r^i r^{ii} = 1; \qquad r^{iii} r^{iv} = 1; \ldots \\
&r^{ii} r^{iii} = r^i; \qquad r^{iv} r^v = r^i, \ldots
\end{aligned}
\tag{3.2.29}
$$

He also says that all the other relations follow from the latter, some of which[29] he used in special cases for $n = 3, 4, 6, 7$, and 11. However, he never stated explicitly (although it follows from his relations) that all the roots of the equation are powers of a (primitive) root.[30]

An important result in this part, which replaces a general statement (owing to Vandermonde's incomplete induction), is the algebraic solution of the equation

$$
x^{11} - 1 = 0,
\tag{3.2.30}
$$

whereas Lagrange[31] stops at the equation of the 5th degree, the solution of which is required in this case, i.e.,

$$
x^5 - x^4 - 4x^3 + 3x^2 + 3x - 1 = 0.
\tag{3.2.31}
$$

The second part of Vandermonde's study is his investigation of permutations. He based this on the study of symmetric functions and on their expression by means of elementary symmetric functions.[32] The terminology he used in studying permutations is subject to the requirements of solving equations. In fact, he studied the functions of the roots but introduced different types of these functions. However, the symmetric

---

[29] For example, he manitained (without deriving, which is easy) that

$$
r^{i2} = r^{iii}, \qquad r^{ii2} = r^{iv} \qquad \text{etc.,}
$$

or

$$
r^i r^{iii} = r^v, \qquad r^i r^{iv} = r^{ii}, \qquad \text{etc.,}
$$

for $n = 7$ (Section X), whereas for $n = 5$ he derived that

$$
r^i r^{iii} = r^{ii}, \qquad r^i r^{iv} = r^{iii}, \qquad \text{etc.,}
$$

or

$$
r^{i2} = r^{iv}, \qquad r^{ii2} = r^{iii}, \qquad \text{atc.,}
$$

which is connected with the choice of roots which is not always the same with Vandermonde.

[30] Vandermonde, without explicitly saying so, sometimes assumes the primitivity of the roots.

[31] Lagrange expressed himself in the sense that Eq. (3.2.31) should be understood as a general algebraic equation of the 5th degree which will be solvable only after a general solution is available.

[32] He did not use this term; his problem was expressing the symmetric functions of the roots by the coefficients of the equation.

functions he considers to be initial types, he "composes" of "partial types". The partial types introduced are really functions of which the values do not change with respect to a given cyclic permutation, but are formed by the given cyclic permutation at the same time. Without discussing the details and meaning of the concepts Vandermonde mentioned and introduced in a comparatively difficult way, this should be sufficient to outline the approach which brought him to the idea of cyclic permutation. He used Roman numerals placed under the permutated symbols, so that

$$(\alpha_{II}\beta_{III}\gamma_{I})$$

indicates three subsequent permutations:

$$\alpha\beta\gamma; \qquad \gamma\alpha\beta; \qquad \beta\gamma\alpha.$$

Using these symbols[33] Vandermonde introduced the composition of permutations of cyclic compositions and again the resolution of cyclic permutations into "smaller" cycles. He actually speaks (Section XXVI) of the product of two "partial types". One may therefore say that he introduced two different operations for his partial types which he also denoted as addition and subtraction. It follows that Vandermonde, even though he had not arrived at theorems of such importance in the theory of permutations or at such a clear and wholesome interpretation as Lagrange, nevertheless stressed the study of permutations in relation to the study of the solution of algebraic equations, introduced certain methods and concepts, as well as symbolic means which played an important role in the development of the theory of permutations later on.

The fact has been mentioned that Vandermonde stressed the arithmetical aspect of solving equations more than Lagrange. This is substantiated by his search for a general expression of the root and by his study of the relations among various expressions. However, he did not go further than the terms current at the time, at best speaking of rational functions (expressions) and of irrationalities. Although he had not reached more substantial results, he was clearly interested in the variety of quantity types. This is substantiated by the topic of his treatise [385b] which was included in the papers of the Paris Academy for 1772 and which, as indicated by its heading, was concerned with "irrational quantities of various orders inclusive of their applications to a circle". However, the means he chose for dealing with this topic had no connection with the algebraic problems being solved at the time. He used special products, in some cases even infinite ones.[34] To conclude, he wrote that the

---

[33] Only the nucleus of Vandermonde's system of symbols has been mentioned in the text; in reality it is much more complicated and constantly connected with the concept of a function of n letters which are permutated.

[34] He considered as the basis the product of elements the first difference of which is equal to unity, i.e., $p(p-1)(p-2)\ldots(p-n+1)$, denoted here by $[p]^n$. This symbol, originally defined for n ($n < p$) natural, is gradually generalized for n being an integer and later also rational. For this purpose he employed the principle which he himself called the principle of continuity (!). He applies the validity and operations of multiplication (and division) to the generalized symbol $[p]^n$.

purpose of the whole study was to show that roots were only a special kind of "irrational" quantity. Therefore, one may get the impression that the impulse for writing the treatise was perhaps a slightly obscure conviction of the particular nature of algebraic quantities with respect to real quantities as a whole, which was no longer anything exceptional at the time.

The third mathematician whom we want to consider here, Eduard Waring, also founded his study on analysis of the work of his predecessors and contemporaries. His studies are abundant in content and have so far been underestimated.[35] His interpretations are not as purposeful as Lagrange's and Vandermonde's. He studied the problems of solving equations from various aspects but he considered his main objective to be the determination of a suitable resolvent. Waring at first determined the degree of the resolvent by elimination. He asked what degree equation could be used to reduce an equation of degree n to an equation of degree m, $n > m$. He reached the conclusion that this would be possible by means of an equation of degree

$$n \, \frac{n-1}{2} \, \frac{n-2}{3} \ldots \frac{n-m+m}{m}.$$

Elsewhere the problem of determining the resolvent led him to study the functions of the roots. He used a concept identical to Lagrange's resolvent and explicitly stated that solution of an equation of degree $n!$ is, in general, required to solve any equation of degree n. Thus he became convinced (in *Scholio* on p. 186) that one would search in vain for the roots of equations of higher degrees using these methods.[36] He tended towards the opinion, which he claimed to be Viète's and which is now proclaimed most often, that one should rely on approximate methods which would yield values closest to the roots.

However, let us revert to the aspects of Waring's study which could have provided incentive for the further evolution of algebra. One of these is the attention Waring paid to the form of the roots and to the relations among the roots. In the same way as other, older authors, he sought simple means (substitution) to adjust roots suitably, e.g., to reduce the given equation to an equation the roots of which would represent powers of the original ones. This, as it were, could eliminate the unpleasant irrationality of the roots. Waring studied the relations among the roots in various forms.

---

[35] This fact had already been presented by Burkhardt ([51] 124, note). Several facts were a hindrance to a greater response to Waring's studies; among others, certainly the incomprehensiveness of the topic, which obscured new ideas in a multitude of theorems and partial examples. Monographs of this type were also more difficult to propagate and were read less than papers published in proceedings of leading academies, especially if the monographs were published in England. The use of Latin, which could have discouraged mathematicians, probably had an unfavourable effect on the historical response.

[36] He had already written in this connection in 1762 ([391], 34): Hac vero methodo aequatio reducens plures habet dimensiones quam data aequatio; frustra esset igitur, per talem reductionem aequationum resolutionem generalem quaerere. At this point, in order to support his opinion, he refers to Newton's *Arithmetica Universalis*.

For example, he considered the question of solving an equation whose roots form an arithmetic sequence, and in the subsequent proposition he considered an analogy of this problem: an equation whose roots form a geometric sequence. In both cases he avoided generalization. As far as Waring is concerned, these problems are only connected indirectly with reciprocal equations for which he proved that a reciprocal equation of degree $2n$ could be solved as an equation of degree n. In his paper of 1762 he had already connected this possibility with the solution of the equation

$$x^n \pm 1 = 0 \tag{3.2.32}$$

for $n = 4$, 5 and 6. Later on he analysed the solvability of Eq. (3.2.32) further, introduced the term primitive root, and expressed all roots in terms of its powers. If n is if the form

$$n = m . r . s . \ldots \tag{3.2.33}$$

where $m, r, s, \ldots$ are prime numbers, Eq. (3.2.32) is reduced to

$$x^m - 1 = 0, \qquad x^r - 1 = 0, \ldots . \tag{3.2.34}$$

However, he did not mention the solvability of Eq. (3.2.32) for any arbitrary n, nor did he discuss the case for $n = 11$.

Waring devoted a good deal of attention to the arithmetical aspect of these problems. He studied various irrationalities. For example, in connection with Eq. (3.2.32) he also studied the roots

$$\sqrt[n]{P}$$

(from the context it follows that P is rational), where he arrived at expressions like

$$\sqrt[m]{(a + \sqrt[r]{(b + \sqrt[s]{(c + \ldots)))}},$$

for n defined by Eq. (3.2.33). There is a whole series of similar cases. They, however, tend to indicate adroitness in calculating but don't lead to more general statements about the various regions of irrationalities, or to a more profound study of the solvability of equations.[37] Nevertheless, Waring was capable of exploiting the maximum which could be achieved in this respect at the time. The arithmetical aspect prompted him to study the mutual relation of polynomials and their divisibility. The polynomial, as a kind of expression, he considered a quantity with which he did not operate quite freely but with which, nevertheless, he calculated. He based his study on Descartes' solution of biquadratic equations or on Newton's search for the divisors of polynomials. However, he only refers very indistinctly to the equation whose given equation is the divisor ([391] 48).

---

[37] One can only propound as a hypothesis that realizing that no progress could be made in the study of irrationalities under the contemporary state of knowledge discouraged the other authors from these deliberations.

In spite of Waring also studying the functions of the roots (especially the symmetric ones) the elements of his study are a much more elementary permutation than Vandermonde's and Lagrange's. Newton's tradition may be seen in the special stress put on the study of the relation between the sum of the roots and the sum of their second, third, etc. powers. Nevertheless, one may observe Waring's fortitude in this respect. For example, he created complicated symmetric functions for the sole purpose of demonstrating the method of expressing them by elementary symmetric functions or sums of the powers of the roots.

This outline of the features of Waring's algebraic study[38] should be sufficiently indicative of the abundance of possible stimuli for further evolution. One could mention others such as the beginnings of the study of the resultant, etc. In the same way the special cases investigated at various times, which make it possible to reach a solution of an algebraic equation of the given type, should be included here. All this forms an extensive study, abundant in ideas, none of which towered high above the achieved standard or drew exceptional attention. In comparison with the work of Lagrange and Vandermonde, for example, even the fact that Waring arrived at the correct determination of resolvents, that he found suitable functions of the roots, etc., appears to constitute no exceptional merit.

This has been an attempt to show how three authors, working in three different countries around the year 1770, proceeded from different viewpoints to reach a new approach to the problems of the theory of equations outstanding at the time.[39] All of them formulated the problems and results with a view to the contemporary state of mathematics and began with an analysis of existing methods of solving equations. The most expressive requirement for a new approach was formulated by Lagrange in his bid for an analysis of the methods used a priori. However, new features could also be observed in the approaches and results of Vandermonde and partly also in Waring. But the terminology the authors used did not change, only a few terms appeared or obtained new significance. Therefore, their studies could be understood well by their contemporaries. However, there was a problem as to what enlightenment they would derive from Lagrange's, Vandermonde's and Waring's studies for their future work and a question whether they could understand the more obscure aspects of their deliberations which signalled further progress.

---

[38] The fact that he included a number of theorems pertaining to the theory of numbers in his *Meditationes algebraicae*, corresponds to the style of Waring's approach to the understanding of algebra and to the extent of the problems investigated.

[39] This trio of mathematicians could be supplemented by the Italian Malfatti and his study [279] of 1771. Using a method of elimination derived from Euler and Bézout, he arrived by calculation of a 5th degree equation at a resolvent of the 6th degree. More detailed information on this study, which had no profound historical effect, was presented by Pierpont [332] p. 33—36. For an outline of the contents as well as for a further historical evaluation, including references to literature, see [55b] 116—117.

3.3 Gauss and Ruffini

All the algebraic studies of the trio of mathematicians who were discussed in the previous section did not affect the evolution of algebra in the same way. Most incentive was provided by Lagrange's work, Lagrange's considerable scientific prestige[1] possibly playing a role in addition to the scientific qualities of his work. Nevertheless, there were very few mathematicians who realized the possibilities these three provided in the half century after they had been published. Even attempts at a mere partial exploitation of these possibilities were an exception. The first to surpass the results published in the years 1770–1772 were K. F. Gauss and P. Ruffini at the very end of the 18th century, i.e., with a delay of more than 25 years. It may be said that these two mathematicians proceeded in two different directions which, however, were based on Lagrange's work to the same extent (as well as on that of others). Whereas Ruffini's name was connected with the proof of the algebraic insolvability of the general algebraic equation of an order higher than four, Gauss deserves credit for his exhaustive analysis of one of the cases in the class of solvable equations. Both these approaches are complementary to each other. In their historical consequences they lead up to the creation of Galois' theory and in this way contributed to the forming of modern algebra.

Although Gauss' influence on the development of algebra is very considerable, his algebraic work is not very extensive. His first published paper, his thesis of 1799 [160], gives proof of the fundamental theorem of algebra and considers the only a topic which could provide a basis for the discussion of the solvability of equations. It is well known fact that Gauss returned to this topic several times later on. However, in most cases the fundamental theorems of algebra exceed the scope of the methods and concepts of algebra, due to the means used.[2] He expressed his doubts about the general solvability of algebraic equations in the first of these proofs: "...post tot tantorum geometrorum labores perexiguam spem superesse, ad resolutionem generalem aequationum algebraicarum unquam parveniendi, ita ut magis magisque verisimile fiat, talem resolutionem omnino esse impossibile et contradictoriam...". To this Gauss added: "Forsan non ita difficile foret, impossibilitatem iam pro quinto gradu omni rigore demonstrare, de qua re alio loco disquisitiones meas fusius

---

[1] The dominating influence of Lagrange's work in comparison to his contemporaries is generally acknowledged. Burkhardt ([51] 131—2) characterized the role of all three clearly: Wenn in späteren Zeiten die Tradition... wesentlich an Lagrange's grosse Abhandlung anknüpft, so verdankt sie das wohl vor allem ihren formellen Vorzügen: der Ausführlichkeit u. Bestimmtheit in der Angabe der leitenden Ideen, der Übersichtlichkeit der Anordnung, der Durchsichtigkeit — am Massstabe der Zeit gemessen — Strenge der Beweisführung. Sieht man nur auf den Inhalt, so wird man Waring und Vandermonde das Zeugnis nicht versagen dürfen, dass auch sie unabhängig von Lagrange und gleichzeitig mit ihm einen grossen Teil seiner Resultate erhalten haben.

[2] Gauss' second proof attempts to avoid this overlapping and to use algebraic means exclusively. As regards the historical inventiveness of this proof, which will be discussed later on, compare Bashmakova's paper [20].

proponam". These words are more than a mere expression of Gauss' doubts concerning the general solvability of algebraic equations; they are also one of Gauss' statements of the extensive results which h e had in mind but which he later failed to publish, nor was a manuscript found in Ga u ss' inheritance with this content.

Nevertheless, Gauss was considering the solvability of equations at this time,[3] and he included at least some of his results in the seventh part to his *Disquisitiones arithmeticae* (1801 [161]). He also used nearly the same words here as two years earlier (Art. 359) to voice his conviction that a general solution of an equation of a higher degree than four was impossible. However, he immediately pointed out that there must exist an infinite number of classes of equations of an arbitrary degree which were solvable. One of these classes of equations is then discussed in the seventh part of the DA which also contains the proof o f the possibility of this solution.[4]

The principal results of the seventh part are generally known. Gauss proved the solvability of the equation

$$x^n - 1 = 0, \tag{3.3.1}$$

for any arbitrary natural n, the method of solution as well as numerical examples, in particular for $n = 17$ and 19, being sh own. The method of solution yielded a remarkable result according to which it was only possible to construct regular n-sided polygons for

$$n = 2^\mu \cdot p_1 \cdot p_2 \cdot \ldots, \tag{3.3.2}$$

where μ is natural and p_i are prime numbers of the form

$$2^{2^\nu} + 1. \tag{3.3.3}$$

for ν natural.

Without repeating the whole of Gauss' train of thought leading up to these results, an attempt will be made to discuss at least the essential moments in this development. An endeavour will be made not to modernize but to preserve the conceptual system with which Gauss worked.

---

[3] It can be seen from his diary that the results published later in the *Disquisitiones arithmeticae* were ready prior to 1799, cf. Gauss [159] Bd X/1. The Disquisitiones arithmeticae are referred to in the text by the abbreviation DA, apart fom the usual reference, and, instead of the page, the appropriate article, i.e., Art., of the original is given.

[4] In other connections Gauss said that he was not interested in an individual, though very important, problem in mathematics, but in the theory from which one might obtain the solution of the problem (this was in connection with Fermat's theorem). A more detailed investigation of Gauss' work will disclose a consistent application of this idea, which characterizes the difference of his results from those of his predecessors. Therefore, one may assum e that Gauss had the same relation to the problem of insolvability of equations of the 5th and higher degrees, and that he considered the study of the classes of solvable equations scientifically more im portant. However, there is no direct proof of this hypothesis, and he probably never reached a ge neralized opinion (i.e. Galois' theory). The equations studied in the seventh section of the DA, howev er, are also attractive in a different way because their study yielded, among other things, a theory which solves old geometrical problems (e.g., the possibility of constructing regular polygons with a ruler and compass), and which provided these equations with a special characteristic name, Kreisteilungsgleichungen.

After making introductory remarks in which he outlined the various mathematical connections of the problem studied,[5] he goes on to prove that it is sufficient to consider the solution of Eq. (3.3.1) in cases where n is an odd prime number, and that it is possible to reduce all the other equations in this way. He formulated the objective of the whole of the seventh section a little later on. He saw it in resolving of the equation

$$\frac{x^n - 1}{x - 1} = x^{n-1} + x^{n-2} + \dots + 1 = 0 \qquad (3.3.4)$$

into more and more factors in such a way that the coefficients of these factors would be determined by equations whose degree would be as low as possible; in this way it would be possible to arrive at the simplest factors or directly at the roots. Therefore, he wanted to prove that Eq. (3.3.4) could be resolved into α factors of degree $\dfrac{n-1}{\alpha}$ for

$$n - 1 = \alpha . \beta . \gamma . \dots (\alpha, \beta, \gamma \dots \text{ prime numbers}),$$

their coefficients being determined by an equation of degree α; these individual factors can again be resolved into β factors of degree $\dfrac{n-1}{\alpha\beta}$ with the help of equations of degree β, etc.

The objective and method described by Gauss are well understood, although one could use other terms today. It was of essential importance that Gauss did not depart terminologically from the customs of the time. He constructed an equation whose coefficients were "derived" from the roots of another equation, which was a frequently used turn. Neither was the resolution of a given polynomial into a product of polynomials anything unusual. Nevertheless, by combining both procedures Gauss expressed the heart of new ideas in a simple and understandable way. First of all, this contains the idea of adjoining roots of an equation to a numerical realm whose elements are the coefficients. The idea is unthinkable without the idea of a numerical realm, or better still, without the idea of various algebraic numerical realms. However, Gauss, without discussing these ideas or giving the merest indication that he had been considering them, goes on to work simply with terms he had used in his description. In this way, for example, he considered the proof of the irreducibility of Eq. (3.3.4), a differentiation

---

[5] It should be pointed out that Gauss drew attention to the analogy between circle graduation and the graduation of a lemniscate, and that he spoke in this connection about a function depending on the integral

$$\int \frac{dx}{(1 - x^4)}$$

He even mentioned a special extensive paper on this transcendental function he was preparing. Compare ([161], Art. 335).

between integers and rational numbers being sufficient for this purpose.[6] In seeking the resolution of Eq. (3.3.4) and in determining the coefficients of the auxiliary equations he stresses the importance of creating a general method of solution, which he illustrates by means of numerical examples. The generality of his method of solution is on creating a general algorithm. The weakness of contemporary terminology became its virtue; in giving instructions for calculation one does not have to worry about the realm to which the elements, one is using belong. One may say that they are still elements of the realm of complex numbers, and the arithmetical operations required for them have already been introduced.

Gauss was aware that the treatment he used with the class of equations studied and which provided him with the solution, was made possible by the special relation between the roots of these equations. The fact that all the roots of the equations (3.3.1) (let it be assumed that n is still an odd prime number for the sake of simplicity) are powers of any one of them

$$\alpha \neq 1, \quad \text{i.e.,} \quad \alpha^1, \alpha^2, \alpha^3, \ldots, \alpha^{n-1}, \quad \alpha^n = 1, \tag{3.3.6}$$

was known much earlier. This statment can be found[7] in Lagrange's studies, it follows from Vandermonde's relations pertaining to the roots of Eq. (3.3.1) which have been mentioned, it was known to Euler as well as others. In contrast to Lagrange and Vandermonde, Gauss went further and he subjected the relations between the roots to further analysis. In this way he had the possibility of forming various subgroups of the roots of the equations. For example, if $n - 1 = e \cdot f$ it is possible to divide $n - 1$

---

[6] Gauss ([161], Art. 341) in fact proved the following theorem: If the function

$$x^{n-1} + x^{n-2} + \ldots + 1$$

is divisible by a function of a lower degree λ

$$\lambda < (n-1), \quad P \equiv x^\lambda + Ax^{\lambda-1} + Bx^{\lambda-2} + \ldots + Kx + L,$$

all the coefficients $A, B, \ldots, L$ cannot be integers.

[7] Of course, not even Lagrange discovered it; Euler and others studied these properties of the roots of Eq. (3.3.1) earlier. However, a detailed comparison should be interesting showing how Gauss treated the roots of Eq. (3.3.1) relative to predecessors. It seems that many of the facts and operations are the same. Similarly to Gauss, Lagrange also proved that as regards the equation (3.3.1) it was sufficient only to consider those with which n was an odd prime number and to which the solution of the others could be reduced. He only solved the equation (3.3.4) by substitution as a reciprocal equation. There apparently existed a limit for him at which he stopped and of which he was probably not aware. Gauss' relation to Vandermonde's work is more complicated. Loewy showed [277] that Vandermonde's solution of Eq. (3.3.1) was fundamentally identical to Gauss'. Regardless of some of the deficiencies in Vandermonde's proofs, there is a considerable difference in the means the two authors used in interpretation, and in lucidity. Perhaps one might describe Vandermonde's interpretation by saying that the author had the correct solution of the problem in mind, but that he was not capable of expressing the essence of the solution and why it was correct, and therefore his explanations remained obscure. This conclusion may be considered to be hypothetical and its verification, by comparing Vandermonde's and Gauss' texts, would call for complementing Loewy's treatise, merely concerned with the mathematical aspect of the problem, by a historical analysis.

roots different from 1 into e groups of f terms, so that the elements of each group are of the form

$$\alpha, \alpha^e, \alpha^{2e}, \ldots, \alpha^{fe-f}. \qquad (3.3.7)$$

Let us avoid the question which Gauss considered very carefully, i.e., how to achieve the independence of these groups, of the selection of the initial root α, and what the other groups look like. They are formed so that they are disjunctive and they have in common each root of Eq. (3.3.1) just once (apart from the root $\alpha_i = 1$). This immediately prompts one to think of the analogy between Gauss' idea and the resolution of Abelian groups into residual classes. The justification for this analogy is substantiated by several facts; first of all, there is the way Gauss exploited his earlier interpretation of congruence, the way in which he freely forms the representatives of the residual classes, and the way he easily turns from the multiplicative relation between the roots to the additive between the coefficients in order to prove conversely that he could also introduce a multiplicative relation between the coefficients. This treatment is sometimes interpreted as proof of the fact that Gauss was aware of the isomorphism among various groups, inclusive of the isomorphism between the additive and multiplicative records of the group.[8] These arguments, the justification of which will be dealt with later on, are perturbed by some of Gauss' own turns. Gauss did not consider "period" (or "aggregate") to be a mere set of elements but also a sum of roots given by a certain rule.[9] Therefore, on the one hand he maintained that a complex of suitably selected "periods" was identical with a set of all the roots,[10]

---

[8] The elements of this new aspect of Gauss' will be discussed later. At this point we will make do with some data concerning Gauss' progress and symbolics. In order to facilitate printing ([161] Art. 342) Gauss did not write the powers of the root α but only the exponent, i.e., $\alpha^\lambda = [\lambda]$. Clearly

$$[0] = 1, \qquad [\lambda] \cdot [\mu] = [\lambda + \mu], \qquad [\lambda]^\nu = [\lambda\nu];$$

$[\lambda]$ and $[\mu]$ are equal or different provided the relation

$$\lambda = \mu \, (\text{mod. } n)$$

is valid or not valid, respectively. Since the numbers $1, g, g^2, \ldots, g^{n-2}$ (for g primitive mod n) are then congruent (with the exception of the order) to the numbers $1, 2, 3, \ldots, n-1$, the roots $[\lambda]$, $[\lambda g], [\lambda g^2], \ldots, [\lambda g^{n-1}]$ (λ an integer, $(\lambda, n) = 1$) are identical (with the exception of the order) to the roots $[1], [2], \ldots, [n-1]$.

[9] Gauss ([161] Art. 343) defined the period as the sum of f roots given in principle in the way indicated in the text and denoted with regard to the description in the previous comment by

$$[\lambda] + [\lambda h] + [\lambda h^2] + \ldots + [\lambda h^{f-1}] = (f, \lambda).$$

Nevertheless, this "complex" cannot always be substituted (as he wrote in his comment to the definition) by a numerical value. This has been substantiated by the subsequent procedure described in the text.

[10] Gauss ([161] Art. 344) maintains, considering the notations he introduced (compare with the foregoing comments) that a "complex" of e periods

$$(f, \lambda), \qquad (f, \lambda g), \qquad (f, \lambda g^2), \ldots, (f, \lambda g^{e-1})$$

is identical to the set of all roots of equation (3.3.1).

however, at the same time ([161] Art. 345) he introduced the product of two periods (f, λ) and (f, μ). If (apart from other conditions) (f, λ) is composed of the roots $[\lambda]$, $[\lambda']$, $[\lambda'']$, ..., it holds that

$$(f, \lambda)(f, \mu) = (f, \lambda + \mu) + (f, \lambda' + \mu)(f, \lambda'' + \mu) + \cdots.$$

The nature of the period as a sum of the roots defined by a certain rule came clearly to the fore when Gauss investigated the relation between the periods and the coefficients of the equations. With a view to the reasons mentioned, it can perhaps be said that Gauss' concept is not unique in this case, in that one could have understood the division of the roots into groups as a method only leading to the determination of the coefficients of the polynomials which divide the original equation. The indications of a different understanding are obscured, among other factors, by the fact that no reason was given in Gauss' interpretation for requiring a different understanding, although it might be possible that Gauss himself was aware that they existed.

Using the means indicated, Gauss was able to resolve the polynomial

$$x^{n-1} + x^{n-2} + \ldots + 1, \tag{3.3.8}$$

into a product of polynomials of lower degrees, i.e., to proceed according to the method he had explained in determining the objectives of the whole section. Among other things, he showed how Eq. (3.3.1) can be solved by means of four quadratic equations for $n = 17$ and as three equations for $n = 19$ of which two were cubic and one quadratic.[11] He already considered these examples to be a mere illustration of the general theory, all the main points of which he had discussed earlier ([161], Art. 352). Therefore, he was of the opinion that the two following points follow from the theory discussed: If one expresses $n - 1$ as a product of prime numbers, it means that one may be able to arrive at an expression for a polynomial of degree $(n - 1)$ in the form of a product of polynomials of lower degrees; in some cases, however, one is perhaps forced to solve an equation whose degree is a prime number higher than 4. For $n = 11$ Lagrange arrived at an equation of the 5th degree and drew the conlusion that there was no known solution to a general equation of the 5th degree so far. Gauss treated this problem in particular beginning with Art. 359. As mentioned above, he spoke of the existence of an infinite number of solvable equations of arbitrary degree. He then maintained that all auxiliary equations which he had em-

---

[11] He calculated the coefficients of the auxiliary equations for the cases mentioned to 10 decimal places, and indicated that different methods could lead to the objective. He sometimes also uses Newton's means for determining the coefficients in calculating symmetric functions (e.g., the sum of the powers of the roots) thus coming close to the treatment of Waring. He also indicated how to calculate the coeficients of the equations into which the equation

$$x^{18} + x^{17} + \ldots + x + 1 = 0$$

is resolved, i.e., the coefficients of three equations of the 6th degree (Art. 348). He also computed the numerical values of the roots for the examples mentioned.

ployed in the whole section, were always solvable or, as he himself proposed, they could
be reduced to "pure" equations, i.e., equations of the form

$$x^n + A = 0. \tag{3.3.9}$$

He immediately went on to say that the investigation of the whole problem would be too
extensive[12] and, therefore, he would only mention the main points which would be
required for the proof of solvability and that he would postpone the detailed treatment,
which this topic deserves, to a later period.

Needless to say, Gauss never undertook the more detailed treatment, and
therefore one must make do with the subsequent pages of the *Disquisitiones arithmeticae*
which contain a mere outline of procedure for Gauss' proof. Like Vandermonde[13]
earlier, Gauss also restricts himself to examples of the procedure in some cases.
However, the nucleus is the proof of the rationality of certain expressions the sources
of which Gauss pointed out correctly. The problem of algebraic realms came markedly
to the fore here when Gauss (e.g., *DA*, Art. 360) considered entire rational functions
of the n-th root of unity. However, as mentioned above, Gauss took these ideas no
further.[14]

The second point on which Gauss commented in the closing paragraphs is the
Euclidean construction of regular polygons and the algebraic expression of the values
of trigonometric functions equivalent to it. As regards the Euclidean construction,
Gauss was satisfied with saying that it was possible for odd prime numbers only,
as we should say, if they were Fermat's prime numbers, and he pointed out (Art. 365)
that not all numbers of the form

$$2^{2^v} + 1 \quad (v \text{ natural})$$

were prime numbers. He also maintained that if $n - 1$ contained other prime numbers
than 2 as the divisor, one would always arrive at equations of higher degrees. He
also said that he was capable of proving precisely that these equations of higher
degrees could certainly not be eliminated or reduced to equations of lower degrees.

---

[12] The German translation ([162] Art. 359) of Gauss' words is: Nichtsdestoweniger ist es sicher,
dass es unzählig viele gemischte Gleichungen jeden Grades giebt, welche eine solche Zurückführung
auf reine Gleichungen gestatten, und wir hoffen, dass es den Geometern nicht unerwünscht sein wird,
wenn wir zeigen, dass unsere Hilfsgleichungen immer hierher gehören. Wegen des grossen Umfanges
dieser Untersuchung aber geben wir an dieser Stelle nur die Hauptmomente an, welche zum Beweise
der Möglichkeit erforderlich sind, und verschieben eine ausführlichere Behandlung, deren dieser
Gegenstand äusserst wert ist, auf eine andere Zeit.

[13] Cf. Loewy [277] p. 191 nn. The principal data on Gauss' procedure are also contained here.

[14] The fragments found in the inheritance, in particular *Weitere Entwicklung der Untersuchungen
über die reinen Gleichungen* (cf. [163]) in which he directly refers to the problems discussed in the
text, represent a continuation in their way. Gauss reacts here (§ 8) to Lagrange's comments on the
seventh section of the *Disquisitiones arithmeticae*. It should suffice to say that Gauss did not introduce
any fundamentally new methods or terms here, and that these are only partial supplements and
changes pertaining to the seventh section.

Unfortunately, the extent of the study, to quote Gauss, did not allow for this proof to be presented; he only mentioned it to stop anyone hoping that he would also be successful with the Euclidean construction in other cases. The subsequent paragraph brings the *Disquitiones arithmeticae* to an end.

Gauss had his difficulties in concluding his work, the printing of which had then extended into the second year. It was possible that the happy creative years in pure mathematics, of which Gauss later spoke, were already gone, and that his interest had begun to turn to other spheres. The main results connected directly with the discussion of the solvability of the equation (3.3.1) were presented clearly and in a form convincing to his contemporaries; they were based on algorithms generally applicable, carefully complemented by calculated examples. It can be said of many of the sections of Gauss' study devoted to this problem that there was nothing more to add to them, at least from the point of view of the time. The seventh section also contains a program of scientific work either for determining the problems which Gauss was already to have but had not published for various reasons, or for problems which ensue directly from Gauss' work (e.g., the calculation and construction of regular polygons). It was possible to begin immediately with treating these problems. The third group resulting from Gauss' studies contains indications of a more profound and more general theory of algebraic equations, in other words, indications of the trend leading up to Galois' theory.[15]

There is a substantial difference between the studies of Gauss and Ruffini. This difference does not concern the problems to which both tend. They differ in particular in the formulation of the problems and in the manner of interpretation. In contrast to Gauss, Ruffini attempted to present a general theory of equations, considering the main objective of his work to be the proof of the insolvability of equations of a degree higher than the fourth.[16] In this sense he also called his principial study of 1799: *Teoria generale delle Equazioni, in cui si dimostra impossibile la soluzione algebraica delle equazioni generali di grado superiore al quarto* [343]. Ruffini returned to his interpretation several times in the subsequent years in order to fill in the gaps in the proofs or to correct inaccuracies, which were pointed out to him in the papers of Abbati and Malfatti. The discussion concerning Ruffini's studies, especially as regards his "proof", shows how difficult it was to master this fundamental problem. Nevertheless, the important thing in Ruffini's study is that he attempted to present

---

[15] I am of the opinion that these three pages of Gauss' study may be found in many of his mathematical papers. Together with the lucidity of the interpretation, the former two were the focus of the admiration of his contemporaries, the third maintaining the long-term incentiveness of his ideas. Galois' response to Gauss stimuli will be discussed again later.

[16] Burkhardt [51], whose treatise is considered to be reliable, published the analysis and evalution of Ruffini's algebraic studies and his discussions about them with his contemporaries in 1892, and even the most recent literature (e.g., Wussing [410]) only repeats Burkhardt's conclusions. As regards the actual data the reader is referred to this paper of Burkhardt, in particular to the data on the discussion of the proof of the insolvability of equations of higher degrees.

a proof and that he displayed sufficient inventiveness in doing so, to enable him to present it in a generally satisfactory form. However, Ruffini's name is not only connected with this proof of the insolvability of the general equation of a degree higher than four but also with the beginnings of the study of the groups of permutations. Whereas attention will be devoted to the latter in connection with the beginnings of the theory of groups later on, the former will be considered rather as a part of the theory of algebraic equations which Ruffini attempted to present. We shall concentrate on his first paper referred to, because the others only tend to bring partial modifications or changes but do not change the basis of the system of terms, nor the whole structure of its build-up of algebra.[17]

Ruffini's extensive study (more than 500 pages of the original edition and over 300 pages of the collected works) has the credit of being the first involved attempt at expounding algebra, or rather the theory of equations, comprehensively, based on the work of Lagrange and thus also on the complex of ideas which can also be found in the work of Waring and Vandermonde.[18] Ruffini acknowledged the authority of Lagrange fully. At the beginning of the preface to his paper of 1799 he wrote that it was in fact Lagrange who inspired him with the essence of the whole proof of the insolvability of the equations of a higher degree than four.[19] As already mentioned, Lagrange also made the statement that the study of the properties of the functions in connection with the permutations of their "variables" is the "metaphysics" proper of the theory of algebraic equations. That is why the study of these functions is subject to the concentrated interest of Ruffini, who arrives at an originally unsatisfactory proof of the non-existence of a function of five letters, having three or four values ([343] Art. 275). However, Ruffini is not satisfied with the proof of the insolvability of equations of a higher order than four and with presenting theorems absolutely necessary for this purpose, but in an endeavour to present a comprehensive interpretation, he exceeded this scope frequently. One would find very little original in these sections of Ruffini's study, rather only a uniform interpretation attempting to make the most of

---

[17] This point of view has yet another methodical advantage for us; it makes it possible to consider Ruffini's theory of algebraic equations to be contemporaneous with Gauss' study, and quite omit a discussion of the possible effect of the algebraic studies dating from the beginning of the 19th century on Ruffini.

[18] Ruffini does not refer to Waring or Vandermonde but it would be a wonder if he had not known Vandermonde's study which was published in the proceedings of the Paris Academy, in particular as he certainly had contact with the Parisian scientific world, at least later on; compare with the correspondence published in his collected papers [342c].

[19] Ruffini's preface begins with the words: La soluzione algebraica delle equazioni generali di grado superiore al quarto è sempre impossibile. Ecco un teorema troppo importante nelle Matematiche, che io credo, se pur non erro, di poter asserire, e di cui la dimostrazione quella si è, che principalmente mi ha spinto alla publicazione del presente volume.

L'immortale de La Grange... ha somministrato il fondamento alla mia dimostrazione: Conveniva dunque premettera a questa, per la maggiore sua intelligenza, un ristretto di simili riflessioni.

The words reflect the contents of the study in which the problem of insolvability and Lagrange's stimuli dominate, the latter as regards method.

permutations. He also interprets the properties of functions in this sense in his first chapter; he considers a function to be an analytical expression and he differentiates among these expressions according to the properties they display when the variables they contain are permutated. Similarly to Lagrange, he also introduces "similar" and "dissimilar" functions and a system of symbols for the further study of the properties of functions with respect to the permutation of the variables. In the chapters concerned with the general properties of equations (Chap. 2 and 3) he expressed, e.g., the fundamental theorem of algebra in a form similar to that used by Bolzano later.[20] Of course, he proved some of the properties of the roots earlier, e.g., the relations between the roots and the coefficients, Newton's formulae for the sum of the powers of the roots, etc. One would also find the theorem pertaining to the number of various possible values of a function of m letters (§ 45, p. 32−33) proved by complete induction, just like the theorems permitting the substitution of a given equation into an equation with roots of different properties, e.g., all roots positive. In the subsequent chapters he devoted considerable attention to the transformations of equations, to elimination, etc. Just this list of the problems treated with the mention of certain theorems (the theorems concerning permutations have been omitted intentionally), indicates that Ruffini kept to traditional topics and to the traditional method of interpretation. Ruffini's book will not be discussed here in detail but only the ideas will be pointed out which were decisive at the time for determining the future evolution of algebra and the new terminology.

In connection with the foregoing considerations the solution of the equation

$$x^n - 1 = 0, \qquad (3.3.1)$$

is among the foremost of these features. Ruffini devoted the tenth chapter of his book, called *Proprietà delle radici della unità* ([343] 128−132), to it. He pointed out that the roots of the equation could be expressed in terms of the powers of a single root different from unity, i.e.,

$$\alpha, \alpha^2, \alpha^3, \ldots, \alpha^n. \qquad (3.3.10)$$

He arrived at the concept of a primitive root, as well as at the equation

$$\alpha^t = \alpha^{kn+r} = \alpha^r, \qquad (3.3.11)$$

in other words at the reduction of the exponent to the least positive residue of the modulo n, to use Gauss' terminology. He also maintained that Eq. (3.3.1) could be reduced to "prime factors" for $n = a \cdot b \cdot c \cdot \ldots$, where a, b, c, ... are prime numbers, i.e., to equations

$$x^a - 1 = 0, \qquad x^b - 1 = 0, \qquad \text{etc.} \qquad (3.3.12)$$

He based the proof of this statement on the properties of the roots. He did not explicitly mention the solution of Eq. (3.3.1) when n is a prime number larger or equal to 11.

---

[20] ([343], 37) Therefore, Burkhardt's rebuke ([51] 136), that he did not give proof of the existence of the roots, is not just. However, the accuracy of Ruffini's proof is another matter. The terms Ruffini used do not have Bolzano's accuracy, in particular continuity is not defined accurately.

Therefore, even as regards the extent of the discussion of this problem he did not exceed the level of Lagrange's interpretation, so that one cannot compare this part of his work with Gauss'.[21]

Another, more general problem important for the progress of algebra at the time, was the problem of numerical realms. Ruffini did not devote himself to this problem particularly. Like many authors of the time (inclusive of his example, Lagrange) his considerations in the realm of complex numbers are current, which is enforced by the nature of the roots of the algebraic equations anyway, and he differentiates among real numbers, irrational, rational numbers and integers. The fundamental (main) realms are the latter two: Ruffini's algebraic equations display coefficients either as rational numbers or integers. This approach is understandable. The cases in which other coefficients occur are very rare, e.g., (asymmetric) functions of the roots (Chap. IX, § 180). The nature of these roots always follows solely from the requirements of calculation relevant to the problem investigated, and Ruffini did not mention it otherwise, nor did he use any of its general descriptions. The functions of the roots did not even outline a different division of numerical realms than the one mentioned. He did consider irrational functions of the roots, but only with a view to the behaviour of these functions when the roots were permutated.[22] One might also say that Ruffini considered an unarithmetical concept of algebra in his interpretation, which was also typical for his paragon, Lagrange.

Ruffini's unarithmetical interpretation was reflected in his study of the irreducibility of equations. As the initial equation had either integer or rational coefficients, the meaning of reducibility was clear. However, since he omitted any more accurate definition of the concept of a numerical realm, the statements he made about the irreducibility of auxiliary equations (resolvents) have little substantiation.[23] Since he did not arrive at the concept of a numerical realm or adjunction, he missed the possibility of the resolution of a polynomial into factors with coefficients belonging to a superior realm or, to be more accurate, he did not consider this problem.

When Ruffini reached the conclusion that a general equation of a degree higher than four was insolvable, he asked himself the question, just like other authors of the time, whether one could possibly arrive at a solution of the equation using the relations among the coefficients or among the roots. He was aware that the contingent solutions as a whole would not create the general algebraic solution called for. However,

---

[21] One cannot say from this chapter of Ruffini's study whether he knew Vandermonde's work; but he certainly did not exploit his results.

[22] As regards this question, also compare with Burkhardt's data [51], p. 136 and especially 158.

[23] Burkhardt ([51], in particular p. 147) made this rebuke in a different connection when he discussed the objections Malfatti had made with respect to Ruffini in 1804. He also drew attention to the fact that no numerical realm had so far been determined, nor was the concept of a general algebraic equation clear. However, Burkhardt showed very precisely that Ruffini was aware of certain possibilities in some respect of defining the realms in another way. However, he was not quite sure of how to treat "additional" irrationalities. Compare Burkhardt ([51] p. 149n), where he also discusses the further evolution of the problem.

he did discuss a number of possibilities in the closing chapters of his study of 1799, and he also reverted to the topic later, in a treatise published in 1802 [344]. As compared to previous authors there is the new feature that the relations among the roots (or coefficients) are used to find suitable functions of the roots which would attain a small number of different values; in this way he was able to decrease the degree of the resolvent. Ruffini also showed that not all the relations among the roots could serve this purpose. It is understandable that he could not arrive at a more general approach in this way but that he rested with investigating various special cases, defined by a certain relation among the roots or coefficients.

If one also considers what has been said of Ruffini's work, its principal and permanent contribution, i.e., the contribution to elaborating the theory of permutations which will be mentioned later on, and if one also gives credit to his repeated attempts at improving the interpretation and the proofs in the subsequent years, one is capable of assessing his historical research. Ruffini introduced important progress in the study of algebraic equations. He answered the doubts about the solvability of algebraic equations of a higher degree than four by presenting proof of the insolvability. In spite of certain deficiences the proof is acceptable so that it could have limited further vain attempts. Historically, it is important that this proof was given using means and ideas published by Lagrange neearly thirty years earlier. An exception is formed by the deliberations concerning the theory of permutations, the earlier realization was not hindered by anything, and no new facts since the time of Lagrange had any effect on Ruffini. With the exception of this region, Ruffini's terms and operations are traditional or at least Lagrangean. If these facts provide problems for historical research, there is, of course, the other aspect of the problem of insolvability of which Ruffini's study also gives some evidence. This important but isolated problem contributed very little to a more substantial change in the algebraic theory of equations in the course of its solution. The small amount of progress in this theory which is due to Ruffini, is also reflected in other parts of his work. Even when he deals with an otherwise very promising problem of determining the classes of solvable equations (of a degree higher than 4) he only arrives at individual isolated cases. There is the question whether these failures (just like the successes represented by the beginnings of the theory of groups) did not link up with the unarithmetical approach which he has in common with Lagrange.

3.4 The response to Lagrange, Vandermonde, Waring, Gauss and Ruffini at the beginning of the 19th century

In this chapter studies made by five mathematicians have so far been discussed, three of whom, Lagrange, Vandermonde and Waring, published their principal algebraic results in the years 1770–1772, whereas the two remaining mathematicians began to publish their algebraic papers in 1799. In spite of the fact that there is an interval of nearly thirty years between their studies, one may speak of a direct intellectual connection; one would also search in vain in this interval for a result which would have influenced Gauss or Ruffini.[1] A further connection with the trend created by the authors mentioned, and its logical continuation and development, can only be found again after several years. First of all, there is Cauchy, who solved one of the main problems of Lagrange's program of the theory of equations in his paper published in 1815, and proved that a function of n letters could not attain more than two and less than n different values if the letters were permutated. He mentions the results which were achieved in the study of similar problems by the Italian school. However, it is not clear which of its studies he knew (cf. 7.2). He did not discuss the consequences of his proof for the solvability of algebraic equations, as if he were only solving a precisely determined, narrow problem of the theory of permutations, without considering the broader connections. He did not go back to this set of problems for a considerable number of years (until 1844). Nevertheless, he did influence some authors in the interval between 1817 and 1844. Abel had already mentioned Cauchy's proof.[2]

It is a well known fact that Abel originally thought that he had solved an equation of the fifth degree, but he soon realized his error, and in this way he arrived at his celebrated proof of insolvability. Apart from Cauchy, he undoubtedly knew Lagrange and Gauss, and he also mentions Ruffini's proof.[3] Following Cauchy's study, the part of the proof belonging to the contemporary theory of permutations may be considered as concluded. Without quoting Vandermonde directly, Abel formulated the task in a similar way. Abel required ([3a] 66) that the general form of the algebraic function be considered and investigated to determine whether it could satisfy a given equation when this function was substituted as the unknown. It is in no way difficult to decipher the meaning of this requirement; it is quite clearly an attempt to express a fact pertaining to field theory. It will be shown later (Chap. 4) what historical role Abel's deliberations played in establishing the concept of a numerical realm. At this point we shall restrict ourselves to stating that Abel clearly interpreted the differences between an entire rational, a rational and an algebraic function, and that he arrived at the classification of algebraic functions. The stressing of the algebraic nature of the elements (numbers) with which he operates, and the determining of the differences among these elements, indicates an arithmetical approach by which Abel

---

[1] As regards Gauss I have in mind the seventh section of the Disquisitiones arithmeticae; as regards Ruffini I have omitted the discussion which affected his further work, some of the places where it has been rendered more accurately and of which Burkhardt wrote [51].

[2] Abel [2] and [3], i.e., in the years 1824 and 1826.

[3] Le premier, et si je ne me trompe, le seul qui avant moi ait cherché à démontrer l'impossibilité de la résolution algébrique des équations générales, est le géomètre Ruffini; mais son mémoire est tellement compliqué qu'il est très difficile de juger de la justesse de son raisonnement. Il me paraît que son raisonnement n'est pas toujours satisfaisant ([8] 218).

differs from Lagrange. One may even feel the strong influence of Gauss in this respect, which however only became outstanding in Abel's algebraic studies following his proof of the insolvability of a general algebraic equation. In the paper *Mémoire sur une classe particulière d'équations résolubles algébriquement* [6] he set himself the task of generalizing the method which would lead to the solution of equations of the form

$$x^n - 1 = 0, \tag{3.4.1}$$

i.e., equations displaying certain relations between the roots. He considered the proof of the following theorem to be one of the most important results ([6] 479): "If all the roots of an equation of an arbitrary degree n can be expressed rationally in terms of one of them, which we shall denote x, and if for any two roots ϑx and $\vartheta_1 x$ it holds

$$\vartheta\vartheta_1 x = \vartheta_1 \vartheta x,$$

that the equation will always be algebraically solvable". He also pointed out that the roots of the equation were then in fact

$$x, \vartheta x, \vartheta^2 x, \vartheta^3 x, \ldots, \vartheta^n x = x,$$

where ϑx is a rational function, and $\vartheta^i x$ is a function of the type ϑx "considered twice, three times, etc.". The purpose of expressing the roots in this way was to stress the connection with Gauss' solution. One should also understand the supplement to the main theorem proved in this sense: "Provided the given equation is irreducible and its degree n is equal to the product

$$\alpha_1^{V_1} \cdot \alpha_2^{V_2} \cdot \alpha_3^{V_3} \cdots \alpha_v^{V_u}$$

where α_i represent different prime numbers, the solution of the given equation can be reduced to the solution of v_1 equation of degree α_1, v_2 equations of degree α_2, etc.". If these words remind one of the purpose of Gauss' seventh section of the DA as expressed im Art. 342, it should be pointed out that a litle further on Abel ([6] 491) himself refers to Gauss' method given in the DA (Art. 359 nn) which he is said to have used. As already mentioned, at this point Gauss maintained that all auxiliary equations, which he used in solving the equation (3.4.1) were solvable, but that lack of space prevented him from considering this problem in detail. Abel's own words are thus proof that he was not only stimulated by Gauss' solution of Eq. (3.4.1) in treating this topic, but that he was also led by its method or by its outline while attempting to generalize Gauss' results. The contents of the study fully agree with these statements of Abel. The ideas in Abel's study were in turn stimuli for Kronecker who considered the so-called Abel equations in particular as a certain class of algebraically solvable equations. As regards Kronecker, a conjunction of the arithmetic, and

---

[4] As regards the concept of Abel's equations compare A. Loewy's comments in the German edition of Abel's study [10]. It is also suitable to remember that Kronecker thought highly of Vandermonde. Cf. preface of Itzigsohn to [385] of 1887 in Berlin, at the time Kronecker worked there.

even numerically theoretical, aspect with the theory of algebraic equations occurs. Thus he began to connect Vandermonde's stimuli[4] and Abel's manner of presenting the problem with Gaussian tradition in his work partly, beginning with 1845, but especially in the fifties of the 19th century. However, let us again briefly return to Abel's studies.

In his paper published in 1829 Abel follows up on the seventh section of the *Disquisitiones arithmeticae*; from a historical point of view it is interesting that although this occurred after more than 25 years, it represented the first follow up which understood Gauss' indications and which generalized them. Abel made more and independent progress in his unfinished studies, publishes after his death[5] *Sur la résolution algébrique des équations* [8]. The dominating term for him still remained the "algebraic function" or the "algebraic expression" with the help of which he arrived at facts containing the essence of Galois' theory and of the study of algebraic realms. As regards method Abel introduced a new and important idea into algebraic deliberations, an idea which is a modification of Lagrange's *a priorism*. Abel also quoted Lagrange in this connection [8]. Basically Abel's new idea is the requirement for the proof of the existence of a solution to an equation, i.e., before trying to find the solution to the equation it is necessary to prove its existence. This led him to a procedure which avoided "the complications of calculating". He wanted to solve two problems in this study:

1. Find all equations of any given degree which were solvable algebraically;
2. Decide whether the given equation was solvable algebraically or not.

Nevertheless, he was aware that he had not produced a complete solution to these problems.

The young French mathematician, Evariste Galois, began to study the problems of classes of equations which were algebraically solvable prior to 1830. He also originally believed that he had solved a 5th degree equation, and he also considered the theory of elliptical functions, in addition to the study of algebra. Their paragons, and one might perhaps say teachers, were also the same. Lagrange's work had considerable influence on Galois, but he received the largest stimulus from Gauss' *Disquisitiones arithmeticae*, which were the key to his discoveries.[6] Galois did not consider the proof of the insolvability of equations of a higher degree than four because he thought the matter sufficiently clear.[7] This fact in itself indicates an important feature of Galois' algebraic

---

[5] For the sake of completeness it should be pointed out that Abel dated his study [6] 29th March 1828, and that it was published in Crelle's *Journal für reine und angewandte Mathematik* the following year; his paper [8] was written in the second half of 1829 and published for the first time in Holmboe's edition of Abel's papers in 1839; cf. note in [1], Tome II, p. 329.

[6] In his study of relatively few pages Galois quoted very little, probably much less than he knew (cf. [153] 23) His repeated quoting of Abel's studies is interesting, in which he also refers ([152[17, 37; [153] 22) to algebraic statements in his paper on elliptical functions published posthumously. He certainly knew Cauchy's work on the theory of permutations, cf. ([153] 40).

[7] As regards Galois'attitude the words he noted in about 1831 are characteristic (cf. [153] 17): C'est aujourd'hui une vérité vulgaire que les équations générales de degré supérieur au 4e ne peuvent se

study: he did not consider old problems even though they may not have been solved quite completely, but he directly follows up the most important results. First of all, he fully understood Gauss' work in its algebraic essence. The outlines of the method as given by Gauss became his base, but he avoided considering their special forms, only expressing that which was general in them. That is why he formed the basis of his considerations from concepts like the irreducibility and the reducibility of equations and adjunction, which would have been impossible without a certain idea of numerical realms, though not quite clearly understood and inaccurately expressed so far. This changed the terminology considerably. Let us mention the difference in understanding of both concepts by Gauss and Galois. Gauss spoke of the irreducibility of equations when the coefficients were rational numbers or integers. The same applies to adjunction for which he had no special term. In contrast, Galois understands the region, closed with respect to "rational" operations, generally, and "in it" the equation is either reducible or irreducible; for this purpose he required no numerical calculations. It is similar to the concept of adjunction which he gave this suitable name. He adjoined the element or elements, which were either the root of the given equation or the root of the auxiliary equation, to the realm whose elements were the coefficients of a given equation and which contained all the elements formed rationally (as follows from the context). Reducibility and adjunction then form the basis for the general study of the solvability of equations, and together with the intuitively understood concept of a realm of elements, they form the conceptual basis of all considerations. However, one must add some more concepts to this basis; the concept of a group and the other concepts pertaining to the theory of groups. These concepts will not be discussed at this point because we shall revert to them in connection with the further reforming of algebra and attempt to determine more accurately the sense in which Galois understood them. At this point attention is only being drawn to them. The conceptual system of Galois is quite different from that of earlier authors, including Abel. In establishing this system (which, it should again be stressed, was not always clearly and accurately determined) Galois tried to grasp the theory of solving equations and, at the same time, the system he was in the process of forming enabled him to explain the fundamentals of the new theory of solving algebraic equations which was later named Galois' theory. The basis for this theory is Galois' understanding of how Gauss could have arrived at the solution of the equation

$$x^n - 1 = 0 \qquad (3.4.1)$$

Gauss brought to the fore the special relation between the roots of Eq. (3.4.1), i.e., that

---

résoudre par radicaux, c'est-à-dire que leurs racines ne peuvent s'exprimer par des fonctions des coefficients qui ne contiendraient pas d'autres irrationnelles que des radicaux.

Cette vérité est devenue vulgaire, en quelque sorte par ouï-dire et quoique la plupart des géomètres en ignorent les démonstrations présentées par Ruffini, Abel, etc., démonstration fondée sur ce qu'une telle solution est déjà impossible au cinquième degré.

It is also worth noticing that Galois had no objections to these proofs although Abel was not satisfied with Ruffini's proof.

all the roots are powers of one root; this aspect of the relation between the roots of Eq. (3.4.1) is also stressed by the special manner of interpretation. In generalizing Gauss' procedure Abel came to understand the reasons for solvability with the more general class of algebraic equations, but he always founded his considerations on the special relations between the roots of these equations. Only Galois grasped the essence of the cases investigated, found the meaning of group attributed to a given equation, and also realized the meaning of an invariant subgroup. In this way he found that solvability in the cases studied by Abel and Gauss followed from the fact that each subgroup of the group of equations investigated was invariant, whereas in other cases, roughly speaking, the existence of invariant subgroups makes it possible to reduce the solution of a given equation to the solution of other equations of a lower degree.[8]

Galois' studies provided the possibility of a more substantial turn not only in the theory of algebraic equations, but also in the whole conception of algebra.[9] It is a well known fact that considerable parts of Galois' preserved studies remained unnoticed in their time, and not until Liouville had published them in 1846 [152] did their elaboration make gradual progress. However, this was at a time when a deeper and more extensive understanding prevailed of Gauss' results as well as Abel's studies. Let us leave these problems till later, and revert to the response the work of Lagrange, Vandermonde, Waring, Gauss and Ruffini drew in the development of algebra prior to the application of the results of Abel and Galois.[10]

At the time considered, the response to mathematical studies was still strongly affected by the degree to which they were known. However, for the present purpose this aspect may be neglected. An attempt will be made at determining the manner in which these studies acted in spheres where they were known. At the same time it is necessary to take into account that Lagrange was one of the world's leading mathematicians in the last three decades of the 18th century, a scientist whose studies, together with the fame of French science which then had a leading role, were penetrating. Moreover, after

---

[8] Of course, Galois' results and the comparison of his ideas with those of Abel, Gauss and others is only outlined very roughly here. Nothing new has been added either; for example, the last statement, comparing Gauss and Galois, is notorious and it has lately been pronounced by Delone [116] and Kochendörfer [238] among others. As regards the appraisal of Galois' work and understanding its contents, apart from Galois' own papers [152, 153, 154] we refer to an historical literature.

[9] In no way can it be said that we have only been drawing on Galois' main algebraic results. One could also speak of a whole series of his revolutionary ideas, e.g., indications of the study of finite fields, of isomorphism or automorphisms, etc. However, these are mostly indications which were only employed much later on in the development of algebra and, therefore, we shall go back to them in other connections.

[10] The formulation in the text already indicates that the time interval in which the response is to be investigated is in no way terminated precisely. Some of Abel's papers were published in Crelle's *Journal* which became intensively read in mathematical circles since its very first year, 1826; the papers on elliptical functions were immediately connected. Practically the whole of Abel's mathematical work had been published by 1839. Nevertheless, it is difficult to speak of the effect of Abel's algebraic papers immediately after they had been published.

the French revolution, Lagrange's ideas received an excellent platform in the schools created by the revolution, in institutions and journals. Gauss' fame was also rapidly propagated from the beginning of the century, and with it also the reading of his princial mathematical work, *Disquisitiones arithmeticae*, which had a profound effect on his contemporaries. Therefore, there existed quite considerable groups in which one might speak of response to these studies. However, no attempt will be made at defining all the points in common of the algebraic studies of the five mathematicians mentioned and of the algebraic treatises of mathematicians who knew them or who one may reasonably assume known them. In investigating the response it should be particularly interesting to determine how these ideas were understood, as well as the methods and terms which indicated the subsequent progress of algebra. In this way, such characteristic problems will come to the fore as the discussion of the solvability of equations of a higher degree than four, the solvability of equations of the type $x^n - 1 = 0$, or of other special types of equations, provided the study of the solvability comes from Gauss or tends towards Galois' theory. The study of the reducibility of equations, the understanding of "rationality", or of the "*a priori*" method were just as important. Only some of the facts leading to further development have been mentioned, with the knowledge that their number could be increased, but also that their interconnections with the various authors differ.

Lagrange's study was quoted immediately after it had been published, which is not surprising with an author of his fame who, moreover, worked at the Berlin Academy at the time, i.e., at one of the few scientific institutions of its kind at the time, and which also published his papers in its proceedings.[11] Not even Bézout could avoid quoting him when he published his study *Théorie générale des équations algébriques* in 1779.[12] In his preface he also stressed, apart from Euler, Lagrange's credit with respect to the development of the general theory of equations without specifying the nature of the credit in any way. In a similar way, at the beginning of the second volume of his book, he also mentioned Vandermonde's study published in the proceedings of the Paris Academy in 1772. Bézout's book, however, is only an improvement on and extension of his ideas already published at the beginning of the sixties. Therefore, as if in response to Lagrange's work, he wrote ([34] Préface II – III): Néanmoins toutes les recherches un peu générales que l'on a faites jusqu'ici sur les équations, se réduisent toutes (si on en excepte seulement les équations du premier degré) à des méthodes pour obtenir le résultat le plus simple de la combinaison de deux équations et deux inconnus .... This conviction then led him to the study of elimination as an instrument to solving equations

---

[11] There is not enough room here to mention in detail all the quotations and responses of Lagrange, Vandermonde and Waring. The purpose of the examples given is rather to characterize the type of response. A very frequent response is a kind of "courtesy" quotation which refers to a celebrated paper, without its content and trend being apparent in the results of the author themselves.

[12] D'Alembert, Laplace and du Séjour reviewed this extensive algebraic study of Bézout's [34] for publishing.

of higher degrees, where he still saw the main difficulty in the imperfection of methods which led to exceedingly complicated computations.[13]

Several years later the *Encyclopédie méthodique* (1784), of three volumes, was published. In it mathematics was treated by d'Alembert, Bossut, Lalande and Condorcet, among others. Under the heading Équations déterminées [104] the author, Condorcet, restricted himself to the results which were reached in the general solution of equations in recent years. He quoted the studies of Euler and Bézout of the sixties, he devoted more detailed attention to the studies of Lagrange and Vandermonde, and also mentioned Waring. As regards Lagrange's work ([104] 664–665) he gave a very brief and therefore perhaps hardly intelligible excerpt. In the conclusion, however, he maintained that only two difficulties remained with regard to the general solution of equations; the length of the computation and the fact that so far the existence of the roots "in a general and finite form" had not been proved. It follows from the context that he summarized two problems in his second comment; on the one hand the fact that no precise proof had so far been given of the fundamental theorem of algebra, and on the other hand certain doubts as to the existence of an algebraic solution to the equation.[14] He indicated, although not very clearly, that Lagrange's approach could lead to the proof of the non-existence of a root in the finite form.

This example proves that the work of Lagrange, Vandermonde and Waring soon drew justifiable attention as the foremost studies of algebraic equations, and that the meaning of Lagrange's endeavour had at least been understood as a whole. However, one would not find a creative analysis of his ideas even in France at the time.[15]

---

[13] As is known, the inventiveness of Bézout in computations eventually led him to a considerable decrease in the degree of the resolvent of an equation of the 5th degree. The futility of attempts at solving equations of higher degrees was stated in Bézout's study, nevertheless one feels that optimism prevails as regards the treatment of the whole problem.

[14] It should be interesting to mention Condorcet's opinion verbatim ([104] 665): Il ne reste donc plus sur la résolution générale des équations que deux difficultés; 1°. La longueur du calcul; 2° qu'il n'est pas rigoureusement démontré qu'une équation déterminée d'un degré quelconque, ait une racine d'une forme générale et finie; le contraire seroit même prouvé si, en suivant la marche indiquée ci-dessus (in Lagrange's interpretation — L.N.), la solution de la proposée, n'étant un nombre premier, se réduisoit à la solution d'une autre équation du degré n, qui n'auroit pas de diviseur rationels, ou, si n n'étoit pas un nombre premier, à une équation d'un degré pour lequel l'équation qui donne les termes sous le radical n, ne se rabaisseroit pas au-dessous du degré n-2, n-3, . . ., 3, 2, 1. Ainsi, dans le cas où la racine n'aurait aucune forme finie possible, la méthode proposée ci-dessus conduira encore à trouver cette impossibilité: C'est donc à diminuer la grande complication des calculs, et à trouver des méthodes qui les abrigent, que les analistes doivent tendre maintenant.

In the conclusion the author referred to his treatise, published in the 5th volume of the proceedings of the Academy in Turin [105].

[15] In the other countries the situation was even worse. Lagrange or the others were known but no more involved interpretations of their results have been found. In Germany this could have been the result of attention having been devoted to other mathematical problems, so that there was nobody who would have studied the solution of algebraic equations in greater detail. Lagrange's papers were known and studied; this is also substantiated by the 1st volume of Hindenburg's *Archiv der reinen und angewandten Mathematik* (Leipzig, 1795), in which Lagrange's work was a stimulus for several articles.

The difference between the knowledge of Lagrange's work, understanding it and elaborating on it became more marked in the last decade of the 18th century. The revolution began to have a positive effect on the development of science in France. The foremost French mathematicians, and among them also Lagrange, began to work in the scientific and pedagogical institutions established, among which the École normale and the École polytechnique had the largest significance for the further progress of science, including mathematics, and their lectures were being published.[16] In this way the "*Séances des écoles normales*" [351] were established, which also contain records of the lectures in algebra presented by Laplace and Lagrange.[17] The lectures at the École normale and the École Polytechnique required new textbooks. In this connection a considerable number of text books, modern in their time, were published, such as Lacroix's *Cours de Mathématique* (Paris 1796—1799) of several volumes, and later Garnier's, Francoer's, Reynaud's and others, most of which went into several editions and had to compete with the repeated editions of older courses, such as Bézout's [35], and also were affected by the progress of including the new results in the reissues.[18] Let us disrupt chronology for the time being and investigate the algebraic parts of these books in their various editions, and only then go back to those of Lagrange's and Laplace's lectures which tie up closely with others of Lagrange's algebraic texts.

Lacroix's *Complément des Élements d'Algèbre* [263], which was re-edited in France six times between the end of the 18th century and the year 1835, and for the seventh time again in 1863,[19] is the most characteristic for the development of algebra at the time.

His papers as well as the numerical solution of equations, published by the Berlin Academy in 1770—1771, or Michelson's translation of Euler's and Lagrange's algebraic papers of 1791 [378] were quoted and highly thought of (p. 65).

[16] Information concerning these lectures as well as the changes in organization of French science after the revolution propagated quickly. For example, Hindenburg's Archiv (p. 384) said in connection with the publication of Lalande's letter containing information on the subjects taught at the École Normale, as early as in 1795: Lagrange and Laplace lecture in mathematics, Haily in physics, Monge in descriptive geometry, etc.

[17] Special stress was put on the clarity and intelligibility of the lecture at the school, which forced the lecturers to deal with the fundamental questions and concepts, on the one hand, and on immediate availability in print, on the other. As the lectures were repeated, their texts were printed over and over again.

[18] The significance of French textbooks and compiled works in mathematics of the end of the 18th and of the first decades of the 19th century deserves a special historical analysis. Textbooks of this standard were not available in other countries (there was nobody and no school to write for), and certainly not in such variety and in so many editions, (rapidly published one after the other) as in France. In France this feature gives the final touches to illustrating the explosion of French mathematics in the decades after the revolution, when the society building up capitalist industrial production, considered specialists educated in mathematics to be important assets to progress in technical standards.

[19] Moreover, Lacroix's *Cours de Mathématique* was published in 1897 and this was its 25th edition. Of course, Lacroix already lectured at the Ecole Polytechnique at the beginning of the century (cf. [351], *Correspondance sur l'école polytechnique* No 1, § 1, p. 3). Bézout's *Cours* was also being published until 1848.

The significance of Lacroix's *Complément* is that it was the first extended textbook of algebra which included Lagrange's results and methodical ideas of the years 1770—1771 in its interpretation of algebra, the author being aware of the results which had been simultaneously presented by Vandermonde and Waring, and taking into account the more recent algebraic treatises.[20] An attempt will now be made at specifying this and showing how Lacroix understands his paragons and what interpretation he bestows on them. However, one should bear in mind the meaning of the *Complément*, which is not an elementary textbook but a set of lessons complementing a fundamental textbook[21] to which they frequently refer.

Lacroix's book, which could perhaps be characterized as a scientific monography, proceeds systematically. In the introduction it defines the concept of a function as an expression composed of quantities, the value of which depends on the said quantities. It repeats the wellknown fact that the roots of an equations can be obtained in general either by approximation or "avec des radicaux" which, however, are quantities depending on the said roots and expressed rationally by means of coefficients of the given equations.[22] Lacroix's conceptual system is in fact restricted to these comments, his interpretation being built on it; in order to avoid being unjust it is necessary to add to this the differentiation among numerical realms which was current at the time. He did not discuss the fundamental theorem of algebra in the introduction, he did not mention other than the expressions already discussed, and he probably considered the term "rational" as intuitively sufficiently clear. One must concede that he did not require specification of the initial terms or introduce others in the subsequent interpretation, and this constitutes, as we shall soon see, the weakness of his interpretations.

After the introductory paragraphs, Lacroix quickly proceeded to the heart of the problems, i.e., to the solvability of equations. Prior to this he only discussed symmetric functions (of roots), he expressed and illustrates the theorem stating that "any rational symmetric algebraic function of the roots of an arbitrary equation" could be expressed by the coefficients of this equation.[23] He then carried on to discuss elimination, to which he aplied the facts concerning symmetric functions, and, among others, he refers to Waring. However, as soon as he reached the solution of equations itself, from the

---

[20] By this is meant the lectures of Lagrange and Laplace at the École Normale, to which we shall return later, and the papers which were published in the first edition of Complément and which Lacroix complemented with their ideas. Lacroix quoted the papers of other authors sufficiently accurately.

[21] *Eléments d'Algèbre à l'usage de l'école centrale des quatre nations* [262], which reached its sixteenth (modified) edition (Paris 374 pp.) in 1836 and its eighteenth unchanged edition eleven years later, published independently, corresponds to Lacroix's course of this fundamental textbook already mentioned.

[22] Quoiqu'on ne puisse obtenir en général les racines d'une équation que par approximation ou avec des radicaux, il y a cependant des quantités qui dépendent de ces racines, et qui s'expriment d'une manière rationnelle au moyen des coefficients de l'équation proposée, ([263 c] p. 1). I am quoting the third edition of 1804 (Paris, p. 315) which I consider to be fundamental and to which the deviations in the other editions are related.

[23] He refers [263 c] to Vandermonde's treatise [383].

equations of the second degree he proceeded according to Lagrange and sought suitable functions of the roots. For example, similarly to his example (p. 30 § 16) he required that a function of the roots be determined *a priori* for a cubic equation, which would only depend on a quadratic equation. His approach is the same as regards equations of the fourth degree. In the conclusion of this part he wrote that Lagrange, who was the originator of this method, pointed out that different functions of the roots could be adopted. However, he maintained that no such function had so far been found for an equation of the fifth degree which would lead to an equation of a lower degree. This formulation represents Lacroix's point of view[24] which is very reserved. Further on (§ 60, p. 116 nn) he expanded on Lagrange and Vandermonde, proving that it was impossible to decrease the degree of the auxiliary equation for an equation of the 5th degree to less than the 6th degree. This led him to the conclusion: C'est une question qui n'est pas encore résolue, que de savoir s'il est possible ou non d'exprimer la racine d'une équation par une fonction composée d'un nombre limité d'expressions radicales d'un degré égal ou inférieur à celui de la proposée.[25] Further on he extended his conclusions: … qu'il est encore douteux que l'on puisse exprimer par un nombre limité d'opérations algébriques, c'est-à-dire, d'additions, de soustractions, de multiplications, de divisions, et d'extractions de racines, généralement, les racines d'une équation quelconque, au moyen de ses coefficients. However, he balanced his doubts as to the existence of an algebraic solution by stating that the value of this solution is illusory as already indicated by the case of the cubic equation, and he repeated Lagrange's words: "On peut assurer d'avance que quand même on parviendrait à résolution généralement de cinquième degré et les suivants, on n'aurait par là que des formules algébriques, précieuses en elles-même, mais-très peu utiles pour la résolution effective et numérique des équations de mêmes degrés, et qui par conséquent, ne dispenseaient pas d'avoir recours aux méthodes arithmétiques.".

This was the point of view and the general way out for Lacroix in 1804. In the edition of 1825 as well as in the subsequent[26] editions Lacroix pointed out that Lagrange made the solution of the 5th degree equation and of equations of higher degrees dependent on the properties of the functions of the roots. It is said that this problem was solved by Ruffini who proved the non-existence of a function of 5 letters which would have 3 or 4 values, and also Cauchy who proved a more general theorem. He also gave

---

[24] He pointed out that Langrange's analysis was much more extensive but that Lacroix was not able to interpret it to its full extent. In the 5th edition of 1825 (p. 44) the summary was complemented by the words: Quelque féconde que soient les principes de cette méthode, elle n'a pu s'étendre aux équations générales des degrés supérieurs au quatrième, parceque la fonction des racines qui sert à les déterminer dépend d'une équation plus élevée que la proposée … as it is said Lagrange had proved in Note 13 to the 2nd edition of the Traité [269]. This note was reprinted verbatim in 1835 ([263 e] 46) and in 1863 ([263 f] 41).

[25] In this connection Lacroix discussed Condorcet's study [105]. He interpreted it in the following sense: if the existence of the root in the form indicated will be proved, the remainder will only be a question of calculation even though this might be tedious.

[26] [263 d] p. 60 nn; [263 e] p. 55 nn; [263 f] p. 62 nn.

Ruffini the credit for trying to prove, beginning with 1798, and gradually (until 1816) improving the proof that an equation of a degree higher than four cannot be solve algebraically. To this he again added: Mais il faut renoncer à obtenir, par un nombre limité d'opérations algébriques généralement indiquées, les racines d'une équation quelconque au moyen de ses coefficients, on doit y avoir d'autant moins de regret, que la forme de ces expressions en rend l'application numérique toujours très longue et quelquefois impossible ...

Some of the features of Lacroix's *Supplements* can be seen from the excerpt mentioned. First of all, they are based on Lagrange's method of solving algebraic equations with all their principal results. Later results are also given, similarly, however, with a certain delay as regards Ruffini's proof of insolvability. This proof justifies Lacroix's scepticism expressed earlier. However, nothing was said of the difficulties of the proof or of its trend of thoughts. Lacroix did not even have the system of terms ready for this. He did not investigate the reducilibity of an equation, and the rationality of an expression did not become outstanding as a problem. The form of the root, which only appeared in the definition of the solvability of the equation, was understood to be intuitively clear, and Lacroix did not even disclose the problems of which Vandermonde was probably aware when searching for a more general form. It seems as if all finesses and difficulties remained without Lacroix's interpretation. This also applied to the opinion concerning the algebraic solution of the equation; from a utilitarian point of view a numerical solution is really more important. However, if Lagrange had reached this opinion as a result of a probably vain and abortive attempt at a further expansion of his theory of algebraic equations, Lacroix's way out was only an uncreative approach which did not see any purpose in the theory of equations itself.[27]

An important item in Lacroix's interpretation has been omitted so far, the analysis of the equation

$$x^n - 1 = 0 \tag{3.4.1}$$

It is understandable that there was no mention of Gauss' *Disquisitiones arithmeticae* in the second edition of 1801, and Lacroix was of the opinion (p. 84) that the solution of the equation (3.4.1) was not known considering the contemporary state of analysis. In this way Lacroix, who was a well-read mathematician, proved that there was nobody who had fully understood Vandermonde's work, in particular his most important results, at the beginning of the 19th century.[28] In the third edition published in 1804, Lacroix paid great tribute (p. 92) to the *Disquisitiones arithmeticae,* calling the study

---

[27] By this it was not intended to deprive Lacroix's work of any credit; his work informed the reader of the new results to a much larger extent than it is possible to report in the present book. It is also more extensive in content, of which only a small excerpt has been adopted. However, this does not alter the fact that he left out of Lagrange's work the most substantial facts which meant progress: doubts as to the contemporary ideas and methods, and the beginnings of the theory of permutations, of which he makes no mention at all. He probably followed Langrange's intentions in the latter, of which mention will be made later.

[28] However, he did quote Vandermonde's paper of 1771 on several occasiosn.

"un ouvrage très-remarquable" in which "M. Gauss a fait que toute équation à deux termes, dont l'exposant est un nombre premier, peut être decomposée rationnellement en d'autres équations dont les degrés sont marqués par le facteur premiers du nombre qui précède d'une unité ce nombre premier". He also mentioned Gauss' examples; for $n = 17$ the equation reduces to four quadratic equations, and for $n = 19$ to two cubic and one quadratic equation. He then went on to mention the resolution of the equation into factors which was supposed to be a reduction making solution possible, but he could not interpret Gauss' procedure in greater detail because he supported his interpretation by the properties of numbers which the reader had no knowledge of then, and only went back to these at the end of his study. In the conclusion to his book he mentioned some of the theorems of the theory of numbers, necessary for Gauss' deliberations, and he also wrote of primitive roots. But he did not attempt to outline Gauss' approach, even in the later editions of his study.

Even though Gauss' investigation did not lead Lacroix to considerations of new possibilities in the theory of equations but only to a mere repetition of the results and known relations between the roots of the equation (3.4.1) this investigation is nevertheless interesting. A note was inserted in the 6th edition of 1835 (and also in subsequent edition) at the end of § 171 in which Lacroix maintained that Eq. (3.4.1) was only a special case of a class of equations, all the roots of which were functions (he did not specify which functions) of one of them. At this point he referred to Libri's study presented to the Paris Academy in 1825, which was not published until 1833 and then only in an altered version, appeared in Crelle's *Journal* [275]. In this way Lacroix arrived at the mention of Abel's results, then published, and of Abel's letter to Legendre of 1828, summarizing his results.[29] To this Lacroix added the following important words: "En 1831, un jeune Français, Evariste Gallois, mort l'année suivante, avait annoncé, dans un Mémoire présenté à l'Académie des Sciences, que "pourqu'une équation irréductible de degré premier soit soluble par radicaux, il faut et il suffit que, deux quelconques des racines étant connues, les autres s'en déduisent rationnellement"; mais ce Mémoire paru à peu près inintelligible aux commissaires de l'examiner".

Therefore, Lacroix's treatise preserved its character, of which mention has already been made, even in this part. It informed the reader of the latest results, their selection[30] indicating that Lacroix was capable of assessing their importance; however,

---

[29] Among other things Abel wrote: J'ai été assez heureux de trouver une règle sûre, à l'aide de laquelle on pourra reconnaître si une équation quelconque proposée est résoluble à l'aide des radicaux, ou non. Un corollaire de ma théorie fait voir que généralement il est impossible de résoudre les équations supérieures au quatrième degré. Lacroix quoted these words according to Crelle's *Journal* (1830, p. 77; cf. also [1], T. 2, p. 279). This letter drew the attention of Galois who, however, claimed the independence of his discoveries; cf. Galois, [153] p. 24.

[30] Lacroix did not quote all treatises and books which were published in France at the time, he only mentioned some. On the other hand, one would search in vain for a more important paper he might have omitted. In the later editions he tried harder to point out foreign production, e.g., in the 5th edition he mentioned the French translation of the *Disquisitiones arithmeticae* and also the papers in Crelle's *Journal*. The attention he devoted to English algebra, especially in the 6th edition, is also interesting. In

it seems that he evaluated their importance rather with a view to the topic of the result than with a view to the method and procedures which spelt progress for algebra. That is why he did not lead the reader to the substance of new methods which might have remained undisclosed to him. For this reason the mental state of algebra, which was formed around the year 1770, remained preserved in Lacroix's book in spite of the abundance of the literature he mentioned.

Other French interpretations of the algebra of the time did not exceed in any way the standard represented by Lacroix's *Compléments*. All of them were strongly influenced by Lagrange, whether immediately by his study of the yeras 1770—1771, or inter-mediately. This can be seen most markedly in the interpretations of the solutions of equations of the 3rd and 4th degrees in which the study of the functions of the roots led to the determination of a suitable resolvent. The functions of the roots and the permutations were then a frequent and accentuated subject. However, this "modernity" was more an external symptom which complemented the traditional manner of inter-pretation as well as the current topic. The fundamental theorem of algebra was most frequently bypassed by pointing out that

$$x^n + Px^{n-1} + Qx^{n+2} + \dots + V = (x - a)(x - b)(x - c) \dots = 0 \quad (3.4.2)$$

held identically[31]. In general, the authors worked in the complex realm, so that the properties of the coefficients only followed from the context,[32] but they gave preference to rational coefficients. It is understandable that an idea of the various functions of the roots appeared, but it mostly followed from the context that the functions considered were rational, and rationality was understood intuitively. As regards Bourdon ([47] 376) the root of an equation is an expression of arbitrary nature (numérique ou algébrique, réelle ou imaginaire), which substituted into an equation nullifies it. In this conceptual sphere considerations of reducibility or irreducibility of a polynomial are meaningless. This does not mean that divisors of polynomials were not studied. On the contrary, Lefébure de Fourcy [272] as well as Reynaud [336] devoted special books to this topic which were then frequently quoted; other interpretations of algebra also have sections devoted to this problem. Bourdon ([47] 382) wrote that even polynomials could be

---

the latter he told of Ivory's entry "*Equations*" [207] in the English encyclopedia, which will be discussed later, and he also mentioned Peacock's "Report" of 1833 [328].

[31] Cf., e.g., Bourdon [47 a] p. 382; Reynaud [337] p. 69. Lacroix complemented his book by a section of the fundamental theorem of algebra in Note A as late as in 1835, where, among other things, he also mentioned Gauss' proofs.

[32] At the point mentioned, Reynaud wrote that the equation could also be constructed with the roots $\sqrt{2}$ and $\sqrt{3}$, so that it would read

$$(x - \sqrt{2})(x - \sqrt{-3}) = x^2 - (\sqrt{2} + \sqrt{-3})x + \sqrt{-6} = 0.$$

To this he added: On voit donc que les coefficients d'une équation peuvent être commensurables ou incommensurables, réels ou imaginaires. Mais nous ne considérerons que les équations dans lesquelles les coefficiens des inconnues seront des nombres... (i.e., integers or rational numbers) ([337] p. 69).

primitive just like prime numbers, but he ascribed to this the meaning that there does not exist a divisor with integral coefficients.[33] This was the way the algebraists reasoned, although they currently stressed the point that an equation was divisible by a binomial $(x - a)$, in which a was the root of the given equation, i.e., generally an expression of arbitrary nature as regards the numerical character.

As already mentioned, Lagrange's method of solving equations, his study of functions of roots, etc., penetrated these traditional interpretations. This then necessarily led the authors to discuss the algebraic solvability of equations. Practically nowhere does one encounter the idea that the proof of insolvability had already been given (or that it could be given). However, the awareness predominates that there had been too many abortive attempts at solving an equation of a higher degree than the fourth, and some authors were led to doubt whether this solution would ever be reached.[34] That is the reason why authors rather refer to the solutions of certain types of equations (e.g., Reynaud [337] p. 302) or to approximate solutions. It was widely held that a general algebraic solution has no value,[35] and that numerical solutions were required for which the approximate method was more convenient.

The second author to influence the interpretation of algebra was Gauss. His solution of the equation

$$x^n - 1 = 0 \qquad (3.4.1)$$

was then frequently referred to[36]. For example, Gauss' *Disquisitiones arithmeticae* were mentioned, beside Lagrange, in the preface to [150]. Nevertheless, the fact that Gauss was frequently misunderstood can be observed. This follows, e.g., from the fact that Reynaud ([337] 267 nn) restricted himself, like Lagrange in the years 1770—1771, to a mere study of the resolution of the exponent n into prime factors with the equation (3.4.1). Garnier ([150] Chap. IX) and others proceeded in a similar way.

It might seem that in interpreting the response, only examples favourable to our hypothesis about the considerable misunderstanding of Lagrange and Gauss are being considered, not to mention others. However, the arguments could be expanded upon by studying articles, as well as algebraic interpretations in other countries. However, it is always possible to rest with examples. Still if one omits England, where the problems

---

[33] Bourdon also thought along these lines in the thirties. Then, of course, the formulations and abundance of thought which Gauss had published in 1801 and which was not understood, becomes prominent.

[34] Bourdon ([47 a] 550) expressed himself as follows: ,,...l'on doute si jamais on pourra parvenir à un résolution...". He declared the same in the editions of his book in the thirties.

[35] Even in the thirties, in the 7th and 8th editions of his book, Bourdon [47 b, 47 c] evaluates the possible algebraic solution of a general equation of an arbitrary degree by saying: "le seul avantage qu'offriraient les résultats serait de confirmer la proposition hypothétique: Toute équation a au moins une racine".

[36] Practically nobody, even in France, remembered the similar results of Vandermonde at the time.

[37] For example, J. Wood [405] in a university textbook, only refers to Waring of his contemporaries. The origin of the English school of algebra will be discussed later.

of algebra, like the whole of mathematics, were considerably sterile at the beginning
of the century, the situation in Germany was no different. Textbooks were mostly of
a low standard, and translations of French books were frequently used.[38] Undoubtedly
the best German work, including an interpretation of algebra, was M. Ohm's *Versuch*
[323]. It devoted considerable attention to the fundamental concepts,[39] but it was
aimed at an interpretation of the traditional system of concepts, logically and as
accurately as possible. This acted so as to cut off all the elements of a more modern
approach to the problems of algebra.

The author divided the solution of equations into two fairly remote sections. In
the first section he dealt with the solution of algebraic equations inclusive of the fourth
degree (pp. 250–381), which, however, represents an exposition of little algebraic
interest. He treated the solution of equations in general in the 2nd part of his study
([323b], 179–259), but some of the sections concerned with the theory of equations
are included in a different connection.[40] However, a certain restriction of the algebraic part in one way prevented Ohm from expressing some stimulating ideas and
using more accurate formulations than his contemporaries.

He considered the solution of equations to be a special part of the theory of
entire functions, i.e., of polynomials. For example, Ohm was aware of the obscurities in
contemporary understanding of polynomials, and therefore he expressed that which
usually remained in between lines with other authors; namely the question about the
nature of the coefficients. That is why he said that the coefficients were "ganz beliebige
allgemeine-numerische Ausdrücke" ([323b] 179) which, as he understood it, means
that they could be real as well as complex numbers. He put the fundamental theorem
of algebra at the head of the theory of solving equations. Since, to use modern terms,
he always considered polynomials in a realm of complex numbers, it necessarily
followed ([323b] 188) that a primitive entire function is represented by the function

$$(x - x_1)$$

where x_1 is a complex number in general. In his further study of the solving of
equations he only arrived at a differentiation of the traditional numerical realms.

Apart from these introductory sections which outlined the manner of his reasoning
and the advantages and disadvantages of his approach to algebraic problems, he
devoted a special chapter to the theory of elimination which did not bring anything

---

[38] A German translation of Francoeur's textbook, which was written in the first decade and which is
predominantly of an elementary nature, was published as late as in 1839. However, the general statement
made in the text deserves analysis, although the historical literature gives no arguments contradicting it.

[39] For details of the nature and significance of Ohm's book see Chap. 4, (2).

[40] For example, the whole chapter on entire symmetric functions including Newton's formulae
(pp. 155—171) was included as part of the theory of functions. Earlier, i.e., before the interpretation of
the solutions of algebraic functions was discussed, mention was made not only of entire functions,
but also of operations with them, including division, which led him to the concept of a "fractional
function".

new. However, when Ohm tried to determine the general form of the root of a general equation, he was incapable of pronouncing anything but trivialities which can simply be expressed by saying that it is a complex number

$$a + bi$$

where a and b are real numbers. This approach in itself indicates that Ohm had not understood the more modern trend in algebra. A more extensive proof of this is given in the section "*Über die Auflösung allgemeiner höherer Gleichungen von jedem Grade*" ([323b] 239−259). In the introduction to it he stated the generally well known fact that all the attempts to present a solution of an equation of degree n were finally successful for $n = 2, 3, 4$, but unsusccessful for $n > 4$. That is why Ohm referred to the papers of Bézout, Euler and Lagrange, and only mentioned some of the methods for the sake of information. In this way Lagrange's interpretation also entered Ohm's work, but in a distorted and simplified form, as if Lagrange had only presented a single proposal for the method of solving equations.[41] He also condemned Ruffini's proof without any analysis whatsoever.[42] Ohm did not refer to any other authors or papers on algebra. His study said nothing of the algebraic results of Abel, or Gauss.

In the thirties Ohm published abbreviated excerpts ([320a, b]) of the original extensive study. In them he restricted himself as late as in 1839 ([320a] Bd I, pp. 35−36) to the stringent statement: Eine allgemeine Auflösung der algebraischen Gleichungen vom fünften und höhern Grade in geschlossener endlicher Form ist bis jetzt noch nicht nachgewiesen worden. Whereas he did not discuss the algebraic solution in any way, he devoted nearly the whole of his attention in the chapter on equations of higher degrees (pp. 166−220) to numerical solutions.

Ohm's opinion of the algebraic solution, published at the end of the thirties, is surprising only at first glance. It can be considered as an attempt to avoid the problem of the algebraic solution. An opinion of this kind would have been understandable if it were of a non-algebraist of the time. It would have corresponded to the earnest opinions formulated at various occasions. Many mathematicians had not given up hope that a solution would be presented in time. Perhaps this was supported by the little experience with proofs of existence (or non-existence). Anyway, it is known that Abel as well as Galois originally thought that they had presented a solution of a general algebraic equation of the 5th degree, but they were soon convinced of their error. A similarly deceptive solution was found by Meier Hirsch who was also quick in recognizing his error, which he corrected in his subsequent paper.[43] It was slightly harder to

---

[41] Besides other methods of solving equations, Ohm did really speak of "Méthode de la Grange", which is what he called Section 544, pp. 241—248.

[42] Der Beweis des Ruffini, dass für höhere Gleichungen, die über den vierten Grad hinausgehen, eine geschlossene (endliche) Formel für den Unbekannten nicht existiere, ist hier abermals weggelassen worden: weil er uns mehr schwache Stellen zu haben, und in der That nichts zu beweissen scheint, so viel Scharfsinn sich auch in ihm entwickelt findet. [323 b] p. 238.

[43] Cf. [202].

convince one of an error in cases where only a proposal was made, but where the method of solving equations of a higher degree than four was not analysed in detail. One of these proposals was made in the paper of M. H. Wronski [406]. Each of these studies in fact omits the result which was the purpose of Lagrange's treatise, i.e., to prove a priori by analysing the method, prior to calculation, that it would fulfill its purpose. Therefore, these authors adopted opinions corresponding to experiments made prior to the year 1770, inclusive of Euler's and Bézout's. Wronski's paper[44] together with a similar paper of Coytier, was subjected to a crushing critique by Gergonne in the third volume of the *Annales* [171, 172], where it was proved convincingly that their method, which anyway led to complicated calculations for equations of the fourth degree, was identical with Bézout's method published for the first time in 1762 [37].

If the attempts at solving equations of higher degrees were not very convincing, the first proofs of the insolvability of equations of degrees higher than the fourth were also accepted with distrust. In the same way as Abel expressed his doubts as to the accuracy and lucidity of Ruffini's proof, Abel's proof was not capable of convincing other mathematicians.[45] The proofs did not even satisfy unprejudiced significant mathematicians like, for example, Bernard Bolzano, who then worked in Prague. With his cooperation the Royal Bohemian Society of Sciences in Prague offered a prize for solving a problem in 1832. In formulating the problem it was written: Ob eine allgemeine Auflösung vollständiger literaler Gleichungen, welche von einem höheren als vierten Grade sind, vermittelst eines endlichen Ausdruckes möglich sei, muss man noch immer als unentschieden betrachten.... The Society then went on to require that either a quite accurate proof of the insolvability of these equations be presented, or that a solution be offered.[46] When the members of the Society (especially Bolzano and Kulik) evaluated the seven treatises presented after the two year limit was up, they finally reached the conclusion that none of them was an answer to the given question.[47]

As regards the confusion which still prevailed in the problem of insolvability of algebraic equations in the thirties, it would be possible to present further data of which only data pertaining to the single year 1833 will be considered. At the time, a seven-page comment on Abel written by the Parisian mathematician G. Libri [276] was published

---

[44] Wronski was certain to have known Lagrange's mathematical work. This is proved by his obstinate (and fundamentally justified) fight against Lagrange's explanation of the fundations of calculus.

[45] Cf. Pierpont, [332], in particular pp. 46—47.

[46] As regards the problem mentioned cf. K. Rychlík [346]. Those who invited the solution of the problem were dissatisfied with Ruffini's proof, which had then recently been published in an abbrieviated form in German in a Viennese journal [345]. As Bolzano's unpublished manuscript notes indicate, Bolzano also made his acquaintance with Abel's proof soon after it had been published.

[47] Among the authors of these competitive studies there were mathematicians from various countries, also E. H. Dirksen, professor and Member of the Berlin Academy, in the proceedings of which his paper was also published in 1836. Although it could not be included in the competition (it was not anonymous), Bolzano pointed out considerable gaps in its proofs in his assessment. Like the other papers, mentioned paper did not solve the given problem.

in Volume LVI of the *Biographie universelle*. The French mathematicians were already capable of recognizing the significance of Abel's mathematical results at the end of Abel's life. As already mentioned, they valued in particular his studies on elliptical functions.[48] This was also the sense in which Libri, also voicing objections to Abel's proof, wrote, that since all reasonings as to the "ambiguity of roots" were unreliable he also considered Abel's conclusion to be uncertain. Mais bientôt il s'occupa de questions plus importantes, Libri praised Abel,[49] and mentioned the graduation of the lemniscate.[50] In spite of all personal acknowledgement and the acknowledgment of the institution of which Libri was a member, Libri's attitude towards Abel's algebraic studies was critical.[51]

Drawing on the examples given, one may also put forward a hypothesis as to why so few mathematicians at the end of the 18th century and even far into the 19th century were capable of affecting the development of the theory of algebraic equations by original contributions, and thus expand on the ideas one is capable of discerning in the studies of Lagrange, Vandermonde, Waring, Ruffini, Gauss, Abel and Galois. The fact that the system of concepts in algebra had practically not changed at all is considered to be of fundamental importance.[52] The results were formulated using the earlier, or one might say traditional, system of concepts. In this way it is also possible to understand Lagrange's interpretation of the years 1770 − 1771; Gauss apparently tried this as well, in the seventh section of his *Disquisitiones arithmeticae*. Everywhere else where the interpretation deflected outside this scope of "comprehensiveness", it was not understood (like Vandermonde) or acknowledged (as was the case with insolvability). Moreover, this state was frozen by textbooks which were written, in particular in France, from the end of the 18th century and which enjoyed numerous but little changed re-editions. In the lavish flow of this production the inextensive and little known published studies of Abel and Galois were incapable of producing

---

[48] I assume this with a view to Abel's correspondence and the response to it. Abel pointed out his results in the theory of algebraic equations even to French mathematicians; he wrote an excerpt of his paper on the insolvability of equations of a higher degree than the fourth, published in Crelle's *Journal* of 1826, for the *Bulletin de Ferrusac*, he also handed the manuscript covering this topic to French institutions, etc., but commendatory response was drawn by the analytical studies, whereas no response was drawn by the algebraic studies.

[49] Abel "...il parvient à démontrer que si la résolution de l'équation algébrique du cinquième degré était possible, il en résulterait une absurdité, dérivée de la multiplicité des racines. Ce genre de démonstration, tiré de la multiplicité des racines, peut ne pas paraître entièrement satisfaisant pour ceux qui connaissent à combien de disputes on a été amené par l'ambiguité des racines, dans la résolution des équations du quatrième dégré. Quoi qu'il en soit, ces recherches resteront comme de beaux théorèmes d'analyse, lors même que la démonstration d'Abel ne serait pas complète. Mais bientôt il s'occupa de questions plus importantes." Libri [276] pp. 4—5.

[50] It is also worth mentioning the data concerning Gauss' response to Abel's proof of insolvability. Cf. ([9] Holst, p. 28; Sylow p. 16).

[51] He thought highly, of course, of his analytical studies.

[52] The subsequent chapter substantiates the fact that this went on long into the 19th century. The statement in the text is, therefore, in advance of the proof.

a turn for the better within the theory of algebraic equations. In order to substantiate this hypothesis let us revert once again to the fate of Lagrange's work, and ask to what extent Lagrange himself influenced the one-sided understanding of his endeavour in his study of 1770—1771.

Mention has already been made of the French revolutionary institutions and of Lagrange's and Laplace's work at the École Normale. Both lectured in algebra (and arithmetic), and one may say that under the given conditions their lectures were modern. They discarded the discussion as to the "justification" of complex quantities and considered real and imaginary quantities as equal in principle.[53] They stressed the general theorems on equations, many of which they presented without proof but with reference to literature. The study of the functions of roots, especially symmetric ones, came to the fore. Even Laplace drew on Lagrange's work and repeated his statements. However, when the solution of equations of the third and fourth degree was being considered ([351] 304nn) he maintained that various methods had been created for its solution but that all had the same basis, i.e., the resolvent is derived from a suitable function of the roots. The search for a suitable function of the roots then becomes "méthode de résoudre les équations de troisième degré", and he quotes Lagrange, Vandermonde and Waring as the authors. The demand for searching " *a priori*" here is therefore only connected with the demand for a suitable function of the roots. He then solved a quadratic, cubic and biquadratic equation easily, using Lagrange's method. In the conclusion ([351] 317—318) he expressed a certain hope, provided by the uniformity of the methods, that a general solution would eventually be reached. Since this has not yet happened, one must be satisfied with the knowledge that even the best of such solutions would be of little use in practical applications, and that approximate methods are still more convenient.[54]

Lagrange lectured on the solution of equations[55] in a similar way, but he expressed his opinion that general solutions have no sense and that one must concentrate on approximate solutions, i.e., numerical ones. His opinion was close to the opinion of his colleague, Laplace, with whose interpretations he connected his own.[56] In analyzing

---

[53] For example, Laplace maintained ([351], Tome II, p. 119): Quoique les quantités imaginaires soient impossibles, cependant leur considération est du plus grand usage dans l'analyse. Souvent les grandeurs réelles se présentent sous la forme de plusieurs imaginaires...

[54] Nous venons d'exposer ce que l'on sait sur la résolution des équations complettes. Les analystes parvinrent bientôt à celle des équations du second, du troisième et du quatrième degré: mais arrivés à ce terme, ils trouvèrent un obstacle que des efforts continués pendant plus de deux siècles, n'ont pu surmonter encore. L'uniformité des méthodes imaginées pour résoudre les équations des degrés inférieurs au cinquième, donnait quelque espoir de les étendre à ce degré; mais toutes les tentatives que l'on a faites pour cet objet, ont été jusqu'à présent infructueuses. Au reste, ce qui doit consoler du peu de succès des recherches de ce genre, c'est que la résolution complette des équations, quoique très-belle par elle-même, serait peu utile dans les applications de l'analyse, dans lesquelles il est toujours plus commode d'employer les approximations.

[55] Cf. [351] Tome III, pp. 276—310, 463—489.

[56] „...le cinquième degré présente une espèce de barrière que les efforts des analystes n'ont pu

approximate solutions Lagrange referred to his own paper on the solution of numerical equations which he had published in Berlin in 1769. At the end of the 18th century Lagrange complemented this treatise substantially and published it under the title *Traité de la résolution des équations numérique* in 1798; its next edition was published in 1808 [269]. These editions and supplements indicate that Lagrange's interest had turned to the numerical solution of equations. He no longer considered algebraic solutions at the time. Lagrange's opinion as to the small usefulness of an algebraic solution can again be seen in his preface of 1808 ([264] Tome 8, p. 15–16). Here he pointed out the casus irreducibile, the analogy of which one has to expect with equations of a higher degree; he also pronounced the following words to which many authors referred later: ...des formules algébriques, précieuses en elles-mêmes, mais très peu utiles pour la résolution effective et numérique des équations.... .

However, in spite of the numerical nature of the study, it does contain sections belonging to the theory of equations or concerned with algebraic solutions. These may be found in particular in the appendices, the so-called Notes. For example, Note X treats the resolving of polynomials. He reached the conclusion that any equation may be resolved into two other equations, their coefficients being real numbers depending only on the real roots of the equation of an odd degree. Even if this conclusion applied to equations, it is not general enough; it indicates that Lagrange was aware of the possibility of resolving the equation in a suitably selected field, to use the present-day term.[57] For our purposes, however, the most important are Lagrange's 13th and 14th notes, of which the former ([269d] 295–327) concerns the algebraic solution of equations, the latter binomial equations ([269d] 328–367).

In the former he gives a brief account of the principal ideas of the first three sections of his study of 1770–1771 with certain important changes in content, which he also refers to and quotes. He also repeated that the objective of the treatise was finding a common base for the solution of equations of the third and fourth degree, i.e., the use of an auxiliary equation, and the "*a priori*" determination of the degree of this

---

encore forcer; et la résolution générale des équations est une des choses qui restent encore à desirer en algèbre. Je dis en algèbre car si dès le troisième degré, l'expressions analytiques des racines est insuffisante pour faire connaître leur valeur numérique dans tous les cas, á plus forte raison le serait-elle dans les degrés supérieurs; et on serait toujours forcé d'avoir recours à d'autres moyens pour déterminer en nombres les valeurs des racines d'une équation donnée; ce qui est en dernier résultat l'objet de la solution de tous les problèmes que les besoins ou la curiosité offrent à résoudre. [351] Tome III, p. 463.

[57] The construction of the coefficients of the divisors of a given polynomial is not without interest in spite of the specificness stemming from the requirement for realness. Lagrange's Note, however, proves that it was possible to proceed from the fundamental numerical realm of rational numbers numerically to various algebraic numbers, and also to exploit some of their properties. The mathematicians of the time would have had to have, however, a sufficient reason for this; which was lacking for the purposes of a general approach. In this case it followed from the numerical solution and, therefore, it also put restrictions on the "form" of the algebraic numbers with which the calculations were carried out. Let us mention later papers, in which the resolution of polynomials was studied separately from other problems of algebra, where the polynomials in question in the various resolutions are ones in a realm of rational numbers.

resolvent, possibly also of its divisors. He included it in the study on the numerical solution of equations because "qu'il en résulte une méthode uniforme pour la résolution des équations des quatre premiers degrés", and because this method was easy to apply to the solution of binomial equations. This, of course, considerably narrows the sense of his older study, but he proceeded generally in the text itself, and considered equations of degree n (where n is natural); $n = 3$ and $n = 4$ only appear as examples. In considering the degree of the resolvent he reached the conclusion that it would in general be higher than n. However, he allows for the possibility that the degree could be decreased, but he expressed himself very sceptical as to finding the decrease a priori.[58] One can thus see, as in the overall stressing of numerical solutions of equations, of which the study is representative, a slightly different opinion in the discussion of the possibilities of a general solution of equations of an arbitrary degree. Lagrange avoided the problem. From this point of view, omitting the problems of the fourth section, i.e., the study of the functions of the roots and of permutations, also indicates that it is not substantial for further progress in the theory of equations, but only a narrowly motivated means of the said theory.

The fact that Lagrange presented an excerpt from his earlier studies has yet another historical significance. Whereas the study of 1770−1771 was only published in the proceedings of the Berlin Academy, its interpretation was published in an extended, independent publication and, together with the remaining editions of the publication, it constitutes a new instruction on how to understand the older study, 35 years after its publication. It is as if he erased the significance of the earlier study.[59]

The new interpretation differs in content as regards two more substantial features from the study of 1770−1771. Lagrange stressed the solution of the binomial equation more and prepared the ground for the following note. Besides, in the conclusion to his 13th note ([269d] 324−327) he praised greatly Vandermonde's treatise of 1771. He stressed the independence and generality of the principle by which Vandermonde eventually reached results similar to Lagrange's by different methods. He also characterized Vandermonde's ideas truthfully, the main purpose of which was to find an analytical expression for the roots of the equation satisfying the conditions mentioned, since the solution depended on this. He was also right in pointing out (p. 326) that the main difficulty was in proving the rationality of certain expressions, i.e. to express them as a function of the coefficients.[60] He considered Vandermonde's method of solution to be more direct than the method he himself explained in the Note and in the treatise of 1770−1771.[61]

---

[58] Il est possible que cette équation puisse être abaissée à un degré moindre, mais c'est de quoi il me parait très difficile, sinnon impossible, de juger à priori ([269d] 307).

[59] Lagrange referred to his older paper several times and he also referred to it in places where he was not able to explain everything, due to lack of space in this Note. But in this way he only stresses the more that the present brief interpretation presented the "nucleus" and "sense" of the older paper.

[60] Lagrange used these vague formulations.

[61] Comme la méthode de Vandermonde découle d'un principe fondé sur la nature des équations,

In Note 14 Lagrange considered equations

$$x^n - 1 = 0. \tag{3.4.1}$$

The stimulus for this note was provided especially by Gauss' *Disquisitiones arithmeticae.* He characterized Gauss' method ([269d] 329) as a procedure for reducing the solution of Eq. (3.4.1.) to the solution of as many particular equations as the number $(n - 1)$ contained primitive factors, their degrees being expressed by the said factors. In the note considered, he wanted to explain how one can obtain a complete solution of Eq. (3.4.1) for a prime-number n without employing any "intermediate" equation.[62] He realized this plan in the following pages. It is understandable that his result as well as the essence of his procedure were identical with Gauss'; he worked with the same irrationalities and thus in fact arrived at the elements of the same algebraic realms. However, he obscured Gauss' method by calculations, the method requiring for the resolving of the equation

$$x^{n-1} + x^{n-2} + \ldots + 1 = 0, \tag{3.4.3}$$

which was irreducible in the realm of rational numbers, the adjunction of the roots of a suitable equation, and he repeated this process, all the equations being "solvable". Therefore, the reader will not find the Gauss' ideas[63] which point to Galois' theory in Lagrange's interpretation of Gauss. However, one should not be unjust to Lagrange; he proceeded in accordance with Gauss' work and he elaborated one of its aspect to its end, i.e., that his results could be presented in a traditional way. Moreover, the procedure was much simpler and Lagrange reached the numerical results in the cases selected much more easily although he partly used Gauss' procedure for this purpose.[64]

In this note Lagrange praised Vandermonde as well. He pointed out that he had presented a solution of the equation

$$x^{11} - 1 = 0, \tag{3.4.4}$$

---

et qu'à cet égard elle est plus directe que celle que nous avons exposée dans cette Note, on peut regarder les résultats communs de ces méthodes sur la résolution générale des équations qui passent le quatrième degré comme des conséquences nécessaires de la théorie générale des équations ([269d] 327).

[62] ... en appliquant les principes de la théorie de M. Gauss à la méthode exposée dans la Note précédente, j'ai reconnu que l'on pouvait obtenir directement la résolution complète de toute équation à deux termes dont le degré est exprimé par un nombre premier, sans passer par aucune équation intermédiaire ni avoir à craindre l'inconvenient qui naît de l'ambiguité des racines. C'est ce que je vais développer dans cette Note ([269d] 330). It is suitable to take notice at this point of Lagrange's fears as to the ununiqueness of roots. Cf. above, comment 49.

[63] And in Galois' papers one can find traces that he was inspired, among other things, by this. This is also the opinion of the more recent historical literature, as pointed out above.

[64] For example, Lagrange worked with Gauss' periods of roots and formed functions of these periods. This is where one can observe that Lagrange mastered Gauss' procedures and ideas in his way, and he presented them to others perhaps in a simpler and more intelligible manner.

in his study of 1771, and he filled in the gaps in the latter's interpretation.[65]

The second edition of Lagrange's *Traité* was also subject to an involved analysis by Poinsot in the journal *Magasin encyclopédique* in 1808. This analysis met with Lagrange's agreement[66] and, therefore, in the third edition of the *Traité* of 1826 it formed a kind of introduction to the whole study. One may assume, therefore, that it agreed with Lagrange's opinions. Here, algebra was divided into three parts: the first is a general theory of equations, the second a general solution[67] of equations, and the third contained numerical solutions. Poinsot agreed with Lagrange and with his arguments that the latter part was most useful, whereas the general solution was of no use for computing the roots. Moreover, it is in no way certain, wrote Poinsot (pp. XII−XVI), that such a solution exists when all the endeavour of the mathematicians was of no avail. The general theory of equations, including general facts about equations and represented by Lagrange's work of 1770−1771 and Vandermonde's of 1771, to which Poinsot also devoted sufficient attention (p. XIII nn), forms the basis for both parts. The generality is represented in this case by the "perfect independence" of the roots. Therefore,

$$x^n - 1 = 0,$$

is a special equation, and Poinsot pointed out that Lagrange reverted to his old ideas in his note 14 (p. V), and he praised the straightforwardness of the method. As regards the actual interpretation, Poinsot kept to Lagrange's text and in fact contributed to the expansion of the form of Lagrange's ideas which were expressed in 1808.

Not long after, in 1810, Delambre [68] published an extensive study on the development of the mathematical sciences from 1789 [115]. The section on algebra is brief and very reserved as regards formulations. Once again the implication was made that Lagrange showed the complexity of the general solution, Ruffini is mentioned, whose objective it was to prove its impossibility, and therefore Lagrange devoted himself (like others) to numerical equations.[69]

---

[65] Vandermonde presented the result without a detailed explanation of the method. Lagrange (p. 355) said of Vandermonde's result: "…et personne, après lui, ne s'est occupé, que je sache, à vérifier ce résultat, qui paraît même être resté ignoré".

[66] Compare with the note to the publication of this analysis in the 3rd edition of the *Traité* [269b].

[67] "…résolution générale, qui consiste à trouver une expression composée des coefficiens de la proposée, et qui, mise au lieu de l'inconnu satisfasse identiquement à cette équation…" ([269 b] p. VI.).

[68] The book [115] has 272 pages and contains the material presented to Napoleon in 1808. Lagrange was characterized in it as one of the foremost French scientists.

[69] Si de la géométrie nous passons à l'algèbre ordinaire, nous trouverons des progrès moins sensibles mais infiniment plus difficiles. Les mémoires de M. Lagrange sur la résolution complète des équations littérales, en réduisant le problème à ses moindres termes, avaient montré combien il est encore difficile. M. Ruffini se proposa de prouver qu'il est impossible. M. Lagrange voulut du moins faciliter la solution des équations numériques…. Apart from Lagrange's credit with regard to the solution of numerical equations, also Budan and Laplace are mentioned. Mention is made of Gauss who" … décomposa depuis

Similar opinions can also be found in articles published, for example in Gergonne's Annales.[70] Lagrange was frequently quoted in them and his method of constructing the resolvent used, but the elements to which his study of 1770−1771 pointed in the development of algebra, no longer appeared. One may assume, therefore, that in his later years (he was 72 when the second edition of the *Traité* was published in 1808), Lagrange himself, together with Laplace, Poinsot, Delambre, etc., formed the opinion that his earlier work did not support more substantial changes in the development of algebra.

---

en facteurs du second degré, des équations dont l'abaissement paraissait impossible: il donna les moyens d'inscrire au cercle, sans employer que la règle et le compas, des polygones dont le nombre premier (de la forme $2^n + 1$) …. These data from p. 8 and 9 [115] were repeated in an altered form on p. 65 and the subsequent; Ruffini is also named (Ruffini s'attacha àprouver directement l'impossibilité…). The nineth volume of the proceedings of the Italian Scientific Society, in which Ruffini's paper [334] was published, is also mentioned.

[70] Lagrange himself quoted Bret's paper (p. 321 nn), which was published in the *Correspondance* (p. 218 nn, Vol. II), and which connects up with Lagrange's lectures at the École Normale. The influence of Lagrange can also be seen in Pilate's, Gergonne's, and Encontre's work, as well as that of others. This whole, with Lagrange's explanations, indicates how the tradition of algebraic interpretations and studies was established, at least in France, which then represented the mathematical vanguard.

CHAPTER 4

STUDY OF THE STRUCTURE OF NUMERICAL REALMS

4.1 Definition and concept of the problem

At the beginning of the 19th century different opinions prevailed as to the definitions of the individual mathematical disciplines and their mutual relations. Nevertheless, it is necessary to consider the arithmetico-algebraic region as a whole until the later years of the 19th century in connection with many problems. Its unity formed a common base in principle in the form it had established much earlier; the object and purpose were the study of quantitative relations expressed by numbers or numerical symbols, the relations or quantities being finite as opposed to infinitesimal calculus. Considering the differences in opinion which were then being expressed as to general problems, it is understandable that no fixed and precise boundaries between the individual spheres existed or could have existed for practical reasons. The use of infinite procedures in the arithmetico-algebraic sphere was not exceptional, however, some opinion remained against the use of methods or results from other disciplines in proofs. Concretely, these objections were concentrated on the use of geometric procedures in proofs of arithmetical and algebraic theorems, which could have been partially a response to the emancipation of the arithmetico-algebraic sphere won earlier.

As already mentioned, numbers and numerical symbols were the fundamental material with which one worked in this sphere and which established its unity.

In the course of the period investigated the problem was in which "numbers" were to be considered, how they were to be defined, and how the numerical realms were to be divided internally. However, the answers to these questions were not given in a vacuum but in context with various attempts, successes and failures within the whole of mathematics. First of all, these problems reflected the earlier tendency towards a substantiated "axiomatic" interpretation of the whole sphere (Section 2). However, the substantial and insubstantial properties of the various numerical realms could only be differentiated when more different realms were known whose comparison should be used to uncover the similarities and the differences in their structures. As regards the tradition in algebra and arithmetic, this was made possible by the expansion of the realm of real numbers, i.e., by excepting complex numbers and by a more varied differentiation of the fields within the realm of real numbers at the beginning of the study of algebraic numbers. Both these important events were occurring around the beginning of the 19th century when they acquired the form of a definite and general acceptance of the realm of complex numbers as a realm within which the whole arithmetico-algebraic sphere operates. This is also when the build-up of the arithmetic of algebraic numbers began. Both processes provided fundamental material for a new study of the structures of numerical realms. The inherent objective of this chapter is to determine how these processes evolved, what form they took, how they conflicted with traditional methods and ideas, and how they influenced the change of those methods and ideas. However, there was another impuse towards the differentiation of numerical realms and towards new possibilities of their concepts in the first half of the 19th century. This was the study of the solution of algebraic equations, which helped create the concept of a field and which also belongs to this chapter because it was a problem of a number field and because it represents, in a way, a climax to the solution of the other problems discussed here.

This chapter is limited to the study of concepts and methods which ensued from numerical realms and the quantitative problems of the traditional sphere of arithmetic and algebra. The period investigated segregates from the development of algebra the elements which perhaps contributed more to the formation of a modern algebra, from a broader historical perspective, than those to which attention has been devoted in this chapter. Namely, the development of the study of "non-numerical" realms will be omitted here although one can see that only through the study of such "untraditional" realms was it possible to arrive at the abstract study of algebraic structures. The forms and influence of the study of these realms are discussed in the next chapter. However, the chapters complement each other. This chapter records the ideas stemming from traditional concepts and searches for new features in them, as well as for further possibilities of development, while the next attempts to indicate the connections between the origin and increasing importance of new concepts whose treatment had in fact exceeded the traditional understanding of the arithmetico-algebraic sphere.

4.2 Continuation of the traditional trend

The fundamental concepts and the problems of approach in the arithmetico-algebraic sphere were mostly interpreted in comprehensive compendia and textbooks, and practically not at all in specialized articles. The number of published textbooks rapidly increases at the end of the 18th century. Thus didactic aspects take precedence, whereas intellectual aspects which were propagated from the middle of the century were repeated. In this way the development of textbook literature more or less embalmed older opinions and multiplied the force of tradition.

Of course, these textbooks also differed in standard, expecially as regards the readers or schools for which they were intended. One cannot say that as a whole they made no attempt to overcome traditional difficulties. However, since they employed traditional means for this purpose, the progress made is negligible. Their methods are based on verbal definitions of the numerical realms and operations. These were then only reliable and satisfactory for the realm of natural numbers. All further attempts at expanding the field of their validity and extending them to other realms under consideration ended in failure.

Nevertheless, as part of this fruitless endeavour, mathematicians reached partial results in certain respects. A stimulating idea, propagated in particular by Kästner's compendium [221], was the idea of "opposite" quantities, of which one was positive and the other negative so that "their sum was equal to zero". The understanding of some of the properties of these operations was similar; for example, the commutativity of operations began to be proved as one of the theorems (cf. [168]). It was not difficult to "prove" commutativity for addition or multiplication of natural numbers; however, in generalizing, difficulties arose from the endeavour to define operations verbally but in a different form. The definition of the multiplication of an irrational number by another irrational number was beyond the capabilities of the time, especially beyond the traditional interpretation. Thus, the failure of any attempt to prove the properties of this multiplication was made more likely.

This situation, which has been described well in literature (cf. [295]), led to a paradox in the foundations of the whole arithmetico-algebraic sphere at the beginning of the 19th century. The important endeavour to provide the whole arithmetico-algebraic sphere with a logically satisfactory explanation according to the axiomatic system of geometry led to the restriction of the numerical realm. This was because serious differences in the introduced definitions of numbers and operations appeared with negative, to say nothing of complex, numbers. In spite of all attempts it proved impossible to find new satisfactory definitions. On the other hand, the parts of algebra which made more substantial progress were based on concepts and procedures which had not been explained sufficiently. New facts were accumulated about them, but insufficient explanations did not prevent them from developing, only from being acknowledged. This sphere provided a further example of how an effort for accuracy became a hindrance to progress in the history of mathematics; in this case, because the conceptual system was practically stabilized by the propagation of Euclid's *Elements*.

However, the fruitless search for ways of overcoming the logical difficulties in the foundations of the arithmetico-algebraic sphere did have one positive result. Apart from giving birth to various new ideas, two of which have been outlined, there appeared a better understanding of the difficulties inherent in trying to present a logically substantiated interpretation, i.e., realizing the unfavourable consequences resulting from the various alternatives.[1] This was the source of the effort to refine the conceptual

---

[1] If one ignores the various alternatives in defining numerical realms and operations within them, which do not mutually differ much and are therefore not especially important, only two contradictory approaches remained. The authors of the time fundamentally had the choice of either introducing certain restrictions into the definitions of operations, or of extending suitably (intentionally or unintentionally) the numerical realms. Accordingly, in the former case one speaks of subtracting a smaller number from a larger one only, and in the latter it is then necessary to introduce negative numbers as equivalent to positive ones. The acceptance of taking the root without restriction necessarily lead to the acknowledgement of $\sqrt{-1}$ again as a number equivalent to the real numbers. To master this

means of describing numerical realms and operations within them. Whereas these efforts were distributed through various textbooks, all of which may be considered unsuccessful on their own and quite as insignificant,[2] a continuation of these efforts and, in a certain sense, their culmination was reached in the only comprehensive attempt at a logically presented interpretation of the arithmetico-algebraic sphere. Its author was the brother of the well-known physicist and professor at the Berlin schools, Martin Ohm. His extensive *Versuch eines vollkommen consequenten Systems der Mathematik* [323] contains an interpretation of the whole of pure mathematics with the exception of geometry. In extent it is an exceptional continuation of the extensive textbook courses of the 17th and 18th centuries, but its didactic purposes were completely suppressed. The geometry was omitted in this comprehensive interpretation for several reasons. Ohm reached the conclusion that mathematics (p. 2) were composed of two parts, i.e., "Zahlenlehre" which was its more general part, and "Grössenlehre" which was a mere application of the theory of numbers to quantities. Spatial quantities, i.e., geometry, and quantities of force (and motion), i.e. kinematics and mechanics, had a special status among these quantities. Thus, Ohm proclaimed himself a consistent advocate of the opinon that the algebraic and arithmetical sphere, in which he included the whole of calculus, was fundamental; geometrical quantities were only a special case. The principal argument which guaranteed the "superiority" of geometry with respect to arithmetic for centuries, i.e., the continuty of geometrical quantities (since arithmetic apparently only operated with certain "numerical" quantities) was not considered particularly. He considered arithmetical quantities to be capable of expressing continuity as well. If Ohm wanted to deal with the general and fundamental parts of mathematics, he could omit geometry; the theorems of which were said to be the consequence of generalities, although he did not specify in which respect.

Elsewhere (in the Preface) he maintained that geometry was only a negligible part of mathematics, in which it became as lost as a drop of water in the sea. Nevertheless, it has an important status which Ohm did not consider to be the result of its subject but of its method. He dealt with the significance of its method not only in the preface to this study but also in a special earlier treatise devoted to considerations of method in mathematics in general. The treatise was published in 1819 under the title *Kritische Beleuchtungen der Mathematik überhaupt und der Euklidischen Geometrie insbesondere* [322].[3] Euclidean geometry as a systematic interpretation of geometrical truths,

---

approach logically and conceptually was beyond the possibilities of the time. This became quite evident with complex numbers. If a quantity is a thing which may be decreased or increased, as all the courses and elementary textbooks repeated again and again, how could one then accept a complex number to be a quantity? However, this situation and these contradictions could have been discerned in authors of the first half of the 19th century as well as with mathematicians of about 1750.

[2] Molodshii [295] in recording conditions in the first half of the 19th century did not in fact find any ideas which differed more substantially from those expressed during the 18th century.

[3] The book is called *Für Mathematiker und Nichtmathematiker als Einleitung in dessen Revision der Mathematik*.

presented by Euclid and others, represents an example for Ohm from which the mathematical method of the time was obliged to stem. At the same time, however, Ohm expressed his deep concern that the Euclidean method had not been sufficiently developed even in geometry. Only Ohm's principal ideas will be pointed out here, without giving their comprehensive interpretation.

Ohm did not agree with Kant's approach to geometry, which stemmed from a philosophical conviction that a spatial conception is given *a priori* (pp. 5–6). He pointed out, on the other hand, that Euclid worked with real spatial and temporal quantities in phenomena (p. 6, § 8a). Elsewhere he also stressed the necessarily empirical nature of geometrical facts, objectivity being the final decision as to the correctness or incorrectness of a statement. He also maintained that the fundamental facts are given (§ 21);[4] in spite of everything, geometry remained in the world of phenomena (§ 22). However, this was not, according to Ohm, in contradiction with the requirement for an accurate and logical build-up from the fundamental theorems (§ 23) on which all other theorems were based. It can be said, therefore, that the real, sensually perceptible world is the foundation of geometry, the structure of which should be axiomatic.[5] He says that the meaning of the fundamental theorems is immediately understood, although his further deliberations did not sufficiently clarify how to reach these fundamental theorems, in a way which would guarantee their unique meaning (also in relation to a single existing model), etc. Nevertheless, Ohm pointed out the deficiencies in the contemporary interpretation of geometry in several examples of definitions and fundamental theorems, as well as in the method of construction.

Even though Ohm called for a reformation of geometry in order to achieve a higher accuracy in its motivation and logical construction, it remained an example for him with regard to the rest of mathematics. He proved "dass unsere Arithmetik, Algebra und was darauf gegründet ist, kann der Schaffen einer Wissenschaft, sondern vielmehr ein Chaos von durcheinander geworfenen, fremdartigen, ohne alle Verbindungen dastehenden, ganz grundlosen und meist unrichtigen Behauptungen zu nennen ist' (§ 20, p. 17). Ohm's study *Versuch* [323], the first edition of which was published in 1825, is devoted to the logical reformation of the whole of mathematics (as regards his approach). It may be possible to find certain nuances, which are not, however, important for our purposes, in his approach to mathematics and in the relations between its individual parts among the studies of 1819, 1825 and 1828. One could also compare

---

[4] "Die Geometrie ist die Wissenschaft, welche die Vergleichung der Raumgrössen zum Gegenstande hat, unter der Voraussetzung, dass uns der Raum und mit ihm die verschiedenen Richtungen, seine drei Dimensionen, mithin auch die Linie, die Fläche, der Körper, ferner das Geradseyn und das Krummseyn der Linien, als Elemente, die keiner weitern Zergliederung fähig, unbedingt gegeben sind, und dass man unter Raumgrösse jede zwischen bestimmten Grenzen enthaltene Linie, Fläche oder Körper verstehe..." ([322], § 21).

[5] At the same time he stressed that it was necessary to deprive mathematics of the "metaphysical properties" of philosophical deliberations, etc., and he also maintained that mathematics was more capable than other sciences of extricating itself from these deliberations (§ 9).

these studies with some of Ohm's other books in which didactic motives were more important and which defined the mathematical objects more clearly for students.[6] However, let us concentrate on the overall nature of Ohm's construction of the arithmetico-algebraic sphere.

He expressly considered the objective of building it up according to Euclid's example.[7] However, he was or the opinion that Kästner had made the most headway in his compendium,[8] but Lagrange and others were also due for credit.

The division of mathematics into "Zahlenlehre" and "Grössenlehre" has already been mentioned above. Ohm considered unnamed integers to be a superior term, he named integers an inferior term; quantities are always named numbers, i.e., numbers of a certain kind, of which the special ones must first be developed into more abstract and more general terms. Proclaiming the concept of a number as a simple given term (p. 1) does not contradict his philosophically realistic opinion, which acknowledges deriving abstract mathematical concepts from opinion (e.g., through sensory means as in geometry). By this he understands an abstract number[9] which we consider to be a distinct and generally accepted fundamental concept.

In the preface he stressed that his objective was not to study the properties of quantities but the properties of operations. In this way a new and fresh idea was incorporated into the study, which could have affected or inspired further progress considerably more. However, in Ohm's interpretation the stressed operations also have a quite special form. Not the operations, but the directly given concept of a number he considered to be the initial concept, from which the properties of the operations then follow.[10]

New and very important is Ohm's procedure by which he defines various parallel numerical realms and the operations introduced in them. He considered abstract

---

[6] For example [320a] and [320b] were intended for technical institutes of a higher level, but also for secondary schools and self-educated people. For example, in the introduction ([320b] 1 nn) he explained that time and space were before us constantly and that they were continuous. Only then did he go on to derive a number. However, he then arrives at a conception of mathematical disciplines in the same sense as in the remaining studies.

[7] The introductory words [323] or "Vorrede" are: "Der Verfasser hat... keine andere Absicht, als für die gesammte Mathematik das zugeben, wodurch Euklid's Elemente für das Studium der Geometrie so wichtig geworden sind."

[8] Cf. [322] p. 63, § 82, where he praised him "dass er die schon vorhandenen mathematischen Wahrheiten mit der grösstmöglichsten Strenge zu erweisen versuchte".

[9] 1. Grundsatz. Die Zahl und mit ihr die absolute (abstrakte) Einheit, wird hier, als uns gegeben angesehen ([323a] p. 7).

[10] He said ([323a] Vorrede, p. VII) literally: "...Eigenschaften der Operationen, welche letztere aus der Betrachtung der Zahl mit Notwendigkeit hervorgehen, und welche Gegensätze und Beziehungen äussern, die in dem ganzen physischen und psychischen Haushalte der Natur allenthalben sich wieder nachweisen lassen". To the quoted excerpt he was adding a note in which he stressed the applicability of mathematical statements even more. Although he never actually said it, even the quotation mentioned indicates that the relations which mathematics prove are abstracted from the outer world and that they have various models in the outer world permitting application.

integers to be given intuitively and directly. One may imagine that these numbers are "composed" in various ways. They may be added, multiplied and raised to a higher power. These are direct operations by which one proceeds from two numbers to a third. It is not even necessary to point out that one can make do with current definitions[11] because the numbers in question are only integers. As opposed to these three so-called "Verbindungsarten", Ohm considered indirect operations which he derived in the following manner: If addition is

$$a + b = c, \qquad (4.2.1)$$

where a and b are the initial numbers and c the resultant number, the reverse operation (which he called subtraction) represents finding a third number, i.e. b or a, corresponding to the pair a and c, or the pair b and c, respectively. He himself said (p. 8) that three direct operations led to determining six indirect ones, but both the operations derived from addition as well as both the operations derived from multiplication, Ohm considered to be identical, without giving further reasons for this.[12] However, the operation of "raising to a higher power", i.e.,

$$a^b = c, \qquad (4.2.2)$$

will yield two indirect operations which Ohm called finding the root and taking the logarithm, depending on whether the result of the operation was a or b, respectively. The indirect operations are defined by the direct ones. In the brief introductory part (pp. 7 − 10), in which the said definitions were given, he briefly defined (§ 7) one of the fundamental terms of his whole interpretation of the arithmetico-algebraic sphere, i.e., the term "expression".[13] This he defined as an arbitrary numerical symbol or as an arbitrary symbol which has the nature of a numerical symbol.

---

[11] Cf. [323a] § 3, pp. 7—8. An attempt was made at paraphrasing Ohm's interpretation in the text. In the introduction he actually wrote: "Zwei Zahlen können aber zunächst auf folgende Arten mit einander verbunden gedacht werden: 1. Man kann sie zu einander addiren, d.h. man kann sich eine Zahl denken, die so viele Einheiten hat, als die beyden gegebenen Zahlen zusammengenommen haben". Other operations were defined in a similar way. It is also worth mentioning that Ohm spoke of imagined numbers.

Ohm's text, which had not been subject to a more detailed historical analysis, in spite of its great celebrity in its time, contains outlines of various trends which played a considerable role in further development. For example, the proposal to study the properties of operations, reminds one of later important ideas of the British school of algebra (which, as regards Ohm, was perturbed by the fact that these operations were derived from the properties of numbers). In a similar way, the future approach of Grassmann reminds one of the attempt at studying operations in general, i.e., what the various operations could be. The analysis of these historical connections, however, is outside the scope of this section.

[12] The commutativity of addition and multiplication is assumed. However, from his own point of view he never made an assumption of this kind; he did maintain that the properties of operations follow immediately from the given concept of abstract (natural) numbers 1, 2, 3, ...

[13] "Jedes Zahlzeichen, und späterhin schon jedes Zeichen, welches die Form eines Zahlzeichens hat, nennen wir Zahlen-Ausdruck oder schlechtweg Ausdruck". The formulation of the definition itself indicates that Ohm was aware that he was slightly in advance of the interpretation; he had not yet introduced any other numerical symbols but natural numbers.

In the subsequent chapters Ohm presented an interpretation of the properties of operations. This is where he applied the term "expression". According to his opinion the difference $a - b$ only represents a real number[14] if $a > b$. However, this does not mean that one cannot calculate with the expression $a - b$ (carry out operations) even if $a \leq b$. He listed the properties of the operations of addition and subtraction very carefully, and he expressed them in terms of equations. He maintained that the operations were unique, and also pointed out the general validity of the equality

$$a + b = b + a. \qquad (4.2.3)$$

Further on he also mentioned the relation

$$(a + b) + c = (a + c) + b = a + (b + c). \qquad (4.2.4)$$

His interpretations are only a little more complicated because he differentiates between equality for addition and subtraction and their various combinations. Without having had a special name for it, he mentioned the associativity and the commutativity of addition as well as a number of other properties of adding whole numbers. Without giving special reasons for his conclusion he listed (§ 14, p. 20) the theorems from which all the others could be derived:

1. $a + b = b + a$,
2. $(a + b) + c = (a + c) + b$,
3. $(a - b) + b = a$,
4. $(a + b) - b = a$,
5. $a - (a - b) = b$,
6. To add to or subtract from the "same" number a "same" number yields the "same" number,
7. If two expressions are equal to a third they are equal to each other.

These rules as kinds of axioms allowed Ohm to derive a number of theorems among which he also introduced the concepts of a positive and negative number (opposed numbers, increasing and decreasing); he also exploited the sign rule hidden in axiom 5.

The purpose here is not to assess the amount of interesting detail to be found in Ohm's text. Let us concentrate on the principal trend. The interpretation of the properties of multiplication and of division, derived from it, is simple. Among other things, Ohm maintained (p. 59) that all the theorems of addition and subtraction also hold when addition is substituted by multiplication and subtraction by division. That is why he intentionally pronounced the parallel (§ 48, p. 63) of the theorem similar to those mentioned above, and maintained that all the others could be derived. Therefore, he established the operations of addition and multiplication as a certain analogy. Without expanding on this analogy, he discussed the distributivity of addition and multiplication in a special chapter, after discussing a larger number of

---

[14] In the spirit of Ohm's realism only natural numbers are "really" numbers; all the rest are only numerical expressions.

theorems, and he included it in the list of axioms of addition and multiplication[15] in the form

$$(a + b) . c = a . c + b . c. \tag{4.2.5}$$

The operations defined, determine the numerical realms, or even better, various kinds of numerical expressions. He named five kinds (§ 111, p. 123):

1. positive integers;
2. positive fractions;
3. negative integers;
4. negative fractions;
5. zero.

He called these five special "numerical forms" summarily "real" numbers, and the numerical forms which could not be reduced to these five kinds were called "imaginary" numbers.[16]

In interpreting the operation of raising to a higher power and the two corresponding reverse operations, he could no longer arrive at an analogy between addition and multiplication. However, he arrived at a system of 21 initial theorems (axioms) from which the others could be derived (§ 150, p. 179). He was also aware that the defintions of the operations were originally only introduced for natural numbers, but he considered it a matter of course that one would not arrive at contradictions if one also applied them to his "real numbers". He "proved" this (Chap. 6) by indicating the validity of the axioms mentioned earlier; nevertheless, this whole section is, at it best, only a deliberation.

In this part of his book he also introduced irrational numbers (§ 163, p. 190). If the root $\sqrt[b]{a}$ is not real, then "diese nur in der Idee lebende, der Wurzel $\sqrt[b]{a}$ gleiche, gebrochene Zahl heisst... eine irrationale Zahl". He also introduced logarithms in a similar way in some cases.

The main features of Ohm's interpretation of the fundamental concepts and operations, on which the whole construction of the arithmetico-algebraic sphere rests, have been pointed out. Drawing on these concepts, Ohm also dealt with the theory of equations, which he considered to be a part of this sphere just like the theory of numbers. However, these sections no longer display new and more stimulating ideas, and are outside the scope of this chapter.

Should one want to include Ohm's book in a broader historical context, one should have to consider three aspects. In the first place, Ohm's book is a climax in the trend which had been developing since the beginning of the 18th century, and which had attempted to present a logical and comprehensive interpretation of the arithmetico-algebraic

---

[15] He summarized these axioms specially in § 82 on p. 100.

[16] "...und alle in der Folge durch das Verallgemeinern der Wurzeln und Logarithmen... noch entstehenden speziellen Zahlformen, welche nicht auf reele gebracht werden können, werden imaginäre Zahlen genannt."

sphere. Ohm mostly reacts to traditional problems by traditional means. However, at some points he was on a higher level than the older representatives of tradition. This includes, e.g., his system of axioms, the stressing of the importance of operations, and expressing their principal properties. That Ohm represents a continuation of the logical development of the traditional trend is also substantiated by the fact that the newer theory of algebra is not reflected in his interpretation. A very marked example of this is his understanding of operations. He only considered operations to be operations with numbers, and that they were only defined for numbers, or, in Ohm's spirit, their properties follow from the concept of natural numbers itself. They do not represent rules of "composition". Although examples were already known at the time, e.g., from Gauss' *Disquisitiones arithmeticae*,[17] the operations Ohm introduced represented operations in known numerical realms. Finally, the third aspect under which it is convenient to judge Ohm's work is its effect on further development. Especially the latter indicates that Ohm's *System der Mathematik* was a conclusion to a trend of development; its subsequent continuation, provided one might speak of a continuation at all, did not bring anything new and creative. Nevertheless, Ohm's book drew a response and favourable appraisal, especially from English mathematicians who were thinking about a new approach to algebra at the beginning of the thirties.[18] The considerable difference between Ohm and them was that they drew on the contemporary progressive tendencies in algebra in establishing their new approach to it, and that they tried to include new discoveries in their system. On the other hand, two of Ohm's ideas could have had a stimulating effect on them. First of all, there was his formalism; he expressed the properties of the operations by formal equations which simultaneously played the role of axioms. Secondly, there was his method in which he tried to extend the validity of these formal equations, originally introduced for the realm of natural numbers, to the realms of rational, real and complex numbers. The essence of this method resembles the so-called principle of permanence.

Whereas Ohm's ideas enjoyed widespread effect from the publication of his *System der Mathematik* and his other texts, a diametrically different fate was in store for the later mathematician who, drawing on the same mathematical stimuli as Ohm, deliberated on the foundations of mathematics and devoted his attention to the arithmetico-algebraic sphere as well. This was the Prague mathematician Bernard Bolzano who worked on his attempt for a comprehensive interpretation of mathematics in the thirties, which he called *Grössenlehre*. This study remained unfinished in manuscript and it could not have had a direct influence on the development of mathe-

---

[17] For example, the composition of quadratic forms or calculations according to an integral modulus; cf. Chap. 5, in particular 5.2. Of course, in the twenties one could have spoken also of operations with geometrical quantities. However, in Ohm's system geometry is only applied to pure mathematics. Here, operations and their properties have not been directly applied to geometrical quantities but numbers were used to give and solve geometrical problems.

[18] Peacock knew and appreciated Ohm's work. Hamilton read his study and other mathematicians also studied it. It is a question about the influence of Ohm on H. Grassmann's Ausdehnungslehre.

matics in the last century.[19] Nevertheless, I consider it suitable to deal with it as with one of the documents of mathematical thinking of the time.

Bolzano's work, in so far as it concerns the arithmetico-algebraic sphere, is concurrent with Ohm's in many ways. Moreover, Bolzano knew Ohm's papers well and he commented them all favourably; in many cases agreement prevails between the authors.[20] Bolzano, like Ohm, drew on the traditional discussion of the principal problems pertinent to the sphere investigated, and both conformingly attribute no significance to modern results in algebra as regards the solution of the problems discussed (again both of them ignore Gauss). Both are also in accordance as regards rejecting Kant's philosophy, together with its consequences for the philosophy of mathematics. Both, in accordance with the period and tradition, consider arithmetic and algebra to be the abstract part of mathematics; the theory of space, concerning geometrical quantities only being a special part of mathematics. One could name a large number of identical opinions held by both authors. Nevertheless, their respective systems are also quite far apart. Philosophical opinions and the philosophical approach led Bolzano to reform Ohm's system and complement it with his own deliberations. Bolzano's opinions regarding mathematical method, which he explained in his paper *Wissenschaftslehre* [40] written in the twenties and which he expanded in an extensive section of the introduction to the *Grössenlehre,* led to stress on the concepts of mathematical objects and of the precise logical determination of relations between concepts. One might possibly say that mathematics became for him a sphere to which he applied his logic, conceived very modernly for the time.[21] This statement is not quite accurate because the philosophical opinions hidden in Bolzano's theory of the objective nature of theorems and truths in themselves (Sätze an sich, Wahrheiten an sich), affected the fundamental mathematical precepts at least as much as the tendency towards a consistent application of logic itself.

As regards mathematics Bolzano seemed to be a realist; mathematical objects really exist. This applies to numbers; however, real numbers (wirkliche Zahlen) are only natural numbers. The remaining numbers were derived as in Ohm's case; they only represented numerical expressions or, in the sense of his philosophy, they were only

---

[19] Only certain parts, particularly interesting mathematically, have been published from this study, provided they had been concluded, e.g., the so-called *Funktionenlehre* [41], *Zahlentheorie* and Bolzano's interpretation of the theory of real numbers [42]. Dr. K. Večerka, who obliged by letting me have substantial parts of the manuscript of the said *Grössenlehre* in typescript, is preparing the edition of Bolzano's mathematical studies (including the manuscripts) as a whole. Bolzano's mathematical manuscripts, with certain exceptions, are in the Vienna Nationalbibliothek.

[20] Whereas Ohm did not quote his paragons or antagonists, Bolzano, in a spirit which reminds one of scholastic quotations, attempted to quote authors who had the same opinion, in spite of his stress on logic and even in cases when the agreement was questionable. As regards Ohm's quotations, it was not a question of proving his own opinions by quoting an authority but of an agreement in opinion in many concrete questions.

[21] As regards Bolzano's mathematical logic, cf. e.g., Berg [28] where other references are also given.

imagined numbers. Some are even contradictory in themselves. To use modern expressions, Bolzano only considered the powers of sets to be real numbers. Since he did not acknowledge an empty set, zero was not a real number, but infinite numbers were real because infinite (enumerable in this case) sets really exist. He considered the sequence of natural numbers to be a sequence of sets of which the first only had a single element and each subsequent, one element more than the previous. This new and really interesting approach to introducing sequences of natural numbers was complemented by a deliberation which was identical with Peano's axioms when given a modern interpretation. However, it only had an existential meaning (for proving the existence of an infinite sequence of natural numbers), and at the same time perhaps a descriptive meaning.[22] Bolzano's introduction of the concept of operations was also very original for the time. Immediately after introducing the concept of a number into his study he included a section which dealt with the functional relations among numbers and which presented Bolzano's modern definition of a function known in other connections. Bolzano then went on to define operations as a special case of a functional relation in which a third element is attributed to two elements (of a set of natural numbers).[23] He explained the manner of attribution first in terms of concept (addition may be interpreted as the integration of disjunctive sets), and then presented the symbolics. Ohm's formal approach is alien to him; he wanted to include everything in concepts to which he attributed terms (words). He considered symbols to be only auxiliary. However, this did not prevent him from stressing specially the fundamental properties of arithmetic operations, i.e., associativity and commutativity. He introduced both forms for the distributivity of addition and multiplication, i.e.,

$$m(a + b) = ma + mb, \qquad (a + b)m = am + bm, \qquad (4.2.6)$$

to which he was prompted by the original differentiation of the multiplier and the multiplicand. He introduced the distributivity even before explaining the commutativity of multiplication.

It may be said that he introduced more extensive numerical realms and operations with their elements with greater care than was usual at the time. At the same time, he considered the interpretation of operations to be more of a technical matter, whereas the definition of numerical realms represented a principal difficulty. This approach then gave rise to his interpretation of rational numbers, as well as to Bolzano's very important attempt at presenting a theory of real numbers [42].

---

[22] Bolzano's interpretation is very complicated, interposed by complementary subsidiary deliberations; he drew on a complicated system of auxiliary concepts which again require the author's special explanation. The foundation on which Bolzano draws is in one of the introductory parts which, on nearly 200 pages of typescript (manuscript 73ª — 188ᵇ), explains that which is proclaimed in the title "*Vorkenntnisse*". He also explained a lot about the concept and properties of series (and sequences), from which he was able to deduce the properties of a natural numerical series. The introduction of natural numbers he included in the well-known paper "*Paradoxien des Unendlichen*" (1852) later (in the forties); this paper was also known to the originator of the theory of sets, Cantor.

[23] Bolzano used the set terminology currently.

Even this very brief outline of the contents, the concepts, and the method of Bolzano's study is sufficient to indicate the novelty in it as well as his scientific prevalence with respect to the period when it was formulated. It is proof of the fact that a logical analysis, founded on suitable, philosophically substantiated concepts, could have had a different trend than Ohm's work, even though originating from traditional problems and stimuli. However, Bolzano's approach is quite distinctive, and one would search in vain for parallels to many of his ideas in the first half of the 19th century, although ideas close to his own played an important role in other sceintific connections later.

Another expressive feature can be added to the characteristic of the traditional trend leading up to the first decades of the last century which have been presented here. All the mathematicians maintained a realistic approach to the subjects of their deliberations; the numbers are defined objectively and the mathematician learns their properties. The extent of this single existing model which ensures extramathematical applicability, however, narrows down until only natural numbers remain from which the others are "derived". The material of the traditional deliberations allows for the expression of the various properties of numbers and operations, as well as for finding the determining and essential properties. But the definitions of numbers and operations still remain a description of the properties intuitively derived from a single model with which the mathematician is confronted. This approach and the whole conceptual system can only be changed by an accumulation of new mathematical experience different from the contemporary. Only this is capable of making further generalization possible. The attempts at presenting comprehensive systems produced by Ohm and Bolzano, and which represent in their way the climax of long years of traditional effort, manifested a fact which frequently plays its role in mathematics, i.e., that logic on its own, leading the logical development, is only capable of indicating the possibilities of a further change, but incapable of effecting this change. For this it lacks reasons and experience which alone can evoke new generalization.

4.3 Arithmetic of algebraic numbers

The knowledge of the properties of algebraic numbers is important for the origin of modern algebra for many reasons. However, two of these are the most significant. First of all, there is the study of algebraic numbers which helped the definition of numerical realms other than those current at the time. Secondly, as these number properties were found even the foremost mathematicians considered them to be paradoxical; this in turn gradually forced them to introduce quite new definitions of the properties of numbers, and therefore also to give known theorems a different form. In this way the study of algebraic numbers contributes to forming the concept of a numerical realm and to creating doubts about the traditional theory of numbers; in this way a transition to studying subjects with "untraditional" properties was being prepared within a traditional sphere.

This significance of the study of algebraic numbers only became clear retrospectively, especially after the seventies, when the concept of a field received a clear-cut form. The study of algebraic numbers itself, however, was born of the actual problems of the theory of numbers and algebra.

The purpose of this section is not to present an outline of the development of the study of algebraic numbers,[1] but to point out the tendencies and moments in the development which were important for the formation of modern algebra. It is also necessary to stress the context in which the tendencies were born, to point out in what form their algebraic meaning was displayed, and what contributions they made to reforming algebra, as well as to investigate how important these were.

As has already been shown in the foregoing chapters, the following numerical realms were identified in the mathematics of the 18th century in spite of all historical peripatecisms and various logical difficulties: positive integers (inclusive of unity), integers (inclusive of zero), rational numbers, all real numbers, and the slightly peripheral realm of complex numbers, which were being used and which included all the other realms. The realms of real numbers were defined intuitively, the complex numbers were justified by computational practice only.

In the 18th century, there were few factors to disturb this image. The division of number into algebraic and transcendental was non-existent; up to then it had no sense in algebra. The concept of transcendental quantities was being established rather more in connection with geometrical problems (Descartes) and with infinitessimal calculus. With regard to the latter, as early as in the eighties of the 17th century Leibniz defined transcendental quantities (also considering a function to be a quantity). In the 18th century one could have found the clearest and most pronounced formulation of the division of functions into algebraic and transcendental in Euler's *Introductio in analysin infinitorum* (1748), however, with no mention of application to number fields [138].

---

[1] The principal factography of the development of the theory of algebraic numbers is known to a large extent. Klein ([234] 320nn) gives the fundamental data on the works of Gauss, Kummer, Kronecker, Dedekind and others; these data have also been assembled by Wieleitner [401]. They can also be found in the historical notes of contemporary authors (e.g., Kronecker [245, 244]). The corresponding sections of Bell's history of mathematics [26] are also very stimulating. He does not go into the factual details, but rather points out a number of historical connections concerned with the research into the theory of algebraic numbers and other contemporary fields of mathematics, i.e. even non-algebraic. As the analysis of these connections has not been considered in the text, the reader is referred to Bell's book all the more, because many of his interpretations have been found acceptable. Just as stimulating is Bell's special study [25] which represents a detailed analysis of Gauss' works. Bell's statement that Gauss did not make an attempt at an autonomous, general arithmetic of an arbitrary field of algebraic numbers is considered to be proved sufficiently. Also Hancock's lecture [195] brings, expecially in its second part, a number of interesting items. The problems treated prior to 1870, remain unto this day a part of the interpretation of the theory of algebraic numbers. Therefore, valuable data, allowing for the assessment of some of the facts, can also be found in some of these interpretations. From this point of view, H. Weyl's paper [399] of 1940, illustrating the author's deep understanding of the contemporary problems, is very important.

The discussion concerning various kinds of real number, however, was initiated by a different problem. The impulse was the question whether the numbers π and e are rational [411]. Several mathematicians took part in the discussion, but it was Euler who managed to prove the negative for e (in 1737, published in 1744), and on this basis, again using analytical means, Lambert[2] proved the irrationality of π and also generally of e^m for m rational. In a letter to von Holland in 1768, Lambert wrote that "... the way in which I proved it can also be extended to the proof that circular and logarithmic quantities *cannot be the roots* of rational equations". Perhaps, this slightly obscure statement is proof that Lambert had begun to consider the problem of the algebraicity of π and e. However, there is no proof that he arrived at positive results with this problem.

Neither was the problem of the reducibility of algebraic equations attractive in the 18th century. It would seem that this problem was trivial from the point of view of the contemporary mathematicians because each equation was reducible[3] to linear factors if the author had a more modern point of view and if he dwelt in the realm of complex numbers; or to linear and quadratic factors, if he remained in the realm of real numbers. This is the way the equation was created, i.e. as a product of these factors.[4] Neither was the problem of reducibility considered in another way in the 18th century. Recently, Bashmacova pointed out Newton's experiment, to use modern terms, of suitably expanding the field of rational numbers to adjoin the square root of a natural number in order to resolve a given polynomial with integral coefficients. This isolated experiment cannot be considered as the origin of the systematic studies of quadratic algebraic numbers.[5]

Newton's experiment can be explained on the basis of contemporary knowledge and procedures. Disregarding the possible discussion as to what extent Newton's genius could or could not intuitively feel the necessity of considering other number fields, especially in contemporary mathematics, one could hardly find reasons which would lead him to the said considerations. Mere computational procedure, currently used

---

[2] As is known, not even Lambert's proof was perfect; it was later supplemented by Legendre. Not until 1844 did Liouville present an exact proof of the existence of transcendental numbers when he also proved a reliable criterion for distinguishing algebraic and transcendental numbers. Hermite presented the proof of transcendentality of e in 1873 and, as regards π, this was proved in 1882 by Lindemann.

[3] As regards the attempts to prove the fundamental theorem of algebra in the 18th century, for details see I. Bashmacova [20]. The author did not consider the wider connection indicated in this paper.

[4] It is the author's opinion that this is the way to understand the long period of disinterest in the proof of the fundamental theorem of algebra, as its statement seemed to be trivial. However, as a polynomial (or an equation) originated as an assembly of factors, it was naïve to try and prove that each polynomial can be expressed as a product of linear or quadratic factors.

[5] I. Bashmacova [22] draws these conclusions from a citation from Newton's *Arithmetica universalis* without considering the context. As the simplest illustration of his concept, Newton interprets the old method of solving quadratic equations by making it into a square. The question is to what extent Newton was considering the reducibility of the equation and whether this was not just a problem of a method of solution.

with irrational numbers, of course, is only sufficient for a similar experiment to occur, at least in isolated cases. Much later than in the 18th century, with other authors who returned to what was more frequent in the 17th century, an extensive analysis of computing with irrational numbers may be found. However, the author has not found with any author but Newton that these computations, including not only quadratic but also biquadratic irrationalities, were applied more extensively to the theory of equations. In the sections about equations these authors mention the reducibility of equations only in the traditional ways.

Thus, the problems of reducibility of equations were left to Gauss, who in his *Disquisitiones arithmeticae* (1801) had significant reasons for studying them. Not until he began to study the insolvability of certain equations, did he need the irreducibility of the investigated equation in proofs. Because the initial realm is that of rational numbers (or integers) this does not mean that he developed the study of algebraic numbers in any way.[6]

The first facts about algebraic numbers, which are not mere transpositions of procedures, known earlier in the indivisible realm of real numbers (inclusive of computing with irrationalities), may be found in the 18th century mainly in investigations concerning the problems of the theory of numbers. From this point of view, especially the last part of Euler's *Algebra*, which is concerned with indeterminate equations, is important. Euler, like other contemporary mathematicians, did not hesitate to use complex numbers without any restrictions whatsoever. In studying the problems of the said part, he passes quite freely into the realm of complex numbers even though he was solving problems in the realm of rational numbers, or rather integers, which were not only formulated but which required a solution in this realm. Without special mention, he considered exclusively complex integers, i.e. number like $a + bi$, where a and b are rational integers. These, of course, do not form the only realm of algebraic numbers with which Euler worked. In modern terms, his considerations dwell quite freely in various quadratic fields, real as well as complex. The paradoxical phenomena which he was bound to encounter in these fields seem to remain rather in the background.

In order to avoid distorting the real purport of Euler's considerations, in which he deals with various algebraic numbers, an attempt will be made to reproduce Euler's procedure; abbreviation and interpretation being carried out with the utmost care, in order to refrain from disrupting its period meaning.[7]

Euler was studying[8] problems of indeterminate equations, based frequently on

---

[6] These results are in the 7th part, concerning the solution of the binomial equation. The irreducibility concerns especially (although the proof and the statement may be understood more generally) the equation $\dfrac{x^n - 1}{x - 1} = x^{n-1} + x^{n-2} + \ldots + x + 1 + 0$. Compare the opinion in the text with Bell's data [25].

[7] It is the author's opinion that in the short note of Bell [25] concerning this particular point of Euler, the danger of modernization is hidden, because he does not point out the special feature of Euler's interpretation.

Fermat's ideas, or on their modifications. Apart from the great Fermat theorem, a number of problems are included concerning the resolution of a power of an integer into a sum of a different power of two numbers, in some cases with additional conditions. The case which is of interest now requires an integral solution of the equation $ax^2 + cy^2 = z^3$. As in other cases,[9] Euler proceeds by resolving the L.H.S. of the equation in the realm of complex numbers (remembering that a and c are rational integers), viz

$$(x\sqrt{a} - y\sqrt{-c}) \cdot (x\sqrt{a} + y\sqrt{-c}) = ax^2 + cy^2 = z^3 \qquad (4.3.1)$$

Hence

$$x\sqrt{a} + x\sqrt{-c} = (p\sqrt{a} + q\sqrt{-c})^3 \qquad (4.3.2)$$

and similarly

$$x\sqrt{a} - y\sqrt{-c} = (p\sqrt{a} - q\sqrt{-c})^3 \qquad (4.3.3)$$

so that

$$ax^2 + cy^2 = (ap^2 + cq^2)^3 \qquad (4.3.4)$$

Only now does Euler consider the question whether he will arrive at integral values for x and y. The answer is obtained by the computation leading to equations

$$x = ap^3 - 3cpq^2, \qquad (4.3.5)$$

$$y = 3ap^2q - cq^3 \qquad (4.3.6)$$

Using this method he solved several examples. One of them is the problem of the integral solution of the equation $x^2 + 2 = z^3$, i.e. for $y = \pm 1$, $a = 1$ and $c = 2$. Equations (4.3.5) and (4.3.6) readily yield the solution $x = 5$ and $z = 3$, which is the only possible integral solution. However, the same procedure does not provide the required solution with equation

$$2x^2 - 5 = z^3 \qquad (4.3.7)$$

i.e. for $a = 2$, $c = -5$ and $y = \pm 1$. In this case Eqs. (4.3.5) and (4.3.6) do not yield the required integral solution; actually, after substituting into Eq. (4.3.6), one obtains

$$\pm 1 = 6p^2q + 5q^2 \qquad (4.3.8)$$

which has no integral solution, nor even a rational solution. Nevertheless, Eq. (4.3.7) is satisfied by $x = 4$ and $z = 3$. Euler supplements this conclusion by saying: "Es ist aber von der grössten Wichtigkeit den Grund zu untersuchen." As he had already encountered similar difficulties earlier,[10] it seems that he was even more aware of the necessity to solve

---

[8] [141] Zweiter Teil, Zweiter Abschnitt.

[9] Euler began with equations of the form $ax^2 + by^2 = r^2$ and through various intermediate steps he finally reached equations like $ax^4 + by^4 = r^4$. The case, presented in the text, is solved generally in Section 188.

[10] He solved, e.g., $x^2 - y^2 = z^3$, substituting the general method for a special procedure; putting $x + y = 2p^3$ and $x - y = 4q^3$, he obtained $z^3 = 8p^3q^3$. He was aware of the extraordinariness of the method.

the difficulties. However, at the same time he was aware that his possibilities were inadequate for the task and, therefore, he rests with the statement that similar difficulties only occur for $c < 0$. Euler's explanation, however, is very obscure.[11] Very interesting indications are contained in Euler's preceding computation (Section 196). It is possible to deduce that the integer 3 has an infinite number of distinct decompositions in the quadratic field $R(\sqrt{10})$, just like the equation

$$r^2 - 10s^2 = \pm 1 \tag{4.3.9}$$

has integral solutions. This last contention is a modern interpretation of Euler's computations. However, it is possible that in the numerical case which he analyzed, Euler at least felt a certain anomaly, which he was not capable of expressing.[12]

Without doubt Euler was confronted with a difficult problem here, and he was aware of it. However, it rested in the realm of computation; without attempting to tackle it theoretically more thoroughly, he left it open to other researchers. One could observe this in Euler's work more than once. Of course, in his computations he practically adopts some of the factors of the arithmetic of algebraic numbers. In computing he strives to remain within the given, to use a modern term, algebraic field. That means that he avoids unnecessary introduction of elements of other fields, intuitively selects algebraic integers, and if he dwells in the $R(i)$ field, he adopts Gaussian integers. Should a or b in $a + bi$ prove to be non-integral, he is guided by the logic of computation and specially stresses it. With a few exceptions, he casually transfers concepts and operations from the rational field to $R(i)$ elements.

The said problems and indications of solutions were so obscured in unfinished, complicated computations for his followers, that none of them took them up in the following decades. It seems that the decisive step forward in the arithmetic of algebraic numbers was made by Gauss who, in the changed atmosphere of new mathematical problems and results, initiated a more or less continuous flow in the study of these problems.

Gauss and Euler in many cases have the same origin,[13] from which they approach algebraic numbers, and this is the same with regard to the realm of complex numbers;

---

[11] He says that difficulties are encounterd "Wenn in der Formel $ax^2 + cy^2$ die Zahl c negativ ist, weil alsdann die Formel $ax^2 + cy^2$ oder diese $x^2 - acy^2$, die mit ihr in einer genauen Verwandtschaft steht, 1 werden kann".

[12] The choice of the example leads to the field mentioned in which the factorization into prime factors is not unique. Euler's procedure is actually as follows: In order to avoid difficulties encountered with the case mentioned, i.e. the impossibility of determining p and q as integers in Eqs. (4.3.5) and (4.3.6) for $y = \pm 1$, $a = 2$, and $c = -5$, he multiplies the appropriate factors $(p \sqrt{2} \pm q \sqrt{5})$ by the factor $f \pm g\sqrt{10}$, where $(f + g\sqrt{10}) \cdot (f - g\sqrt{10}) = f^2 - 10s^2$. As, according to him, $f^2 - 10s^2 = \pm 1$, it has an infinite solution (e.g., 3, 1; 19, 6; 37, 117; etc.), then $(f + g\sqrt{10})(p\sqrt{2} \pm q\sqrt{5})$ (with the exception of a further constant integer) is the divisor of three. The last conclusion, the consequences of which Euler suggests should be exploited for solving the original case, was not explicitly drawn by Euler.

[13] Gauss knew Euler's work and in his youth he studied Euler's introduction to algebra; however, it does not seem that he was directly stimulated by it. There is no evidence for this. Compare Bell [25].

for solving problems of the theory of numbers, formulated in rational integers and requiring integral solutions, both authors found it necessary to transfer to another realm, most frequently the complex realm. Euler was also aware of this.[14] However, the problems which lead the authors to expand the field of investigation of the theory of numbers are different. Gauss, moreover, arrives at formulations for the basic definitions and theorems, from which the subsequent development of the study of the whole field was enlarged.

The principal problem from which Gauss' investigation of algebraic numbers developed was the theory of biquadratic residues. It is a fact that in Gauss' work, especially in his earlier papers, there were a number of ideas which could have lead to the arithmetic of algebraic numbers. It was as if Gauss avoided this aspect of his deliberations. This is the case in *Disquisitiones arithmeticae*, inclusive of the part on solving the binomial equation, in which Gauss works in a most matter-of-fact way with complex numbers. He does not even make use of the possibilities provided by the theory of higher congruencies.[15] At this time he only mentions cubic and biquadratic residues. Of course, in *Disquitiones arithmeticae*, besides the fundamental, more general definitions of residues, he discusses only quadratic residues in detail.

In the same way that he presented several proofs of the law of reciprocity of quadratic residues, Gauss also looked for similar general und fundamental theorems for the residues of higher powers. He himself said[16] that he started studying cubic and biquadratic residues in 1805. However, it seems that he was always encountering great difficulties.[17] This is also substantiated by the fact that, apart from a number of com-

---

[14] Die hier gebrauchte Methode ist um so merkwürdiger, da wir durch Hilfe irrationaler und sogar imaginärer Formeln Auflösungen gefunden haben, für die einzig und allein rationale und sogar ganze Zahlen erfordert wurden. [141] § 191.

[15] An interesting analysis of these possibilities is presented by Bell [25].

[16] Gauss ([164] 67) states that in comparison to the theory of quadratic residues, i.e. with the most beautiful part of higher arithmetic "...longe vero altioris indaginis est theoria residuorum cubicorum et biquadraticorum".

[17] In a report on one of his proofs of the quadratic law of reciprocity in *Göttingische gelehrte Anzeigen* of 1817 ([159] Bd II, p. 161) he expains it in these words: "Seit dem Jahre 1805 hatte er (der Verfasser) nemlich angefangen, sich mit den Theorien der cubischen und biquadratischen Reste zu beschäftigen, welche noch weit reichhaltiger und interessanter sind, als die Theorie der quadratischen Reste. Es zeigten sich bei jenen Untersuchungen dieselben Erscheinungen wie bei der letztern, nur gleichsam mit vergrössertem Massstabe. Durch Induction, sobald nur der rechte. Weg dazu eingeschlagen war, fanden sich sogleich eine Anzahl höchst einfacher Theoreme, die jene Theorien ganz erschöpften, mit den für die quadratischen Reste geltenden Lehrsätzen eine überraschende Aehnlichkeit haben, und namentlich auch zu dem Fundamentaltheorem das Gegenstück darbieten. Allein die Schwierigkeiten, für jene Lehrsätze ganz befriedigende Beweise zu finden, zeigten sich hier noch viel grösser, und erst nach vielen, eine ziemliche Reihe von Jahren hindurch fortgesetzten Versuchen ist es dem Verfasser endlich gelungen, sein Ziel zu erreichen. Die grosse Analogie der Lehrsätze selbst, bei den quadratischen und bei den höhern Resten, liess vermuthen, dass es auch analoge Beweise für jene und diese geben müsse; allein die zuerst für die quadratischen Reste gefundenen Beweisarten vertrugen gar keine Anwendung auf die höhern Reste, und gerade dieser Umstand war der Beweggrund, für jene immer noch andere neue Beweise aufzusuchen".

ments which mostly remained in manuscripts,[18] he presented the first part of his *Theoria residuorum biquadraticorum* for publication in 1825, its direct continuation in 1831, while the second continuation never appeared. Gauss wrote about his intentions in a letter to Dirichlet on 13th September 1826; whereas the first part was already in press "... die Hauptmaterien für das Übrige sowie für die ähnliche Theorie der cubischen Reste ist, obgleich noch wenig davon ordentlich zu Papier gebracht ist, im Wesentlichen als abgemacht zu betrachten".[19]

If we compare the first and the second parts of Gauss' theory of biquadratic residues, we find there is a material difference between them, of which Gauss himself was aware. The first part is devoted more to special theorems, partial results and proofs. The second part contains general theorems, including the biquadratic law of reciprocity. The reason for this procedure, so unusual for Gauss (he himself has said that he already had the results) may be found in the method which he used for the general proof, i.e. in the transition into the complex realm which required the establishment of the arithmetic of so-called Gauss' complex integers. That he was earlier well aware of the necessity of this procedure is proved by his comment at the beginning of the first part, saying that for the general theory it will be necessary to extend *campus Arithmeticae Sublimoris* suitably. He also indicates the way in which this should be brought about: in a way similar to the theory of the roots of unity.[20]

---

[18] The author has in mind mainly fragments which the publisher assembled into a whole with the title *Zur Theorie der biquadratischen Reste I—VI*, published in ([159] Bd. II, pp. 313—74). In the notes (idem p. 375) Schering mentions that fragments I and II of 1811 (or shortly afterwards), and fragments III—VI are in a separate volume and, with the exception of the last, they are not of a much later date than the first two.

[19] In the letter mentioned, Gauss also writes that instead of a second volume of "*Disquisitiones arithmeticae*" he must be satisfied with the publication of individual papers; he stresses that this is not a continuation of the old problems, but that he is concerned with newones Cf. [121 Bd. II, p. 375.

[20] Gauss' *Campus, Feld* in German, should be understood as a realm of elements in which Gauss is working. In order to illustrate Gauss' own opinions, let us quote the appropriate sections: "... mox vero comperimus, principia Arithmeticae hactenus usitata ad theoriam generalem stabiliendam neutiquam sufficere, quin potius hanc necessario postulare, ut campus Arithmeticae Sublimioris infinities quasi promoveatur, quod quomodo intelligendum sit, in continuatione harum disquisitionum clarissime elucebit.... Quum iam ad promulgationem harum lucubrationum accingamur, a theoria residuorum biquadraticorum initium faciemus, et quidem in hac prima commentatione disquisitiones eas explicabimus, quas iam cis campum Arithmeticae ampliatum absolvere licuit, quae illuc viam quasi sternunt, simulque theoriae divisionis circuli quaedam nova incrementa adiungunt". In [159] Bd. II, pp. 67—8. In the introduction of the first part of the book (idem p. 165nn) he says that already in 1805 he had worked out the substantial parts of the theory of cubic and biquadratic residues, but that he was stopped by other work from elaborating at least a part. "Der Anfang ist jetzt mit der Theorie der biquadratischen Reste gemacht, die der Theorie der quadratischen Reste näher verwandt ist, als die den cubischen. Inzwischen ist die gegenwärtige Abhandlung noch keinesweges dazu bestimmt, der überaus reichhaltigen Gegenstand zu erschöpfen. Die Entwicklung der allgemeinen Theorie, welche eine ganz eigenthümliche Erweiterung des Feldes der höhern Arithmetik erfordert, bleibt vielmehr der künftigen Fortsetzung vorbehalten, während in diese erste Abhandlung diejenigen Untersuchungen aufgenommen sind, welche sich ohne eine solche Erweiterung vollständig darstellen liessen".

However, what Gauss had still not stated was that the whole subsequent procedure is in close connection with establishing the arithmetic of algebraic numbers. It seems that Gauss had made a number of experiments and partial discoveries in his youth, but that he was not satisfied with them to such an extent as to include them in his published papers.[21] In the same way that some authors had earlier studied the forms $x^2 + ay^2 + bz^2 + abw^2$ and eventually ended up by studying the properties of the elements in the corresponding quadratic field,[22] Gauss studied[23] the equation $x^3 + ny^3 + n^2z^3 - 3nxyz = \pm 1$, where the L.H.S. is the norm of the number $x + vy + v^2z$ in a cubic field, defined by the equation $v^3 = n$, n being a rational integer. More examples could be presented of this kind, which Gauss encountered not only in studying the theory of residues but also in his attempts at solving the great Fermat theorem or in studying quadratic forms. He frequently dwelt very deftly in a suitably chosen algebraic field, but simultaneously this field was somehow defined within the considerations and calculus implicitly and, just as with the 18th century authors, it was only a tool, never being specially investigated eo ipso. Therefore, there is no proof that Gauss had considered the problem whether it was also possible to apply the properties of arithmetic from the rational realm to the expanded "realm of higher arithmetic". It is, therefore, questionable to what extent he was aware that the theorem about the unique decomposition into prime factors need not hold in every algebraic realm. He avoided any positive verdict whatsoever and was satisfied with using Euclid's algorithm (which is only a sufficient, but not necessary, condition for the existence of a unique decomposition). No general statements or definitions can be found; to use modern terms, he always considers a given simple algebraic (quadratic or cubic) field. This is telling, if one considers that later authors searched in vain in Gauss' work for a definition, e.g., of an algebraic integer. One can only conclude that Gauss was very careful in his verdicts and, perhaps, that is why he hesitated with his publications, why he elaborated them in greater detail and why he left out all that would lead him into uninvestigated territory.[24] It is possible that the reasons for which Gauss hesitated in publishing his paper also containing the theory of complex numbers, were only in the anticipated arithmetic of algebraic numbers.

The interpretation of the so-called Gauss integers was considered by Gauss to be a significant component of the second part of his *Theoria residuorum biquadraticorum*[25]

---

[21] Bell's analysis, which tends to this conclusion, the author considers to be absolutely convincing. In the following the author makes use of the factual arguments from Gauss' work, used by Bell.

[22] This mainly concerns Euler, but also Lagrange; compare [264] T. 2, pp. 527—32, T. 7 pp. 170—9.

[23] Cf. [159] Bd. 8, p. 5nn.

[24] Gauss' letters from 1816 ([159] Bd. X_1, pp. 75—6) are interesting from this point of view; in them, in connection with biquadratic residues, he writes of a theory belonging to those, "...wo man nicht voraussehen kann, inwiefern es gelingen wird, dunkel vorschwebende Ziele zu erreichen". The probability of success is about 1 to 1000. He at least thinks of some of the difficulties as overcome. However, he indicates that a full-scale success could lead to the solution of the great Fermat theorem.

[25] This is also supported by the fact that the larger part of his report in *Göttingische gelehrte Anzeigen* is devoted to this subject. Cf. [166].

In a remark at the beginning he indicated that the theory of cubic residues can be construed in a similar manner if one initially considers numbers like $a + bh$, where h is the complex root of the equation $h^3 - 1 = 0$, viz. $h = -\dfrac{1}{2} + \sqrt{\dfrac{3}{4}}\,i$. He adds: "... et perinde theoria residuorum potestatum altiorum introductionem aliarum quantitatum imaginarium postulabit",[26] which is the only more general indication which could have immediately influenced his successors.

In the interpretation itself Gauss introduced a number of terms and derived a number of principal theorems. The initial definition determines that the subjects of the investigation are numbers of the form $a + bi$, where i is said to be "quantitatem imaginariam ... atque a, b indefinite omnes numeros reales integros inter $+\infty$ et $-\infty$". He calls these numbers complex integers, which include integers (if $b = 0$), as well as purely imaginary numbers (if $a = 0$). This is the way Gauss understands the expansion of the said "*campus arithmeticae*".[27] Although the subsequent interpretation is not systematic, at various points theorems may be found which eventually compose a wholesome description of the whole realm. Gauss defined four units here, four conjugate numbers, a norm. He introduced a theorem stating that the norm of a product is equal to the product of the norms; he considered unity to be a number, the norm of which is equal to one, etc. He defines a composite number as a number which may be expressed as the product of two "non-trivial" divisors. On the other hand, a complex prime is a number which has no non-trivial decomposition. It was found that a number which is a prime in the rational field need not be a prime in the complex. He also presented a proof of the uniqueness of the decomposition. The procedure applied is as follows: If A, B, C, ... are prime factors and if it holds (with the exception of unities) that $M = A^{\alpha}B^{\beta}C^{\gamma}$ ..., and if P is another complex prime factor, a, b, c, ..., m, p being the norms of these numbers, then for $P \mid M$ it must hold that $p \mid m$. However, as a, b, c, ... are either primes or their squares, and the principal theorem holds in the realm of rational integers, it is possible to prove that $p \mid m$ does not hold and, therefore, there is no P satisfying $P \mid M$. Gauss then points out that the principal theorem of the uniqueness of the decomposition follows from the latter.

Gauss' expansion of the "arithmetic field" was followed up nearly simultaneously. One may, therefore, assume that similar thoughts were being developed in the work of other mathematicians as well. It is quite understandable that the first to react (almost immediately) was C. Lejeune Dirichlet, the man who was the foremost continuator in developing Gauss' theory of numbers. As early as September 1832 he wrote a treatise

---

[26] Note to Section 30 ([165] 102).

[27] A little later (idem p. 104) he introduces division and finding the root (addition, subtraction and multiplication he considers self-evident), but he does not exclude the possibility that he might pass outside the realm of integers. He also says that many definitions remain valid if a and b are rational or even irrational numbers.

[28] [165] § 37, pp. 107—9.

following up directly on Gauss' ideas.[29] He considers the principal idea to be the discovery of the analogy between the theorems on rational integers and the complex integres, defined by Gauss. He wants to exploit this analogy in his own work. But as if wanting to prove that at the same time he is considering a much more substantial expansion of the realm, which is being considered also in the theory of numbers, he suggests in a footnote that instead of considering expressions in the form of $t + ui$, one should consider much more general forms, i.e. $t + u\sqrt{a}$, where a has no quadratic divisor. His hypothesis that even then the theorems are analogous and that they can be proved in a similar way,[30] is not clear. However, in his paper he only considers Gauss' integers. But perhaps being under the influence of the idea of analogy, in a number of papers he introduces theorems, proved in the rational realm, to Gauss' integers, inclusive of the quadratic law of reciprocity, the study of quadratic forms, as well as his theorem on the existence of an infinite number of primes in an arithmetic sequence.[31] However, with a view to the latter, it was not necessary to develop the concept basis contained in the second part of Gauss' work, and he is satisfied with a mere repetition of Gauss' fundamental definitions and statements. This is the case with his paper of 1842,[32] in which, apart from other terms, he explains clearly Euclid's algorithm for Gauss' complex integers.[33] The following theorem is based on it: If m and m_1 are relative primes and the product $m . n$ is divisible by m_1, then $m_1 \mid n$. He is fully aware of the significange of the theorem and he derived the uniqueness of the factorization into prime factors from it. The exploitation of these terms for studying other algebraic realms is being indicated at the time only very indirectly. This must be sought in close connection especially with problems of the theory of numbers. In considering them, Dirichlet prior to 1832 used expressions like $a + b\sqrt{D}$ (a and b are rational integers, D a rational integer indivisible by a square),[34] but for the reader it is only a numerical expression in the realm of real or complex numbers; no more

---

[29] Démonstration d'une propriété analogue à la loi de réciprocité qui existe entre deux nombres premiers quelconques, [125].

[30] On peut, au lieu des expressions de la forme $t + u\sqrt{-1}$, considérer celles de la forme plus générale $t + u\sqrt{a}$, a étant sans diviseur carré. Les expressions de ce genre, considérées sous le même point de vue, donnent lieu à des théorèmes analogues à celui qui fait l'objet de ce mémoire et susceptibles d'une démonstration toute semblable (Dirichlet [125] p. 175). The question is whether Dirichlet did not think at the time that the arithmetic of integers in every algebraic field was similar to the arithmetic of rational integers.

[31] Cf. [121], Bd. I, p. 503—32.

[32] Cf. [128]. In shorter form the concepts are also explained in the paper [127] read on 27th May 1841 in the Berlin Academy.

[33] He says about him ([128], p. 541): "Le théorème que nous venons de démontrer, étant entièrement semblable à celui qui dans la théorie ordinaire sert de base à toutes les recherches sur les nombres en tant qu'ils sont divisible les uns par les autres, décomposable en facteurs simples, etc., on en tirera les mêmes conséquences pour la théorie des nombres complexes."

[34] Dirichlet [121], Bd. I. In many cases the problem is connected with quadratic forms or directly with Pellian equation or Fermat's Theorem.

than is said by the mathematicians of the 18th century. The series of papers following after 1840[35] may be considered within the scope of the more or less traditional ideas. He also stresses, e.g., the connection with Lagrange's supplements to Euler's algebra. Among other problems, he also studied generally an example which will now be presented for a special case of $n = 3$. Considering an equation $s^3 + as^2 + bs + c = 0$, (a, b and c are rational integers) the roots of which are α_1, α_2 and α_3, Dirichlet studied expressions like $x + \alpha y + \alpha^2 z$ (x, y and z are again rational integers) and he investigates the value, e.g., of the expression

$$(x_1 + \alpha_1 y_1 + \alpha_1^2 z_1)(x_2 + \alpha_2 y_2 + \alpha_2^2 z_2)(x_3 + \alpha_3 y_3 + \alpha_3^2 z_3), \qquad (4.3.10)$$

i.e. whether x, y and z can be selected so as to render Eq. (4.3.10) equal to 1.

It is now difficult to say how these papers, published in German and French, were understood by other mathematicians. In 1846 Dirichlet himself calls his contribution *"Zur Theorie der complexen Einheiten"* [130], and one is justified in interpreting this paper and the papers closely related to it as an endeavour to study units in higher algebraic fields.[36] In this case Dirichlet arrives at a simple determination of the number of these units. However, the study of "units" was at the time also strongly influenced by another part of Gauss' work, of which it has been said that it did not inspire Gauss himself to study aglebraic numbers, i.e. from the theory of roots of unity, which was published in *Disquisitiones arithmeticae*. Apart from the effect it had on Dirichlet, it also clearly stimulated other mathematicians, especially Jacobi and Eisenstein, but also Kronecker, to name just a few. However, their papers rather tended towards the development of the theory of equations and the theory of algebraic fields, which is to be discussed in other sections. Let us return to the problems of algebraic numbers, as regards their effect on the changes in understanding the field of arithmetic and algebra, i.e. from the point of view of their arithmetic.

The results discussed so far have only brought such changes as were compatible with traditional general ideas in the knowledge of contemporary mathematicians. With a view to computational practice there was nothing untoward in expanding the region of considerations in the theory of numbers into the complex realm, selected by Gauss. Euler proceeded in his computations in the elementary course in the same way much earlier. For equations to have complex roots was current practice and the situation could not have been otherwise, even with binomial equations. The definition of Gauss'

---

[35] Especially beginning with the letter to Liouville ([129], p. 619nn), however, there are also indications in earlier papers. It is necessary to point out that in the years, which were in a certain sense decisive (1842—6), there is a gap as regards publications in Dirichlet's work.

[36] Dirichlet's work was also understood in this way. Klein ([234] p. 99) stresses among Dirichlet's merits "die Inangriffnahme der Theorie der höheren algebraischen Zahlen (von 1840) an". However, if the concept of unity is ascribed to Dirichlet, in the sense of an algebraic integer, which satisfies equation $x^n + ax^{n-1} + bx^{n-2} + \ldots \pm 1 = 0$, with integral coefficients, the author is afraid that this already Dedekindian concept could hardly be found in an explicit form in Dirichlet's papers.

integers was "natural", it corresponded to current concepts and procedures in the rational realm. The existence of more units did not draw exceptional attention and complex conjugate numbers were current in the theory of equations. The theory of equations was also used to prepare the study of units of higher fields than quadratic, since Dirichlet drew on Lagrange's work of 1770−1. Also in these papers the terms, definitions and many results were only a generalization or a simple modification of similar ones in the rational field. There are only indirect indications that some mathematicians were aware of the difficulties which they would be faced with in the course of the further study of algebraic numbers. The measure of this awareness is hard to assess. However, in the thirties the hitherto reliable foundations, supported by the second part of Gauss' work on biquadratic residues, suddenly began to give way. Terms and statements which could not be transposed from the rational realm to the newly studied objects began to come forward. That Gauss' integers expand the realm of rational integers, so that a rational prime could have a divisor among Gauss' integers, was evident. The definition of a prime did not change. Factorization into primes (with the exception of unities and order) remained unique. However, these concepts could be carried no further into the new regions. This became most apparent in connection with the uniqueness of resolution. Neither Cauchy nor Kummer were initially aware of the trap which had been set.[37]

It is now hard to say who was the first to point out the non-uniqueness of the factorization into prime factors. It seems that doubts were the strongest amongst Gauss' followers. An important role is accorded to Jacobi, a great admirer of Gauss (whose attitude towards Jacobi was so cool that Jacobi later mentioned Gauss' disinterest in his result as regards the considered region).[38] It is also certain that Jacobi publically pointed out that a rational prime p can be factorized in different ways into a product of different primes in a suitably chosen realm of algebraic numbers,[39] not later than the

---

[37] Kummer "proved" even the great Fermat Theorem, but Dirichlet, to whom he gave the proof to read, pointed out the incorrectness of the assumption of the uniqueness of the decomposition in the algebraic fields, the elements of which he used. This was said to have happened in 1843, i.e. immediately before the principal work of Kummer on algebraic numbers. Cf., e.g., [195], p. 31.

[38] For example, Jacobi ([213] p. 166; treatise of 1837) mentions that certain theorems were communicated to Gauss more than 10 years ago, elsewhere (pp. 171—2) he indicates that as early as in 1827 he considered numbers of the form $a + b\varrho$ (ϱ is the root of the equation $x^2 + x + 1 = 0$) in connection with cubic residues, etc.

[39] He states this specifically of the realm formed by the λ-th roots of unity. He stated this result on 16th May 1839 in the Berlin Academy; it was shortly published in Crelle's Journal [212] and in 1843 it was published by Liouville in the 8th volume of his *Journal de Mathématique pures et appliquées* under the title "Sur les nombres premiers complexes que l'on doit considérer dans la théorie des résidue de cinquième, huitième et douzième puissance " (he points this out on p. 171 of the same volume). It would be interesting to compare this work of his with the treatise referred to in the preceding note. In the latter he mainly considers the problem of exploiting the *n*-th roots of unity for studying *n*-th residues, also for other cases than for $n \leq 4$. He said that as soon as the corresponding laws of reciprocity for $n = 5$ and 8 achieve the required completeness, he would present them to the Academy (he read the treatise in the Academy in 1837). He points out that, like Gauss, he wanted to

beginning of 1839. This simple statement, which was probably also reached independently by other mathematicians,[40] soon drew attention. This result had a profound effect on further studies of the arithmetic of algebraic numbers, especially as regards Kummer's effort. In his paper of 1844 *De numeris complexis, qui radicibus unitatis et numeris integris realibus constant,*[41] the profound effect of the discovery of the non-uniqueness of the resolution into prime factors can be felt. In the introduction he mentions of Jacobi's discoveries "…Quod idem numerus primus pluribus modis diversis in factores duos diffinditur, et quod producta certa ex iis factoribus formata per alios factores divisibiles fiunt, neque tamen hi ipsi factores cum illis compensari possunt, res maximi momenti, indicat hos factores non esse primos sed compositos."[42] A little further on in this paper (p. 202) he expressed his sorrow over the disruption of the uniqueness of his interpretation and outlines his program for a further effort: "Maxime dolendum videtur, quod haec numerorum realium virtus, ut in factores primos dissolvi possint, qui pro eodem numero semper iidem sint, non eadem est numerorum complexorum, quae si esset, tota haec doctrina, quae magnis adhuc difficultatibus laborat, facile absolvi et ad finem perduci posset. Eam ipsam ob causam numeri complexi, quos hic tractamus, imperfecti esse videntur, et dubium inde oriri posset, utrum hi numeri complexis ceteris qui fingi possint praeferendi, an alii quaerendi essent, qui in hac re fundamentali analogiam cum numeris integris realibus servarent."

He does not realize his program in this paper, however, but he studies the realms created from the rational realm by adjunction the *n*-th roots of one. He proceeds generally, and determines the unities and the relations between them. Thus he carries on in the processes worked out not long before by Jacobi and Dirichlet. At the same time he states a number of interesting theorems, important for further development. Among other things, he points out a certain parallelism between the factorization of element *p* in the given realm with the decomposition of a polynomial above the realm in which the factorization of *p* is being carried out.

The requirement to transpose the arithmetic of integers, its terms and procedures, to a maximum extent to the newly studied objects, is expressed clearly as a program, which

---

avoid the factorization into prime factors in the proofs, as used by Legendre, (p. 172), but he does not actually mention the non-uniqueness of the factorization.

[40] It is possible that it could have been Dirichlet (cf. Note 37). Of course, he expresses this statement quite clearly later. It is possible that he repeated it in his lectures on the theory of numbers. Later (1863) Dirichlet's lectures of 1855—58, published by Dedekind [122] contain a note (Section 16) at the end of the part about divisibility, saying that the theorems considered need not hold for numbers of the form $t + u\sqrt{-a}$ ($a > 0$, different from a square), and he gives the example $15 = 3 . 5 = (2 + \sqrt{-11})(2 - \sqrt{-11})$. In this connection he says: "Der Grund dieser interessanten Erscheinnung liegt allein darin, dass es bei den Zahlen dieser Form nicht mehr gelingt, einen nach einer endlichen Anzahl von Operationen abschliessenden Algorithmus zur Auffindung der gemeinschaftlichen Divisoren zweier Zahlen zu bilden".

[41] Quoted according to its publication in *Journal de Mathématiques* (Cf. [261]).

[42] He quotes Jacobi's paper, presented at the Berlin Academy on 16th May 1839 ([261] p. 185).

in the cases of Dirichlet and Jacobi tended to remain in between lines.[43] At the same time two possibilities of further progress are indicated in this work. One of them is further research into the fields of algebraic numbers. A large number of limitations still existed in this respect. Mostly (with exceptions) only complex fields were studied, which usually followed from the adjunction of complex roots of the binomial equation $x^n - 1 = 0$. Here also n was considered to be a prime. These limitations were later removed. This is greatly due to Kronecker. One of his first papers, his thesis of 1845 *De unitatibus complexis*, dedicated to Kummer, removed many of these limitations. It represents the beginning of Kronecker's systematic research into algebraic fields, which he summarizes in his paper of 1882. In the latter, also dedicated to Kummer, he considers it suitable to again publish his paper of 1845, as if to stress the consistancy of his effort. However, as this work is connected with other problems of the theory of equations and all the more with the study of the structure of algebraic fields, it will be discussed in other connections.[44]

The second possibility, which was perhaps first persued by Kummer, endeavoured to save the definitions and principal theorems originally valid only for rational integers. In this case it was necessary, at all costs, to change the system of concepts. Kummer became aware of this in 1844 at the latest. From a quotation of Kummer's introduction, already mentioned, it follows that one cannot transpose completely even the concept of a prime factor (prime number). Kummer was attracted basically by the following fact. From the realm of rational integers the following theorem, also mentioned earlier, is known: If p is a prime and if $p \mid m = n_1 \cdot n_2$, $p \nmid n_1 \Rightarrow p \mid n_2$. If, however, there are at least two distinct decompositions in the given realm of algebraic numbers of the given number $a = b_1 \cdot b_2 = c_1 \cdot c_2$, $b_1 \mid c_2$ does not necessarily follow from $b_1 \nmid c_1$. Two conclusions were now possible. Provided b_1 and b_2 have no other but trivial divisors, they are not prime numbers as in the rational realm and they must be considered as composite numbers. In the realm of algebraic integers, where various factorizations into "prime factors" are possible, the said theorem from the realm of rational integers does not hold. In his paper of 1844 Kummer decided for the first alternative,[45] thus indicating the trend of his further work. A year later, in March of 1845, he lectured in the Berlin Academy on the outline of this theory of complex numbers, and its final interpretation is dated in September of 1846. These papers (together with Kronecker's study of 1845)

---

[43] Bell mentions that the requirement for maximum analogy between the realm of rational integers and integers in the realms of algebraic numbers was not due to Gauss, but to his followers. However, this fact can be interpreted from various points of view. Possibly this is evidence of Gauss' carefulness. Gauss, however, tends to consider only realms and theorems where the analogy holds. On the other hand, it may be proclaimed a requirement only, when it does not appear to be a matter of course, i.e. at the time when difficulties with its application are encountered.

[44] Cf. below, Chap. 4, (5).

[45] It has been mentioned above that Dirichlet, in connection with the proof of the uniqueness of the factorization into primes of Gauss' integers, gives special reasons for the importance of this theorem. Kummer, therefore, even in this follows his predecessor.

represent a turning point in the development of the theory of algebraic numbers.[46]

Kummer proceeds as follows. Let λ be a rational odd prime and α_i the roots of equation ($\alpha_i \neq 1$)

$$x^\lambda - 1 = 0. \tag{4.3.11}$$

The complex integers, of the realm defined by Eq. (4.3.11), then have the form of

$$f(\alpha) = a_0 + a_1\alpha + a_2\alpha^2 + ... + a_{\lambda-1}\alpha^{\lambda-1}, \tag{4.3.12}$$

where a_i are rational integers. In resolving an arbitrary complex integer into factors (in a given realm) Kummer defines, apart from actual complex factors which divided the number considered, also other factors (further irreducible) in such a way as to preserve the uniqueness of the factorization into prime factors in the set of factors, as well as the theorem mentioned from which Kummer initially proceeded. Then even an apparent prime factor, which has no non-trivial divisors in its limited realm, can have other factors. The factors added to a given interval are called ideal primes by Kummer (*ideale Primfaktoren* as opposed to *wirkliche Primfaktoren*). These objects have been artificially construed by Kummer in order to establish the validity of the said theorem. Kummer justifies his action by the fact that it is basically not exceptional in mathematics. In resolving polynomials, in order to obtain linear factors which are prime, unreal elements were also introduced — the imaginary unit. The procedure is similar in geometry. This is the reason for Kummer's not very suitable notation of these factors as "ideal".

The construction of these elements is naturally very complicated and not quite illustrative. The definition itself, adhering only to the most necessary explanatory remarks, is roughly as follows: Consider λ a rational odd prime number and $\lambda - 1 = e.f.$ The roots of Eq. (4.3.11) α_i can then be resolved into e periods of f elements, similarly as Gauss had done in his *Disquisitiones arithmeticae* in the chapter on the theory of binomial equation. Let the sum of the elements of these periods be denoted by $\eta, \eta_1, \eta_2, ..., \eta_{e-1}$. Let $\psi(\eta)$ be an arbitrary complex integer, formed linearly from the sum of the elements of e periods, having the property that its norm is divisible by the rational prime q, but indivisible by q^2 ($q^f \equiv 1 \bmod \lambda$). Considering a suitable choice of u, let $\psi(u) = 0 \bmod q$. Let $\psi(\eta) = \psi(\eta_1) . \psi(\eta_2). ... .\psi(\eta_{e-1})$. Then an arbitrary complex number $f(\alpha)$, whose product $f(\alpha) . \psi(\eta_r)$ is divisible by q, contains an ideal factor of the prime element q, a factor which is attributed to period η_r. If $f[(\alpha) . [\psi(\eta_r)]^\mu$ is divisible by q^μ, but $f(\alpha) . [\psi(\eta_r)]^{\mu+1}$ is not divisible by $q^{\mu+1}$, f(α) contains the said ideal prime factor μ times only.

From this definition[47] Kummer then derived a number of theorems on the intro-

---

[46] Both papers mentioned were published by Kummer together in the 35th volume of the *Crelle Journal für die reine u. angewandte Mathematik* under the titles "Zur Theorie der complexem Zahlen" (pp. 319—326) [258] and "Über die Zerlegung der aus Wurzeln der Einheit gebildeten complexen Zahlen in ihre Primfactoren" (pp. 327—67) [259]. L. E. Dickson ([118], Vol. II, p. XIX) then concludes: "...The theory of algebraic numbers was really born in 1847".

[47] With the exception of slight differences, Kummer presents this definition in both papers mentioned in the same way. Cf. ([258] 322) and ([259] 342—3).

duced prime factors; he proved that theorems, which he wanted to preserve even in expanded realms, were valid. Among other things he proved that every number resolves only into a finite number of prime factors, that the product of two or more complex numbers has the same prime factors as the factors of the product together, that two complex numbers having the same ideal prime factors can only differ by a complex unit; moreover, every complex number has the property that there always exists an integral power of this number which is an actual complex number, etc. The set of these results makes it possible for Kummer to state that the sorrowful sigh of 1844 is now unnecessary and that the program he had then outlined was fulfilled. In April 1847 Kummer writes in this sense to Liouville: "Quant à la proposition élémentaire, qu'un nombre composé ne peut être decomposé en facteurs premiers que d'une seule manière, je puis vous assurer qu'elle n'a pas lieu généralement tant qu'il s'agit de nombre complexes de la forme $T(= a_0 + a_1\alpha + a_2\alpha^2 + ...)$, mais qu'on peut sauver en introduissant un nouveau genre de nombre que j'ai appelé nombre complexe idéal."[48]

In spite of the fact that Kummer's theory was explained in a complicated form, it drew sufficient attention and admiration. The main fact was that apparently the validity of the analogy between the realm of rational integers and the fields of other algebraic integers had been saved. Nevertheless, the fundamental idea of this analogy did not have a more significant direct continuation. It seems that it soon became apparent at what cost the analogy had been preserved. The complicated and unillustrative theory of Kummer did not become a live theory. As will be shown further on, progress was achieved rather by searching for other ways and means. However, this does not on any account mean that it was immediately and totally forgotten. On the contrary, it was the subject of interest even outside the circle of German mathematicians. Even much later Kummer considers ideal complex numbers, which also applies to Kronecker, the two being followed up by Dedekind.[49] Much later it still provided the possibility for establishing fields analogous to the field of rational numbers, initiated by Kummer, with sufficient stimuli and space for various interpretations, as may be found, e.g., in the papers of Klein and Hilbert.[50] Kummer's theory had the deepest effect on Kronecker and Dedekind, who were, of course, able to find stimuli for their own progress in Kummer's ideas.

---

[48] [260], p. 136.

[49] Cf. Kronecker's paper ([255] p. 273) of 1860, connecting up with Kummer's paper of the same year.

[50] For a short indication of the problem cf. Klein ([234] 321—22). Klein interprets Kummer's theory basically in the sense that the theory actually represents an adjunction of further elements, the same purpose being reached by different adjunctions. For example, for $6 = 3 . 2 = 1 + \sqrt{-5}) (1 — — \sqrt{-5})$ one may adjoin $\sqrt{2}$, so that $2 = \sqrt{2} . \sqrt{2}$, $3 = \dfrac{1 + \sqrt{-5}}{\sqrt{2}} . \dfrac{1 - \sqrt{-5}}{\sqrt{2}}$, or i (as suggested by Hilbert), so that $2 = (1 + i)(1 - i)$, $3 = \dfrac{1 + \sqrt{-5}}{1 + i} . \dfrac{1 - \sqrt{-5}}{1 - i}$. Basically these inter-

With a view to further development of the arithmetic of algebraic numbers, Kummer's influence on Dedekind seems to be most interesting. Unfortunately, there are no direct facts available concerning this influence. Although Dedekind's ideals were maturing as of the beginning of the sixties, their first interpretation was only included in the *X*-th supplement to the 2nd edition of Dirichlet's *Vorlesungen über Zahlentheorie* of 1871.

The said supplement has a special name, *Über die Composition der binären quadratischen Formen*, but Dedekind himself points out (Section 159) that this theory only forms a special case of the theory of forms of the *n*-th degree with n variables, which must be considered from a wider aspect. In this way he forms a transition to the first systematic interpretation of the field theory, in which the facts of the theory of numbers are, of course, appropriately stressed. In the introduction he highly appreciates Kummer's merits, especially as regards the papers published in 1847 which brought a deeper insight into the correct nature of algebraic numbers.[51] Similarly ([122] p. 455) he points out other of Kummer's papers of the fifties and their main results. If the problems of the field theory are left for further interpretation, the arithmetic of algebraic numbers as developed by Dedekind, and its connection with Kummer's, are rather more interesting at the moment.

Dedekind points out ([122] 440) a known fact, i.e. in the realm of all algebraic numbers one cannot define a prime number in such a way that it has no different divisors, apart from itself and unity: there is no such number. On the other hand, one may define a relative prime number having the following property: different integers α and β are called relative prime numbers, if every number divisible by α and β is also divisible by the product $\alpha . \beta$.[52] He also rejects, as he says, the continuation of the analogy with rational numbers in studying the factorization of an integer into prime factors. In the realm of all algebraic numbers it has no sense with fields created by the "adjunction of a finite number of elements (i.e. with finite fields, as they were then called), a phenomenon appears with an infinite number of them" as regards the non-uniqueness of the resolution. He thus reverts to Kummer's discovery ([122] 450) and he does not consider an unresolvable number necessarily to be a prime number, and "wir suchen daher für den wahren Primzahlcharakter ein kräftiges Kriterium als diese unzulängliche Unzerlegbarkeit aufzustellen, ähnlich wie früher bei dem Begriffe der relativen Primzahl". He is aware that his approach is in many ways similar to Kummer's and, therefore, he

---

pretations require a further suitable expansion of the field considered. In this way Kummer's opinions are not only modernized with a view to further development, but also modified. Kummer never mentions his ideal prime factors in the said papers in this way. A substantially different and, it seems, closer interpretation can be found in Hancock's book *"Foundations of the Theory of Algebraic Numbers"*, *Vol. I, Introduction to the General Theory*, New York 1931, p. 334 nn.

[51] "...endlich hat Kummer durch die Schöpfung der idealen Zahlen einen neuen Weg betreten, welcher nicht nur zu einer tiefern Einsicht in die Natur der algebraischen Zahlen führt", [122] Section 159, p. 424.

[52] It seems that it is not possible to use the definition usual in the field of rational numbers, as the highest common divisor is not defined (and it cannot be defined).

compares his train of thought with the latter's.[53] On the following pages (Section 163n) he develops his theory of ideals, the terms introduced here forming to this day part of the interpretation of algebra. As the reader is aware, in Dedekind's theory the relations among numbers are substituted by relations among numerical sets, attributed in a certain natural way to certain numbers. The concepts of ideal, principal ideal, and prime ideal are defined for an arbitrary algebraic field, or more accurately, for integers of an arbitrary algebraic field. An ideal is defined as follows: Let o represent the set of integers of field Ω. System a, containing an infinite number of numbers, contained in o, is then called an *ideal* if the following two conditions are satisfied: 1. the sum and difference of two arbitrary numbers from a is again a number from a; 2. each product of an arbitrary number from a and of an arbitrary number from o is again a number from a. The ideal, defined in this manner, determines the resolution of set o into disjunctive classes. Using concepts, Dedekind could then express and prove theorems analogous to the theorems on the decomposition of rational integers into primes, these theorems being a natural expansion of the usual theorems, i.e. for the field of rational numbers they are identical with the theorems of divisibility.[54] Of course Dedekind's theory was also the origin of a number of other significant theorems on the number of classes, number of equivalent ideals, etc. A discussion of these problem would necessarily exceed the scope of the problem in hand, materially and temporally.[55]

---

[53] This comparison is interesting also for Dedekind's interpretation of the nucleus of Kummer's theory ([122] 451): "Ist μ keine Primzahl (und auch keine Einheit), existiren also zwei durch μ nicht teilbare Zahlen η, ϱ deren Produkt $\eta\varrho$ durch μ teilbar ist, so schreiten wir zu einer Zerlegung von μ in wirkliche oder *ideale*, d.h. fingirte Faktoren. Giebt es nämlich in σ einen grössten gemeinschaftlichen Theiler ν der beiden Zahlen η und $\mu = \nu \cdot \mu'$, der Art, dass die Quotienten $\eta : \nu$ und $\mu : \nu$ relative Primzahlen sind, so ist μ in die beiden Faktoren ν und μ' zerlegt, von denen keiner eine Einheit ist, weil weder ϱ, noch η durch μ theilbar ist. Der Faktor μ' ist wesentlich dadurch bestimmt, dass alle Wurzeln α' der Congruenz $\eta\alpha' \equiv 0 \pmod{\mu}$ durch μ' teilbar sind (z. B. auch $\alpha' = \varrho$), und dass ebenso jede durch μ' theilbare Zahl α' der vorstehenden Congruenz genügt. Umgekehrt, gibt es in σ eine Zahl μ', welche in allen Wurzeln α' der Congruenz $\eta\alpha' \equiv 0 \pmod{\mu}$ und nur in diesen aufgeht, so ist auch μ teilbar durch μ', und der Quotient $\nu = \mu : \mu'$ ist der grösste gemeinschaftliche Theiler der beiden Zahlen η und μ.

Aber es kann sehr wohl der Fall eintreten, dass in σ keine solche Zahl μ' zu finden ist; als nun diese Erscheinung (bei den aus Einheitswurzeln gebildeten Zahlen) Kummer entgegentrat, so kam er auf den glücklichen Gedanken, trotzdem eine solche Zahl μ' zu fingiren und dieselbe als *ideale Zahl* einzuführen; die *Theilbarkeit* einer Zahl α' durch diese ideale Zahl μ' besteht lediglich darin, dass α' eine Wurzel der Congruenz $\eta\alpha' \equiv 0 \pmod{\mu}$ ist, und da diese idealen Zahlen in der Folge immer nur als Theiler oder Moduln auftreten, so hat diese Art ihrer Einführung durchaus keine Bedenken. Allein die Befürchtung, dass die unmittelbare Übertragung der bei den *wirklichen* Zahlen üblichen Benennungen auf die idealen Zahlen im Anfang leicht Mistrauen gegen die Sicherheit der Beweisführung einflössen könnte, veranlasst uns, die Untersuchung dadurch in ein anderes Gewand einzukleiden, dass wir immer ganze *Systeme* von wirklichen Zahlen betrachten."

[54] Dedekind in the introduction to the 2nd edition, however, says of this part: "Der Aufbau der Theorie in § 163 befriedigt mich selbst zwar noch nicht vollständig" ([122] viii) and points out that he had given it this form only after long deliberation, when he had built up the theory of higher congruences rather under Galois' influence 10 years ago.

[55] Kummer's and Dedekind's theory have their further mathematical connections, which, like

Without desiring to analyze the development of the arithmetic of algebraic numbers in detail,[56] several partial conclusions follow from the facts mentioned above. The arithmetic of algebraic numbers was created within the frame of, so to speak, the innermost traditions of arithmetic, from the arithmetic of rational integers. Supported by computational techniques, the mathematicians of the 18th century expanded the region of argumentation for the requirements of this realm, intuitively transposing the properties of the rational realm into more general realms. This mostly occurred in studying indeterminate equations, quadratic and higher residues, forms and the great Fermat theorem; singular paradoxical phenomena long remaining unnoticed. Even though some of the mathematicians in the twenties of the 19th century conjectured more than one may find in their work, a real turn towards research into algebraic numbers was initiated by Gauss' paper on biquadratic residues (2nd half of 1831), which required an appropriate expansion of the "field of higher arithmetic" for studying higher residues. This was followed up in subsequent years by several mathematicians (Jacobi, Dirichlet, Eisenstein, but also by less well knowns like Lebesgue and others) who worked under the direct influence of Gauss' paper. They necessarily encountered paradoxes which probably promted Gauss to adopt an attitude of considerable reserve. Thus, inside the traditional realm, in which one currently computed and which had the semblance of such security, one had to start anew and think again about the initial definitions and principal theorems. The concept of a prime, the highest common divisor, resolution into primes, but also the concept of an integer, unity, and conjugate numbers had to be defined again. The depth of the shock is well expressed by Kummer's words, which became a symbol and a program. For the present purposes the fact is important that in the forties it has become clear, and this was proved by papers published in both the leading mathematical journals of the time, as well as in other publications, that the time of the hitherto illustrative and understandable definitions and statements is past and, if order is to be preserved in the arithmetical and algebraic region, one must adopt a different approach. An example of this was Kummer's work on his "ideal prime factors". If the period following the publication of the second part of Gauss' work on biquadratic residues (1831) up to the time of Kummer's work of 1847 is a transitive period of grasping the difficulties and the necessity for change, a period of developing a new theory follows which is crystallized into a clear form by Dedekind's approach to

---

Kronecker's stimuli, have not been mentioned, as this is only concerned with certain problems of historical development. Many of these connections can be found by the reader in a very historical stimulating work by Weyl [399] in which, apart from comparing the results of all three mathematicians mentioned and their assessment from a modern point of view, one may also find a critique of Dedekind's theory of ideals. Just as a matter of interest the author would like to point out its critique due to Klein ([234] 323).

[56] This development would provide material for a much more extensive analytical research in which it would not, however, be possible, as it has been done here to a maximum extent, to neglect the generating theory of algebraic fields.

the arithmetic of algebraic numbers.[57] The analogy of the arithmetic of algebraic integers with the arithmetic of rational integers is preserved to a maximum extent, the new theory is a "natural" expansion of traditional theories but the objects which have been defined are as if from another world for a mathematician of the thirties. And not only for him. Even Felix Klein, a mathematician then connected with the new mathematics, complains that Dedekind's terminology, used for the numerical aggregates he defined, "aller Anschaulichkeit entbehrt".

4.4 Complex numbers

In the 18th century complex numbers were already being accepted as quite a matter of course. Not only were more computations being carried out which penetrated into the complex realm, but there were also more and more general theorems, proofs and deliberations which were being presented in general about "numbers", i.e., in the complex realm. The use of complex numbers thus penetrated into various spheres of mathematical analysis. In this connection d'Alembert's solution of the differential equation describing the vibrations of a string, the discussion on the significance of the logarithm of a complex argument, Euler's power series development, etc., were generally known. However, the sphere in which the concept of complex numbers was born is algebra. The existence of complex roots which had begun to be accepted, at least in between lines, in the 16th century, was still being assumed in algebraic deliberation of the 18th century. On its own it did not evoke discussion.[1] The realm of complex numbers represented a natural expansion of the realm of real numbers. It was born in a similar way to that of the negative numbers. Since the negative numbers allow for the definition of the operation of subtraction without restriction, in a similar way complex numbers make it possible to carry out the operation of taking the root without restriction. Both make it possible to formulate theorems simply and generally.

---

[57] These words do not mean an underestimation of Kronecker, of whom more will be said in another connection. The author is aware that the fact has become apparent here that the separation of the arithmetic of algebraic numbers and the theory of algebraic fields from the fifties is very artificial and that it leads to an apparent underestimation of certain facts of development.

[1] We have been constantly speaking of complex numbers but their concept had not been established in the 18th century. At the time one might perhaps have spoken of imaginary numbers, an imaginary part of a certain expression, etc. For details see Tropfke [379]. It is also suitable to point out that the symbol for the imaginary unit which Euler used, only began to spread slowly in the course of the whole period considered. As late as in 1847 Cauchy still wrote ([82] 87): "...le signe symbolique $\sqrt{-1}$ auquel les géomètres allemands substituent la lettre i...". It was only in the middle of the century under the influence of the expansion of German algebra (but also as a result of Galois' studies) that the symbol $i = \sqrt{-1}$ was generally accepted. The details concerning the development of terminology and symbolics, however, have practically no importance for the problems we are investigating, and therefore we shall not consider them.

As has already been mentioned above several times, difficulties arose in deliberations on complex numbers only at the point where it was necessary to define the complex number itself. The situation was similar to that in calculus where mathematicians of the time were not able to define the concept of a differential in a satisfactory way. On the contrary, each contemporary definition evoked justified criticism which invariably pointed out the discrepancies to which work with objects defined in this way must have led. This was also the case with complex numbers; however, with complex numbers and differentials alike, computations were being carried out on an ever increasing scale. Moreover, their use made it possible to reach important and truthful results. Thus, successful mathematical practice clashed with the insufficient foundations of both parts of mathematics.

Complex numbers gave rise to doubt because they were handled according to the same rules and customs as the other numbers, while at the same time their nature was beyond the realm of numbers. They did not even have the natural property of the other numbers which was attributed to all quantities; namely, they did not increase or decrease. Further, it was not clear what one should understand by them; as the mathematicians of the time complained, they were not illustrative. Negative quantities could be considered to be a debit, the opposite direction, etc., but no such model was known for complex numbers.

As soon as these and other difficulties became known, one necessarily encountered them in all interpretations or textbooks of arithmetic and algebra which included at least a part on the solution of a general quadratic equation, if roots or even computing with complex numbers were not the subject. The awareness of this state of affairs led even serious authors at the beginning of the 19th century to attempt to exclude complex numbers as such from the spheres of arithmetic and algebra. However, most authors realized that this did not solve anything because contemporary mathematics could no longer work without complex numbers. Therefore, mathematicians were attempting to produce a new concept of the whole realm of numbers from the end of the 18th century. This endeavour had not been uniquely successful even towards the end of the sixties of the 19th century.[2] Nevertheless, a number of concepts had been presented which led to some new general approaches, and which contributed to the change in appearance of the arithmetico-algebraic sphere as well. Summarizing the attempts at establishing a uniform mastery of the realms of numbers, including the complex realm, one can observe three main groups, although some studies represent specific transitions and join the ideas of various groups into a whole. The main group of papers, also the most extensive in number, contained studies which sought to anchor the theory of complex numbers in a geometric interpretation. The second group wanted to build up a "more general"

---

[2] Hankel ([196] 71—2) wrote as late as in 1867: Was aber die imaginäre Einheit eigentlich sei, ob eine mögliche oder unmögliche Grösse, eine Zahl, ein an sich nichts bezeichnendes Symbol, ein Affectszeichen, oder was sonst — mit einem Worte, die wahre Metaphysik der Imaginären liegt in den meisten Darstellungen bis heute sehr in Argen und eine Begriffe — und Sprachverwirrung... kehrt nur allzuoft wieder.

algebra especially by stressing a formal approach. Finally, the third chose special operations and means which would provide a logically satisfactory interpretation (or construction) of objects equivalent to complex numbers.

The geometric interpretation of elements of the arithmetico-algebraic sphere had a profound mathematical tradition. The quantities which algebra or general arithmetic dealt with were not only numbers, as we have seen above, but also line segments and geometric formations in general. The theory of negative numbers found one of its models in geometry, with Newton as well as with later authors.[3] As regards the geometric interpretation of complex numbers (in this case imaginary numbers only) one would probably find the beginnings of the tradition with Wallis. The root of a negative number was interpreted as the mean geometric proportional between the positive and negative number; if these numbers were represented by line segments of opposite directions, their mean geometric proportional would be perpendicular to this direction.[4] Several attempts at a geometrical expression for complex numbers can be found in the 18th century. The discussion between Kühn and others, which was published in the proceedings of the Petersburg Academy and which was also entered into by Euler, has a special place. Kühn himself [256] eventually rejected the possibility of geometric interpretation, but the idea of geometrical representation was something which was always iminent.[5] The next step towards the geometric interpretation of complex numbers was only made by Caspar Wessel at the end of the 18th century, and he was followed, but independently of his results, by a whole series of mathematicians who produced similar results.[6]

---

[3] Cf. Newton ([310] p. 9) who gives as an example of negative quantities debit, reverse motion or a line segment. In this case, however, he did not differentiate between a negative line segment and subtracting a line segment: If AB goes from left to right it is in this case considered to be positive, so that the reverse line segment BC is negative because, as Newton put its, "it execution diminishes the line segment AB to AC".

Similarly Karsten ([225] 211) later (1786) spoke of reverse line segments (or directions). However, at this time, thanks to Kästner's study the general interpretation of "reverse" quantities had also become quite frequent. Cf. also Klügel's paper of 1795 [236].

[4] Gauss also referred to this deliberation later. However, the interpretation which Karsten gave it in the paper mentioned ([225] 229) is also interesting. He considered the following: If AD and DE are positive, the mean geometrical proportional can be constructed as known from elementary geometry. If AD is positive and DB negative, a similar construction is impossible. From this he also drew the conclusion that the root of a negative number is not "geometric" but only an algebraic quantity. However, he admitted (p. 282) that "algebraic calculus" was undoubtedly truthful. On the basis of a different argument Klügel also arrived at the rejection of complex numbers ([236] 480). He wrote: "Die unmöglichen Grössen sind eigentlich nur Zeichen eines Wiederspruchs in den individuellen Bedingungen einer Aufgabe. Man behandelt sie als wahre Grössen, um den algebraischen Sätzen die völligste Allgemeinheit zu geben. Aus den unmöglichen Grössen entstehen mögliche da, wo in der allgemeinen Rechnung keine irrationale Grössen blieben."

[5] More about these attempts can be found in Cantor [55b]. A more detailed analysis especially of Kühn's and Euler's opinions is presented in Molodshi's book ([295] 190—208). However, one must admit that the geometric representation of complex numbers did not get a satisfactory form in the 18th century (with the exception of Wessel of whom mention will be made later).

[6] F. D. Kramar's paper of 1963 [241] on vector calculus at the end of the 18th and beginning of

Caspar Wessel was a practical surveyor and his only mathematical paper probably connected up with his interests; his paper was presented to the Danish Academy in 1797 and published in 1799 under the title *Om Directionens analytiske Betegning* ... However, the paper was forgotten and had no further influence on the development.[7] As the title of the paper indicates, the author approached the problem of geometric representation of complex numbers from the opposite point of view from the preceding authors. He himself formulated the objective of his study at its beginning (p. 3); he required that an analytical expression be found which would be capable of expressing in a single equation the length and direction of a line segment. One may say, therefore, that he was looking for a specific geometrical calculus, the requirement of which was also expressed by Leibniz. He introduced four different units for a plan: $+1, -1, +\varepsilon, -\varepsilon$, which represented the four senses along two mutually perpendicular directions, and with the help of which he could express the magnitude and position of the line segments. Only in the course of computing with these quantities[8] did he derive that

$$\varepsilon = \sqrt{-1}$$

Wessel's interpretation of the initial concepts and calculus operations is very clear although adequate terminology is frequently lacking. However, this did not prevent him from realizing the importance of commutativity and distributivity with the operations introduced. In spite of the fact that Wessel's procedure might seem very simple and

---

the 19th century is very stimulating. It not only presents very valuable data concerning the problems discussed here, but it also presents important hypothetical statements on the general state of the problems at the time. This treatise as well as the introductory parts of Crowe's book [108] provide the fundamental bibliographic data.

[7] Its fate was certainly affected unfavourably by various factors, among others the language in which it was written, the fact that the author was unknown, etc. More recent literature points out that the proceedings of the Danish Academy were certainly studied by Abel (cf. Kramar [241] 243). However, one will find no indication in Abel's work of Wessel's treatise having attracted him; one must also admit that he was not interested in the problem, that he did not even react to similar concepts put forward by authors of the beginning of the 19th century, and that he probably tended towards a different, purely algebraic, interpretation of complex numbers. Wessel's treatise was only discovered by mathematical historians at the end of the 19th century, and a French translation of it was published at the 100th anniversary: *Essai sur la représentation analytique de la direction*, Copenhague 1897, [398]. The fundamental data and analyses are contained in the introductions by H. Valentiner and T. N. Thiele to this translation.

[8] Wessel's procedure basically is this: If $+1$ denotes a unit in one direction, $+\varepsilon$ a unit perpendicular to it, -1 a unit opposite to it and $-\varepsilon$ opposite to $+\varepsilon$ with respect to $+1$, the other units display a deflection of 90°, 180° and 280°, respectively. Let the multiplication of the units correspond to the addition of these deflections. For example,

$$(+1)(+\varepsilon) = +\varepsilon, \qquad (-1)(-1) = +1, \qquad \text{etc.}$$

Since

$$(+\varepsilon)(+\varepsilon) = -1,$$

one may put

$$\varepsilon = \sqrt{-1}.$$

matter of course because of its intelligibility, it represented at its time a revolutionary concept. From his predecessors he differs in the first place by not just presenting a geometric expression for complex numbers, but also the graphical significance of the operations to which they were subject. At the same time he also opened up the way for progress along several lines. First of all, one can observe elements of vectorial calculus;[9] Wessel stressed the numerical (algorithmic) aspect, whereas with other authors around the turn of the 18th and 19th centuries, probably beginning with Carnot, the geometrical approach based on "intuition" is more prominent. With Wessel, therefore, one might find the beginnings of vector traction in the form of algebraic elements.

However, Wessel did not just adhere to the algebraic expression of the magnitude and direction of line segments in a plane. He also tried to generalize his procedure for three-dimensional space. Thiele showed at the end of the last century (and his statements have since been repeated in historical studies without substantial change) that Wessel was close to the concept of quaternions. However, one must concede that his method of expressing line segments in space was not very clear. He introduced another imaginary unit η for which he showed that

$$\eta = \sqrt{-1},$$

and that it expressed the direction perpendicular to the plane represented by directions ± 1 and $\pm \varepsilon$. Using a system of axioms he introduced in principle a special operation which he denoted as ,, ; this he did not call multiplication directly, but he said that, in part, it corresponded to multiplication. He was also aware that this operation was not distributive, and from some of his notes it follows that he did not actually consider it to be associative ([398] 30). From the equations Wessel gave it is even possible to see that he did not consider the operation to be commutative.[10] However, he did not say so explicitly and at the point were he analyzed the rotation of a three-dimensional body his own approach caused him difficulties. But Wessel did not give any examples from mechanics which were for him, a practical surveyor, outside the scope of his interests. On the other hand, he did analyze examples of expressing polynomials in a plane and on a sphere.

Therefore, Wessel progressed in his study of the new calculus to the implicit knowledge of unexpected properties of operations which were in contradiction with the generalizations of ancient laws of arithmetic.[11] His formulations are, of course, frequently imperfect, and it is not clear to what extent he was aware of the importance

---

[9] For details of this aspect of the whole development at the time see Kramar [241] and Crowe [108].

[10] From them it does actually follow that

$$\eta,,\varepsilon = \eta \qquad \text{and} \qquad \varepsilon,,\eta = \varepsilon.$$

Cf. the analysis of this procedure with Kramar ([241] 239).

[11] Kramar ([241] 243) thought that most mathematicians of the time were not prepared to take a progressive step like Wessel and therefore they were neither capable of understanding, nor, one may also say, elaborating his ideas. It is possible that this fact played a considerable role. However, if one accepts that his treatise might have been read on a larger scale, there still remains the important

of his results. However, one must give Wessel credit for the fact that the new calculus was, in a way, one of his objectives. He himself indicated this at the beginning of his paper (p. 3). He wrote that contemporary algebraic operations enabled one to deal with the positive and negative directions, but for his objectives (the calculus of magnitude and direction of line segments) he said it was necessary to extend the definition of the algebraic operations in a suitable way which was not in contradiction to the usual. In this way it was necessary to proceed from operations with abstract numbers only, to operations which would also include operations with oriented line segments.

As a result of such an ambitious program, which was something quite exceptional in its time (which makes one the more sorry that it had remained unnoticed), the objective of finding a geometric representation of complex numbers got pushed into the background. The latter was a mere result or part of the means of his deliberations, but not the program itself. Nevertheless, Wessel was aware even of this aspect of his results. This is substantiated by his remark that, among other things, he sought a method which would do away with "impossible" computations. However, this endeavour was much more important to the authors of the first decades of the 19th century.

The discovery of the geometric representation of complex numbers, or to use stronger words, of the arithmetic of complex numbers, has for a long time been attributed to the French mathematician Argand,[12] who published his paper "*Essai sur une manière de représenter les quantités imaginaires dans les constructions géométriques*" in 1806. However, even this paper, published in Paris, might have been forgotten were it not for a discussion of the geometric interpretation of complex numbers in the *Annales de mathématiques pures et appliquées*, which was generated by an article of J. F. Français in the years 1813—15, and in which Argand, referring to his book, and also the editors of the journal, J. D. Gergonne and Servois, took part. It also published Lacroix's mention of an article by the French mathematician Buée, resident in England, who had published a paper on the same topic in *Philosophical Transactions* 1806 (Cf. [12b]). Although the well-read mathematician was sure to have heard of these papers, it is not clear to what extent the ideas expressed in them were generally known. That they were not generally known is indicated by the fact that similar opinions were later expressed again by other mathematicians like C. V. Mourey (1828) in France and J. Warren (1828, 1829) in England.[13]

---

question, why these ideas should have been expanded upon. With a view to the geometry of the time, even Wessel's calculus was much more complicated and obscure than the current parts of elementary geometry with which the solution of the said problems was much easier.

[12] This is what H. Hankel, an expert on this problem even from the historical point of view, thought as late as in 1867. However, he expressed himself ([196] 82) very carefully: "...wenn sich nicht eine Abhandlung früheren Datums beibringen lässt, so ist Argand der wahre Begründer der Darstellung der Complexen in der Ebene."

[13] This is the way Hankel classifies their studies ([196] 82); Kramar ([241] 276) also pointed out that it was not known that, e.g., Warren knew Argand's studies, although he argued (in the Edinburgh Review in 1808) against some of Buée's opinions.

In his study of 1806 Argand started with old deliberations on proportions. He asked what quantity (quantité) x satisfied the proportion

$$(+1) : (+x) = (+x) : (-1).$$

This proportion is similar to proportions

$$(+1) : (+1) = (-1) : (-1) \quad \text{or} \quad (+1) : (-1) = (-1) : (+1)$$

but it is not satisfied by any positive or negative number (nombre positif ou négatif). Argand therefore sought a different real quantity which would satisfy the proportion. He pointed out the similarity to looking for the "reality" of negative numbers. As Argand pointed out, negative numbers were also "imaginary". With many quantities they have no meaning[14] in themselves, the meaning may only be given them. Once they have acquired a real meaning they become real quantities, just as real as positive numbers. He adopted the same procedure for complex quantities. If the line segment $\overline{KA}$ represents the positive unit $+1$, $\overline{KI}$ the negative, the perpendicular line segment $\overline{KE}$ represents[15] the quantity $+\sqrt{-1}$. He added: En effet, la direction de $\overline{KA}$ est, à l'égard de la direction de $\overline{KE}$, ce que cette dernière est à l'égard de la direction de $\overline{KI}$. The same can then be said of the direction $\overline{KN}$ which may be expressed by $-\sqrt{-1}$. He used this to construct other directions, like the mean proportional between directions already known. In this way he arrived at the expression for complex numbers which he denoted by

$$(\pm a \pm b \sqrt{-1}).$$

Their parts may be called real and imaginary, but he maintained that this was not accurate because the quantities in question were real, i.e., they really existed. However, there is another possible way of denoting complex quantities: $m + n\sqrt{-1}$ could be written as $m \sim n$, and $m - n\sqrt{-1}$ as $m \backsim n$. He used these notations sometimes; he adopted the operations from the arithmetic of complex numbers, he did not introduce spatial deliberations, and he did not consider the peculiarities of the operations.[16] The largest part of his study is taken up by geometric deliberations in which he frequently drew on determinations using trigonometrical functions, i.e., of the type $\cos a + i \sin a$, etc. In the final section it appears that he wanted to stress his uncertainty; the operations of addition and multiplication which he introduced were only based on the principles of induction "qui ne possèdent pas un degré suffisant d'évidence", and therefore they were only a hypothesis which must first be proved ([12b] 60).

---

[14] He said, e.g., that —1 franc, —2 francs, etc. are imaginary quantities which can only be given a meaning, by which he means debit (pp. 3—4).

[15] Argand differentiated between the line segment KA, which he only considered to be a length (an absolute quantity, la grandeur absolue), and an oriented line segment $\overline{KA}$, the direction of which is from K to A.

Argand's text, some of the main points of which have been briefly outlined, seems to indicate that Argand, at least in the year 1806, clearly pronounced and published the concept of the geometric interpretation of complex numbers, including the interpretation of their arithmetic, in which he also saw the possibility of ridding them of the unjust indication of "imaginariness". But in the sphere of algebra he did not exploit this concept any further. No mention was made of treating line segments of a three-dimensional space in the same way.

In 1813 Français published similar ideas, pointing out that he had been stimulated by Legendre's letter to his brother which referred to interesting deliberations of an unknown mathematician ([147] 71; [12b] 74). Français was of the opinion that his own paper in Gergonne's *Annales* would make the mathematician, who was soon proved in the *Annales* to be Argand, publish his ideas. Therefore, Français' geometric interpretation of complex numbers is very similar to that of Argand. It is also based on the same proportion and expressly maintains that the so-called imaginary quantities were just as real as positive and negative quantities.[17] He considered the introduction of four mutually perpendicular units $\pm 1, \pm \sqrt{-1}$ to be more convenient than the same notation for rectangular co-ordinates. He made no mention of operations or three-dimensional problems.[18]

Argand was quick to react to Français' article with reference to his previous paper. But in the subsequent exchange of opinions there was perhaps only one new element, i.e., the attempt, as Français put it, to extend the theory of imaginary quantities to the geometry of three dimensions. Français [147] had already tried to use the same imaginary unit, which he employed for the multiplication of line segments in the plane,[19]

---

[16] Argand interpreted the operations of addition and multiplication geometrically in the current manner. Argand's interpretation was described, e.g., by Kramar ([241] 267—8). It is known that Carnot introduced the term complex number in the present sense, and Argand the modulus $\sqrt{(a^2 + b^2)}$ for the quantity $a \pm b \sqrt{-1}$, which expresses the absolute magnitude of the line segment $a \pm b \sqrt{-1}$, in a later treatise.

[17] Français presented the following statement as a theorem ([12b] 67) which he also proved with a pseudo-proof: Les quantités imaginaires de la forme $\pm a \sqrt{-1}$ représentant, en Géométrie de position, des perpendiculaires à l'axe des abscisses; et réciproquement, les perpendiculaires à l'axe des abscisses sont des imaginaires de la même forme. As one of the consequences of this theorems he mentioned (p. 68): Les quantités dites imaginaires sont donc tout aussi réelles que les quantités positives et les quantités négatives, et n'en diffèrent que par leur position, qui est perpendiculaire à celle de ces dernières.

[18] From the final remark of Français and from the remarks of Gergonne it seems to follow that the idea of geometrical interpretation of complex numbers was then already becoming more current. However, the authors were only very little aware of the importance of the interpretation. For them it was more like an interesting idea, as characterized by Legendre.

[19] We have omitted the attempts of these authors to prove the fundamental theorem of algebra on the basis of the geometric interpretation of complex numbers; cf. Hankel ([196] 94). A certain degree of progress may perhaps be found in these studies as regards their purely geometrical aspect, in the concept of the vector and the vector calculus.

for rotating line segments in space. At this point Argand, Gergonne and Français had to admit that they were incapable of solving the given problems satisfactorily. It may also be said that, in their endeavour to master three-dimensional problems, methods of trigonometry prevailed, which indicates that they themselves had not grasped the possibilities of the method whose foundations they had laid. It follows that they did not develop research into the operations. Therefore in many respects their results were only of secondary significance with respect to those of Wessel.

Fresh blood was introduced into the discussion by Servois' letter. In it scepticism prevailed. The dominating tone being: What good is all this? Without expressly saying so, he tended towards the opinion that $a \sqrt{-1}$ was just a quantity (grandeur).[20] The important part of Servois' letter was not his own positive opinions, but the objections which he raised. These uncovered the weak points in the formulations and opinions of the other participants in the discussion, in particular in the manner in which they recorded and explained the relation between complex numbers and their geometrical representation. Servois considered the determination of the relation to be "inutile, erronée". He saw no reason why the mean geometrical proportional should be perpendicular to the quantities $+a$ and $-a$, and he did not understand why the whole geometrical interpretation, as Argand had put it, should be substantiated by a large number of applications *a posteriori*.[21]

Servois' criticism provoked Argand especially to order his opinions and to try to express them more clearly. Right in the introduction to his answer he wrote: The new theory of "imaginary" quantities has two different and independent objectives; "elle tend, premièrement à donner une signification intelligible à des expressions qu'on était forcé d'admettre dans l'Analyse, mais qu'on n'avait pas cru jusqu'ici pouvoir rapporter à aucune quantité connue et évaluable". The second objective can be expressed as an attempt to create a new calculus suitable for geometry.[22] As this second objective is not important for the present purposes, only the first will be considered. It is interesting that Argand was not aware that this attribution was "given by definition",

---

[20] That is why he probably mentioned the book of A. Suremain-Missery [364] which stresses, as we shall point out further on, the purely formal point of view. Gergonne, who considered the expression $\pm a \sqrt{-1}$ to be a negation of the quantity (grandeur), argues with Servois' opinion mentioned in the text in the note on p. 102. However, if he said it was "un être de raison", whereas $+a$ and $-a$ were "deux grandeurs effectives", Gergonne again indicates the significance of the real interpretation of imaginary quantities.

[21] In this connection Servois ([12b] 103) criticized Argand's obscure statement according to which computations with complex numbers may be substituted by computations with other symbols, the result of which is "le simple emploi d'une notation particulière". To this Servois added: "Pour moi, j'avoue que je ne vois encore, dans cette notation, qu'un masque géométrique appliqué sur des formes analytiques dont l'usage immédiat me semble plus simple et plus expéditif." Gergonne commented on Servois' statement by asking whether Servois was at all capable of seeing the sense in the endeavour to rid algebra "de ces formes inintelligibles et mystérieuses", by which he indicated one of the most important motives of the whole endeavour.

[22] The expression of this second objective remained very obscure with Argand; cf. ([12b] 112).

that he adopted Servois' platform, and that he was ready to discuss the justifiability of the proof that the direction corresponding to $\sqrt{+1}$ was perpendicular to that corresponding to $\sqrt{-1}$. He considered this discussion to be permanently opened. However, on the other hand, he was aware that the meaning of many expressions in mathematics was given by definition. He himself (p. 113) gave the example that the meaning of the negative exponent was defined (given) by the equation

$$a^{-n} = \frac{1}{a^n}$$

(*n* natural). As soon as the units were defined, Argand was of the opinion that one could only introduce by definition the concept of identity of the magnitude and position between two line segments. Only then (p. 117−8) did he try to find the meaning of the operations of addition and multiplication of line segments in a way which in fact represented (he was still referring to line segments) an obscured geometrical interpretation of the results of adding and multiplying complex numbers.[23] At the same time he pointed out that these operations, geometrically speaking, were independent of the meaning which they might acquire in the new theory. Only then did he reach the kernel of his argumentation which, however, he expresses almost marginally. Directed line segments (les lignes dirigées) fully represent numbers of the form

$$a + b\sqrt{-1},$$

and they are equally capable of increasing, decreasing, multiplying and dividing. Concrete quantities represent abstract numbers but, on the other hand, abstract numbers cannot represent concrete quantities.[24]

The meaning of the whole endeavour could not be elaborated in detail by Argand and the others with sufficient accuracy. They did not known how to express and prove the isomorphism between complex numbers with their arithmetic of line segments on the one hand, and geometry (at least in the plane with suitable defined operations), on the other. But they touched on the problem of representing an abstract system, and they looked for "realness" in it. They also outlined a method which would justify "des êtres de raison" in mathematics, to use Gergonne's words.

Argand's presentation terminated the discussion in Gergonne's *Ànnalcs*, but, as pointed out earlier, new "original" interpretations of the geometric approach to

---

[23] In his brief explanation he did not consider the details and the special cases. He then arrived at the expression $(n + m\sqrt{-1})(n - m\sqrt{-1})$ and then at $\sqrt{(n^2 + m^2)}$, which he considered to be the absolute magnitude of the quantity. He called it the modulus (p. 122) and he presented a number of theorems on it.

[24] Argand ([12b] 118) remarked: Les lignes dirigées seront donc les symboles des nombres $a + b\sqrt{-1}$. Comme ces nombres, elles seront susceptibles d'augmentation, diminution, multiplication, division, etc.; elles les suivront, pour ainsi dire, dans toutes leurs fonctions; en un mot, elles les représenteront complètement. Ainsi, dans cette manière de voir, des quantités concrètes représenteront des nombre abstraits; mais les nombres abstraits ne pourront réciproquement représenter les quantités concrètes.

complex numbers appeared from time to time. It seems that the awareness of the possibilities of this approach was widespread, although it had its opponents among the foremost mathematicians.[25] Just as in many other respects, Gauss' acknowledgement of the geometric interpretation of complex numbers had a great effect on other mathematicians. It is true that indications of this could be found in several of Gauss' studies, beginning with his studies around the turn of the century.[26] Gauss devoted a considerable part of his annoucement of the second part of his study on biquadratic residues, published in the *Göttingische gelehrte Anzeigen* in 1831, to the explanation of the geometrical interpretation of complex numbers.[27] Here, he in fact defended the extension of the "field of arithmetic" mentioned in the previous section. In the treatise itself he kept to the arithmetical aspect but he refused as completely erroneous the objection that investigations based on the arithmetic of complex numbers were unobjective. Gauss countered these objections ([166] 174) as follows: "Im Gegentheil ist die Arithmetik der complexen Zahlen der anschaulichsten Versinnlichung fähig ...". Roughly this explanation follows: On a straight line one can plot the same distance to left and right of a point, and the end points of these distances can be denoted by $\pm 1, \pm 2, \ldots$. In a plane on both sides of the original straight line one can draw parallels to it with points similarly arranged. The original point, marked 0, has neighbours ± 1 on the original straight line, and $\pm i$ on the parallel straight lines. The points on the parallels divide the whole plane into squares of equal area. These points then in fact represent complex integers. This is sufficient for Gauss to maintain: Bei dieser Darstellung wird die Ausführung der arithmetischen Operationen in Beziehung auf die complexen Grössen, die Congruenz, die Bildung eines vollständigen Systems incongruenter Zahlen für einen gegebenen Modulus u. s. f. einer Versinnlichung fähig, die nichts zu wünschen übrig lässt. Gauss then complemented his concept with several comments on the next three pages. He was of the opinion that the realness of negative numbers had been acknowledged sufficiently, but as regards complex numbers (or as he said, imaginary numbers) the acceptance was not general, as they were still being considered as antithetic to real quantities. Computations with them seemed to be "wie ein an sich inhaltsleeres Zeichenspiel, denn man ein denkbares Substrat unbedingt abspricht" ([166] 175). Even Gauss then considered $+i$ to be the "mittlere Proportionalgrösse" between $+1$ and -1 and therefore corresponding to the symbol $\sqrt{-1}$. In the conclusion he followed up by saying that $+1$, -1 and $\sqrt{-1}$ were not actually the positive, negative and imaginary units, respectively, but rather the direct, opposite and lateral units. He said that all the difficulties would then disappear.

Gauss' interpretation was given a more extensive breakdown in order to be able, as far

---

[25] One of them was probably for a long time Cauchy, as we shall try and prove below.

[26] Gauss himself ([166] 175) mentioned the traces of the geometrical interpretation in the proof of the fundamental theorem of algebra of 1799.

[27] In the study on biquadratic residues [165] Gauss considered the arithmetic of complex integers, but he assumed them to be fully justified, and he did not mention their geometrical interpretation at all.

as possible, to document that as regards ideas there was really nothing new to be had. He secured the "realness" of complex numbers but he derived nothing of the geometric interpretation of the arithmetic of complex numbers, being satisfied with the mere statement that it was clear.[28] At the same time his interpretation had its considerable advantages. Geometric deliberations became redundant, and he spoke of no geometrical calculus with line segments or vectors. He had a realm of complex numbers and he pointed out their representation, including the representation of their arithmetic without trying to derive it excessively.[29] In this way he opened up possibilities for further studies which, based on the given foundation, would necessarily represent a summary of quite matter of course statements. Studies of this kind were also published. There was no doubt about the significance of the geometrical interpretation, and it had its application in mathematical calculus, as well as making its contributions to geometry. However it seem that at the moment when the geometric representation of the arithmetic of complex numbers in Gauss' plane had become clear as such, further deliberations regarding it had no more effect on changes in the approach to algebra. However, the geometrical interpretation was not the only possible approach to the study of the "nature" of complex numbers.

From the end of the 18th century some mathematicians narrowly connected the deliberations on the nature of complex numbers with the endeavour to reach a more conclusive concept of the whole of algebra. They were in fact looking for the sense and justification of the symbolic language of algebra, together with those of its elements which were impossible in the sphere of arithmetic.[30] It is interesting that many authors strongly voiced the philosophical opinions of Condillac with his stress on empirically given phenomena. Abstractions, especially mathematical ones, were then understood in a nominalistic sense to be mere symbols which together form a language, giving evidence of given empirical summaries. In this sense A. Suremain-Missery quoted Condillac in the conclusion of his extensive study on the purely algebraic theory of imaginary quantities of 1801: "L'Algèbre est une langue qui n'a pas encore de grammaire; et la métaphysique peut seule lui en donner une" ([364] 299). In this spirit one can also find strong elements in the treatise of Buée "*Mémoire sur les Quantités imaginaires*" [49]

---

[28] Cf. the very important comment which was made, e.g., by Crowe ([108] 9); "...Gaus himself did not accept the geometrical representation of complex numbers as a sufficient justification for them", and he referred to [159] Bd. X/2, pp. 55—7.

[29] I do not know to what extent Gauss knew the earlier papers on the same topic, some of which were mentioned above. Nevertheless, it is possible to find many similarities in the initial presentation of the problem, as well as in the rejection of "imaginarity". All this may be the result of the same intellectual stimuli and not of connection. One should also notice that Gauss, like his predecessors, only outlined the "real" interpretation, with Gauss even sensuousness guarantees a kind of verity. Gauss saw "isomorphisms" in this but did not point out why it was valid, and in this also he is like his predecessors.

[30] From this point of view the attempt to find a geometric representation for complex numbers touches on the said endeavour in the text. It formed a special case, in its way, of a certain more general trend.

which was founded on Carnot's *Géométrie de Position* and which also presented a geometrical interpretation of complex numbers.[31] He also tried to grasp what this interpretation was about. He maintained (and in this respect he differed from Argand and the others) that deliberations on the magnitude of the line segment concerned arithmetic, and those on the direction, geometry. By connecting up the operations, an arithmetico-geometrical operation must be created. This opinion then led Buée to criticism of the contemporary concept of algebra. He maintained that algebra could not represent "universal arithmetic" but that one must consider it to be "une langue mathématique" (pp. 25, 43, 85). With this approach to algebra one might also include in it the study of arithmetico-algebraic operations. But Buée was satisfied with this general formulation and was incapable of expanding it in detail. He did not even give any suggestion for the study of operations in algebra approached in this way.[32] However, it is certain that there are anticipations here of the fact that the connection between algebra and the arithmetic of integers and rational numbers provided too narow a scope for its development.[33]

One of those who in principle held on to Buée's attitude was Cauchy. This mathematician, with an abundance of ideas, papers and points of view, expressed himself in various ways with respect to the nature of complex numbers. In a somewhat incidental way he explained his formal "symbolic" approach to these problems in the *Cours d'analyse*. His formulation sounds even paradoxical: En analyse, on appelle expression symbolique ou symbole toute combinaison des signes algébriques qui ne signifie rien par elle-même, ou à laquelle on attribue une valeur différente de celle que elle doit naturellement avoir. On nomme de même équations symboliques toutes celles qui, prises à la lettre et interprétés d'après les conventions généralement établies, sont inexactes ou n'ont pas de sens, mais desquelles on peut déduire des résultats exacts, en modifiant et altérant selon des règles fixes ou ces équations elles-mêmes, ou les symboles qui elles renferment.... Parmi les expressions ou équations symboliques dont la considération est de quelque importance en analyse, on doit surtout distinguer celles que l'on a nommées imaginaires.[34] Perhaps Cauchy wanted to express a certain

---

[31] In fact Buée said nothing of the interpretation of the arithmetic of complex numbers. On this, and on this only, is based the justified critical opinion of Hankel, who wrote in the part on geometrial representation ([196] 82): Einen höheren Standpunkt als seine Vorgänger nimmt auch Buée in seinem *Mémoire sur les quantités imaginaires* (Phil. Trans. für 1806, p. 23—88) nicht ein.

[32] Kramar [242] pointed out the deviation from the Newtonian concept of algebra as a universal arithmetic which occurred at the turn of the century. However, he had difficulties in finding proof for his statements; it rather seems that he overestimated the profoundness of this deviation. As regards the concepts and definitions of mathematical disciplines, including algebra, non-uniformity and fluctuations between various opinions always prevailed.

[33] In this connection cf. Chap. 2, (1), where the various possibilities of interpreting Newton's ideas of universal arithmetic are pointed out.

[34] Hankel ([196] 14) severly criticized this part. I am of the opinion that his criticism was justified only with a view to logics, but not with regard to history as perhaps may be seen from the subsequent text.

idea of an abstract mathematical theory in these words, in which elements represent symbols and the relations between them are expressed in terms of equations which have no meaning on their own. It is certainly not possible to interpret with current theories like, e.g., complex numbers or $\sqrt{-1}$ within the scope of current arithmetic. Nevertheless, the rules for operations with these symbols and symbolic equations are given, and Cauchy, without specifying what they were defined by, indicated that they were adopted from the "current" sphere with which the symbolic relations tie up and in which it was possible to interpret results.[35] Cauchy did not expand his ideas in this direction later, but in 1847 he published two very different experiments.[36] In both of them he tried to present the expression $\sqrt{-1}$, as well as the relevant operations, so that its paradoxical nature as an "impossible" quantity would disappear altogether. In his study[37] he tried to eliminate

$$\sqrt{-1} = i$$

by turning from deliberations regarding equations (equalities) to equivalences of modulo $(i^2 + 1)$. It was found that for the realm of real quantities the equalities remained valid. As long as he considered the simpler "expressions[38] he met no difficulties. He considered each expression to be linear and of the form $A + Bi$ (i.e., modulo $i^2 + 1$) but in order to arrive at it he would have had to assume that each function had could be expanded into a power series in terms of i (he considered i to be "indeterminate").

This study, which was stimulated, among other things, by Kummer's study of congruences, was only an intermezzo in Cauchy's opinion. In his second study he inclined towards the geometrical interpretation of complex numbers. However, one must not forget that these studies go back to the middle of the century, when mathematical knowledge had already made considerable progress relative to the first decade

---

[35] In this sense Cauchy's words actually encourage to adopt the opinion of the English school of algebra, concretely to its principle of permanence. Cf. Chap. 6, (2) below. However, with Cauchy one can also find exactly opposite opinions, e.g., *Exercise d'analyse* IV, p. 309, in which he referred to, among other things, the *Cours d'analyse algébrique*.

[36] In the fourth volume of the *Exercise d'analyse et de physique mathématique* (Paris 1847) he published the treatise [82] and [83], which connects with several other papers on "geometrical quantities".

[37] Cauchy ([82] 94) formulated his opinion in the following words: Dans la théorie des équivalences algèbrique substituée à la théorie des imaginaires, la lettre i cessera de représenter le signe symbolique $\sqrt{-1}$, que nous répudieront complètement, et que nous pouvons abandonner sans regret, puisqu'on ne saurait dire ce que signifie ce prétendu signe, ni quel sens on doit lui attribuer. Au contraire, nous représenterons par la lettre i une quantité réelle, mais indéterminée; et, en substituant le signe $\smile$ au signe $=$, nous transformerons ce qu'on appelait une équation imaginaire en une équivalence algébrique, relative à la variable i et au diviseur $i^2 + 1$.

[38] Cauchy considered polynomials, expressions of the type

$$f(i) = \sum_{n=1}^{\infty} a_n i^n,$$

$$\cos x + i \sin x = e^{ix},$$

etc.

of the century. This could also be clearly seen in the concept of a vector, which Cauchy could use as a matter of course with a considerable degree of freedom. He considered a vector[39] to be a geometrical quantity for which he defined the operations of addition, subtraction, multiplication, division, raising to the power, and taking the root. Whereas the geometrical assumptions remained more or less between the lines,[40] one of the determining ideas was that the definitions were to be "natural". Cauchy himself expressed this ([83] 159) with the requirement that the concept of a geometrical quantity should include algebraic quantities as a special case, and therefore also the special case of algebraic quantities, i.e., arithmetical quantities or numbers. Although he considered geometrical quantities in some of the other papers of the collection referred to (for example, he also "applied" them to functions of the complex variable), he did not produce anything new.

Cauchy's opinions, in which he gradually acknowledged the various trends in explaining complex quantities, not only indicate his personal indecisiveness but also the real situation which then prevailed in the approach to complex numbers. Both the trends mentioned, one of which used the idea of the geometrical model while the other stressed the symbolic and formal approach, provided material for further ideas in the logical construction of algebra. It would be possible to bring in even more evidence in favour of either trend, which would indicate the breadth of both trends as well as their differences. As regards the fundamental issue, i.e., how the trends contributed to change in the approach to algebra, nothing more was outstanding.[41] Of course,

---

[39] It is not necessary to analyse the significance of Cauchy's ideas of a vector here; he called it "le rayon vecteur", and differentiates it from a positioned line segment which expresses it (e.g., in adding vectors), but it is always positioned, has a length, a beginning and an end. Of course, two "vectors" are equal if they have the same sense and magnitude. I should like to make the marginal comment that in the introduction to the study ([83] 157—8) he referred to the papers of Buée and Argand of 1806, later studies of Argand, Français, Faure, Mourey, Vallès, etc.; he mentioned H. D. Truel as well as the stimuli of Saint-Venant. This proves how mathematicians were aware of the extent of the discussion at the time, of which we have only mentioned some contributions. Cauchy also only mentioned authors from the circle of French mathematicians. In this connection Cauchy also mentioned his opinions of 1821 when he had wanted to present an accurate theory of imaginary expressions and equations by considering them to be symbolical. "Mais, après de nouvelles et mûres réflexions, le meilleur parti à prendre me paraît être d'abandonner entièrement l'usage du signe $\sqrt{-1}$, et de remplacer la théorie des expressions imaginaires par la théorie des quantités que j'appellerai géométrique..." ([83] 157—8).

[40] Each geometrical quantity, according to Cauchy, has a magnitude (la valuer numérique) and a direction (le module). He determined the direction in some cases as the angle from an axis and the corresponding sine or cosine. For example, he defined multiplication as follows ([83] 161): "Ce que nous nommerons le produit de plusieurs quantités géométriques, ce sera une nouvelle quantité géométrique qui aura pour module le produit de leurs modules, et pour argument la somme de leurs arguments." Although the definitions do not expressly limit the defined object to a plane, they do not exceed it. No new elements appear in the properties of the operations.

[41] For example, the general acceptance of the idea of Gaussian plane gradually became more and more a matter of course. This also provided the development of functions of the complex variable with illustrativeness, but this no longer had any significance for the progress of algebra. Neither did

Cauchy's first experiment of 1847 does not really fit in with the mentioned trends, but rather represents one idea on how to overcome the difficulties in explaining complex numbers through the new techniques of algebra.

W. R. Hamilton's experiment of 1833−5 is the most interesting in terms of method and actual effect. Like other British mathematicians, Hamilton was looking for an approach to algebra which would enable him to build up a logically defensible system. Therefore, he considered it suitable with regard to establishing his approach to complex numbers, to explain his approach to algebra and then to demonstrate this new approach on the example provided by complex numbers. In this way an extensive treatise was born, the "Theory of conjugate functions, or algebraic couples; with a preliminary and elementary essay on algebra as the science on pure time".[42] As regards his approach to algebra, his papers reflect his discussion with the other British mathematicians (mainly with Peacock at the beginning of the thirties). He rejected the opinion that algebra was a kind of language with its symbols and relations between symbols, an idea which was proclaimed by Buée as early as 1806, or opinions which stressed the formal determination of "the rules of Algebra".[43] He wanted to develop algebra as a science. He considered it to be a different sphere from geometry, to which it was not subject. Therefore, the geometric expression of complex numbers was not correct. In these deliberations he proclaimed himself for the "orthodox" Euclidean tradition. He considered geometry to be Euclid's *Elements*,[44] a science of space. In a similar way he required that algebra be a science of a certain subject. Affected by Kant's philosophy, he maintained that algebra was a science about time which is given in our intuition, since geometry was a science about space which was also given us intuitively. It is a science about pure (one might say, abstract) time. That is why he considered algebra to be a real science, studying its subject, and having its own single model.

The main purpose of the treatise was not to explain the science of pure time itself, but to prove the vitality of a discipline defined in this way. He wanted to demonstrate this specifically with the theories of negative quantities and complex numbers, and to overcome the difficulties which he said manifested themselves in other interpretations. However, Hamilton's philosophical approach also led to difficulties. For

---

formal calculations with the n-th roots of unity, the study of their primitivity, etc., contribute anything. Here one can also read in between lines the geometric representation, given by Gauss, which followed from the relation between the solution of the equation $x^n - 1 = 0$ and that of the binomial equation. Cf. Chap. 6 below. The concept of vectors and operations with them, as well as the symbolical (formal) concept of algebra are outside the scope of this chapter and they will be treated elsewhere.

[42] [182]; indications of the fundamental ideas can also be found in [183]. Hamilton explained his opinions and how he arrived at them in the preface to *Lectures on Quaternions* of 1853, as well [192].

[43] It is convenient to consider his arguments and point of view more in connection with building up the concept of algebra in the English school (Chap. 6, (2)). Only the fundamental outlines which will allow for the interpretation of Hamilton's opinions are given here.

[44] Hamilton in fact explicitly said ([182] 4): No candid and intelligent person can doubt the truth of the chief properties of Parallel Lines, as set forth by Euclid in his Elements two thousand years ago; though he may well desire to see them treated in a clearer and better method.

example, the determination of the equality of two time intervals of "pure" time had a certain weakness. However, the time which passed had a natural orientation and therefore a positive and negative time interval is defined naturally, ensuing from the arrangement of moments of time (points). On the basis of the equality of time intervals he introduced integers, and by dividing the time interval, rational numbers.[45] He had certain difficulties in defining the operations and the incommensurable numbers for which current verbal definitions were expressed only in terms of a different language, with different terms and turns. However, even in this case, the existence of a model which describes, but which could not help with the definition of the product of negative numbers, could be a certain help. These explanations took up nearly four fifths of his study. Only the remaining fifth was devoted to complex numbers.[46] The interpretation was based on the concepts and procedures used in the first part. He introduced ordered couples of elements; at first, couples of time points (moments he called them), and then couples of time intervals, and finally number couples. But this was not important. Hamilton then introduced number couples in order to preserve "the analogy" between the theory of couples and the theory of singles. He used the following relations:

$$(b_1, b_2) + (a_1, a_2) = (b_1 + a_1, b_2 + a_2), \tag{4.4.1}$$

$$(b_1, b_2) - (a_1, a_2) = (b_1 - a_1, b_2 - a_2), \tag{4.4.2}$$

$$(b_1, b_2)(a_1, a_2) = (b_1, b_2) \times (a_1, a_2) = (b_1 a_1 - b_2 a_2, b_2 a_1 + b_1 a_2), \tag{4.4.3}$$

$$\frac{(b_1 b_2)}{(a_1 a_2)} = \left(\frac{b_1 a_1 + b_2 a_2}{a_1^2 + a_2^2}, \frac{b_2 a_1 - b_1 a_2}{a_1^2 + a_2^2} \right). \tag{4.4.4}$$

Having introduced these definitions Hamilton thought it his duty to counter the possible objection that these were quite arbitrary by pointing out that they followed from previous statements in the treatise.[47] He then went on to point out that addition and subtraction, like multiplication and division, represent mutually inverse operations, and that the following equalities hold:

$$(b_1, b_2) + (a_1, a_2) = (a_1, a_2) + (b_1, b_2), \tag{4.4.5}$$

---

[45] However, Hamilton's approach is much more complicated, with an endeavour to achieve maximum accuracy and discussions of many examples. For example, he discussed the difference between "ordinal names", i.e., first, second, etc., and "cardinal names", i.e., one, two, etc.

[46] ([182] 76—96). This part of the study was first presented to the Irish Academy on June 1st, 1835. It is called the Theory of Conjugate Functions of Algebraic Couples; On Couples of Moments, and of Steps in Time.

[47] Were these definitions even altogether arbitrary they would at least not contradict each other, nor the earlier principles of Algebra. It would be possible to draw legitimate conclusions by rigorous mathematical reasoning from premises thus arbitrarily assumed. But persons who have read with attention the foregoing remarks of this theory, and have compared them with the Preliminary Essay, will see that these definitions are really not arbitrarily chosen. Though others might have been assumed, no others would be equally proper. ([182] 83)

$$(b_1, b_2) \times (a_1, a_2) = (a_1, a_2) \times (b_1, b_2), \tag{4.4.6}$$

$$(b_1, b_2)\,[(a_1', a_2') + (a_1, a_2)] = (b_1, b_2)\,(a_1', a_2') + (b_1, b_2)\,(a_1, a_2), \tag{4.4.7}$$

which in modern terms represents the commutativity of addition and multiplication, and distributivity. He then introduced for any arbitrary real number[48] a the relation $a = (a, 0)$. He then maintained that the symbols of numbers might be interchanged with the symbols of number couples, if one considers each "simple" number equivalent to a number couple. Thus 1 or (1,0) is a primary unit, (0,1) a secondary unit, and the couple $(0, 0)$ is zero. One of the main results of the subsequent section which was devoted to the powers of number couples[49] were the powers of units, which he presented without proof (which would easily follow anyway from the definition of multiplication) in the following form:

$$(a_1, 0)^i = a_1^i(1, 0)^i, \qquad (0, a_2)^i = a_2^i(0, 1)^i \tag{4.4.8}$$

(for any arbitrary real i). Then

$$(1, 0)^i = (1, 0); \qquad (0,1)^{4k-1} = (0, -1); \tag{4.4.9}$$

$$(0, 1)^{4k-3} = (0, 1); \qquad (0, 1)^{4k} - (1, 0); \tag{4.4.10}$$

$$(0, 1)^{4k-2} = (-1, 0); \qquad (k \text{ an integer}). \tag{4.4.11}$$

A little later, in analysing the n-th powers of unity, where he did not even indicate the possibility of computing them, he almost incidentally considered the case of $(-1, 0)^{\frac{1}{2}}$. Here he differentiated between two roots:

$$(-1, 0)^{\frac{1}{2}} = (0, 1) \qquad \text{and} \qquad (-1, 0)^{\frac{1}{2}} = (0, -1) \tag{4.4.12}$$

According to the foregoing, the former root may be expressed by

$$\sqrt{(-1, 0)} = \sqrt{-1} = (0, 1) \tag{4.4.13}$$

This then represents the essence of Hamilton's celebrated introduction of operations with ordered pairs of real numbers. If one disregards the philosophical background, the terminology, complexity, and muliplicity of deliberations connected with Hamilton's opinions his results sound very modern, quite independently of the intentions of the author. Hamilton intended that his algebra should describe an illustrative and intuitively given model. However, the definition of number couples and of the relevant operations can be understood as an axiomatically constructed system whose representation is the arithmetic of complex numbers, including also the arithmetic of real numbers. Further progress did not bring a development of the philosophical approach,

---

[48] Of course, he did not use this terminology.

[49] The title describes the contents more accurately: *On the Powering of a Number-Couple by a Single Whole Number* (pp. 84—96).

which Hamilton proclaimed even later.[50] This idea is perhaps exclusively connected with his interpretation of number couples and the relevant operations. As already indicated, this interpretation could be taken as an axiomatically given system represented by complex numbers. These concepts were developed later. Besides this, Hamilton's studies provided the basis for the transition from couples to triples, etc., i.e. to vector calculus, and, in its consequences, to linear algebra in general (although this was not supported by Hamilton's statements of the thirties).

Hamilton's studies are the climax of the endeavour to master the theory of complex numbers in the first half of the 19th century. The problem seen by the mathematicians of the 18th century turned out to be exceptionally fruitful. Its difficulty and its elementary formulations stimulated the search for various means of building up a logically correct, or at least reliable, arithmetico-algebraic sphere. The discovery of the geometric interpretation of complex numbers, which found most advocates and, in fact, general acknowledgement, again wanted to subject the arithmetico-algebraic sphere to geometry (Cauchy, nature of geometric quantities), which was still the only illustrative model in the minds of mathematicians, if understood in the Euclidean spirit. The weaker alternative at least provided complex numbers with a certain possibility of becoming illustrative (Gauss). Perhaps there were attempts in between lines to justify them by finding a geometrical interpretation and thus rid them of contradiction, attempts which the authors could not guarantee by a purely logical construction. In this spirit one must understand in particular, e.g., the words of Buée who maintained that algebra was only in fact a language; the meaning, truthfulness and justification of the symbols of this language were provided only by the interpretation. In spite of the obscurity of the statements which tend towards this objective, one must see the indications of a future, new approach to algebra as an abstract axiomatic system. It is paradoxical that one can consider Hamilton's couples to be such a system, even though only partial, when

---

[50] For example, in the introductions to his later papers on quaternions. In the preface to the "*Lectures on Quaternions*" (1853) he also discusses the way he arrived at his point of view. He remembered (p. 123) the comment in Cauchy's *Cours d'Analyse* (Paris 1821): "Toute équation imaginaire n'est que la représentation symbolique de deux équations entre quantités réelles," which he had already known earlier. Nevertheless, Cauchy's idea did not lead him to a new approach to complex numbers to which he is said to have been led from a different direction, rather with a view to the philosophical aspect. This later statement, which gave credit to Cauchy's stimulativeness, though not exploited, should be complemented by Hamilton's words pronounced in 1834 ([183] 97) when he spoke of a certain mathematical incentive. The latter was represented by the so-called Cauchy—Riemann equations which had been presented by d'Alembert in 1752. If the function of a complex variable $\Phi(x + y \sqrt{-1}) = u + v \sqrt{-1}$, and if the relations

$$\frac{\partial u}{\partial x} = \frac{\partial v}{\partial y} \text{ and } \frac{\partial u}{\partial y} = -\frac{\partial v}{\partial x}$$

hold, these relations, which are both real between real functions, could possibly have been produced by similar algebraic relations. It is said that this analogy then led Hamilton to the concept of number couples.

the author himself wanted to consider algebra as a science of a subject intuitively just as given as the traditional Euclidean geometry.

In connection with the search for a satisfactory interpretation of complex numbers, i.e., in connection with the traditional problems of arithmetic and algebra, a new concept of the whole of algebra began to emerge. Further stimuli and perhaps also more time were necessary for its development. This was also the case with the theories which appeared in this connection, among which the foremost are vector calculus and quaternions. These, however, already represent newly created spheres of research.

4.5 Number fields

As has already been said, the concept of a field as one of the fundamental algebraic structures was introduced into mathematics and intentionally used, practically in the present sense of the word, especially in the work of Kronecker and Dedekind, only in the seventies. However, not even then did it appear in its abstract form, but only as a number field. Not until the subsequent decades, when the Galois theory had been mastered terminologically and when other stimuli had also had their effect, did the abstract understanding of field begin to prevail. Its accurate definition by means of a system of postulates only dates to the turn of the 19th and the beginning of the 20th century.[1] Thus the introduction and exploitation of this concept lies beyond the limits of the period being investigated. Nevertheless, forming of the concept of a field was closely connected with the development of algebra from the very beginning of the 19th century.

The concept of a field was born by a process which was similar to the generation of other concepts of algebraic structures. These concepts were not created suddenly by a single act, but were formed in the course of several stages which differed, especially in nature, although they sometimes overlapped in time. Only after long years of study of the individual properties of a certain structure in special cases of it (stimuli for creating a new concept which might have contained the understanding of the new structure in a historically conditional form) did it have a distinct effect. The new concept, without appropriate definition, was then used intuitively for a long time in solving various problems. The concepts themselves remained obscure, too specialized, and lacking in distinct relations to other concepts. Only after having acquired a certain amount of "experience" with this and other new concepts in connection with the elaboration of definite theories was the concept "introduced" as equal with the others and defined relatively as accurate. Not only the concept of a field met this fate (the other circumstances and connections of which are being omitted for the time being) in the 19th century. The concept of the group also developed in this way. The development of the

---

[1] Bell ([26] 213) dates the origin of this system to 1903; it is said that even then it required further complementing (definition of equality).

theory of equations or the construction of Galois' theory were unthinkable without these. However, whereas the concept of a group contributed substantially new elements[2] to the traditional sphere of arithmetic and algebra, the situation was quite different with the concept of a field. A substantial property of the current realms of arithmetic, in which traditional deliberations were conducted, i.e., the realms of rational, real and complex numbers, was that they were fields, which means realms with two laws of composition. Moreover, these operations of addition and multiplication, which displayed reverse operations in the given field, i.e., subtraction and division, were also connected through the law of distributivity. The main trend which led to the birth of the concept of a field was in fact the result of a further differentiation of these realms, into some kind of refined differentiation. Historically speaking, there were two approaches to this trend. There was mathematics approached from the aspect of arithmetics, on the one hand, and mathematics from the aspect of algebra, on the other.

In arithmetic, its calculations and mainly its "higher" part, i.e., the theory of numbers, gave birth to deliberations on algebraic numbers. However, there is a question as to what extent the differentiation of the various "kinds" of algebraic numbers itself led to the concept of a field and to deliberations based on this concept.

The latter approach, due to which the concept of a field developed, was the development of the theory of solving equations. With time the mutual connection of these approaches, which contemporaries only anticipated, has become clear.

Let us attempt to find the main features which characterized the generation of the concept of a field and of the whole field theory[3] in the studies of the mathematicians from the beginning of the 19th century. The knowledge that the roots of equations need not always be rational numbers did not lead to the generation of the concept of a field, in fact not even to the more refined differentiation of the realms of complex numbers. Calculations with irrational or complex numbers agreed with the current understanding of the fundamental realms of numbers. A substantial step was made only when it was found that by "adding" the root of a given algebraic equation to integers or rational numbers (which the coefficients of the equation were as well) a set of numbers was created in which certain algebraic equations might "behave" differently. It is necessary to understand the state of the problem at least as indeterminately as it was understood in the 18th century. As mentioned above, it is probable[4] that with Newton it was only a question of the method of calculation, which was then capable of resolving a polynomial

---

[2] From the very beginning it also appears in his "non-quantitative" representation in the theory of substitutions. Cf. in particular Chap. 7, (2).

[3] Historical references agree on the fundamental fact that the concept of a number field was introduced by Dedekind and Kronecker beginning with the seventies. It will later be shown that there is no unique agreement as regards the other facts, which is due to this concept existing in an obscure form, without a suitable term to describe it. In many instances it can only be discovered retrospectively, contemporaries did not see it.

[4] Chap. 4, (3).

with integral coefficients into polynomials with irrational ones, using current means. However, the problem was slightly different in Lagrange's known and important study of 1770−1. He was still only differentiating among four realms of numbers, i.e., integers, rational, real, and complex numbers. However, he did deal more extensively with the properties of (rational) functions of coefficients and roots of a given equation, and it is common knowledge that the relations between these functions formed the initial points of his analysis. He made use of the property known earlier; that the symmetric functions of the roots can be expressed in terms of (rational) functions of the coefficients. But, at the same time, he also studied the rational functions of the roots which attained different values when the roots were permutated. In this form he also expressed some of the theories of the theory of groups, although he only spoke of substitutions, and although one can hardly speak of a theory of groups at the time. Nevertheless, his deliberations on the rational functions of roots also implicitly contain certain facts about algebraic field, to use modern terms. The "values" of the rational functions of the roots of a given equation in fact form a certain realm of elements. The "values" of the rational functions of the values of the functions of the roots are not outside the scope of this realm of elements. One may also interpret Lagrange's explanation as that this realm of elements contains rational numbers or even integers, because the function of the roots may be symmetric. Besides this, it is also possible to find in Lagrange a slightly obscured claim that it is sufficient to "prendre la racine simple x" instead of the rational function of the roots of a given equation to reach the same "objectives". In this one might possibly see indications of the statement that the resultant realm may be formed by adjoining a single root instead of adjoining more roots.[5]

The fact that it is impossible to find even the smallest direct indication of a realm of numbers (apart from the four fundamental ones mentioned) makes any more modern interpretation of Lagrange's words unjustifiable, provided one does not want to substitute for the interpretation a modern analysis of these problems. The values of the mentioned functions are elements of one of the four known realms, but for Lagrange the problem of "being able to express" them by a suitably selected function is the only substantial one. With Lagrange there is no reason for understanding the problem otherwise, and therefore there is also no reason for considering algebraic numbers and their realms, which would refine the division into the four fundamental ones. There are not even indications of a conceptual change.

The next to substantially influence the process of generation of the field theory was K. F. Gauss. As already mentioned above, Gauss took no substantial step towards

---

[5] Lagrange's latter deliberation is in the said paper [267] § 103—4. As regards the modern interpretation which Langrange's deliberations were given in the text, the French edition of the Encyclopédie des sciences mathématiques ([135] I/2, p. 240, Note 21) says: "...Un cas particulier du même fait est même implicitement contenu dans J. L. Lagrange, *Réflexions sur la résolution algébrique des équations...*". A considerable degree of reserve can be seen in this evaluation of the said section of Langrange's paper.

introducing algebraic numbers, or towards the study of the properties of their realms. His contribution was more indirect. He considered two groups of topics: his proofs of the fundamental theorem of algebra, and the study of the binomial equation.

Gauss devoted his doctor's thesis (1799) to the fundamental theorem of algebra and he returned to the topic several times with new proofs. These proofs, like similar proofs of his contemporaries, all required analytical means as well (e.g., continuity of a function). However, for us Gauss' second proof of 1815, which tried to avoid these extra-algebraical means, is the most interesting.[6] Gauss' approach is based on the idea of creating algebraic expressions, so that the product of linear or quadratic factors suitably derived from them would be equivalent to a given polynomial. He proved the existence of the said expressions by complete induction, according to the degree of the polynomial.[7] Gauss' principal idea suggests the procedure which Kronecker used in the field theory later;[8] he did not assume the existence of a field containing the field of rational numbers, to use modern terminology, or, as was more accurately Gauss' case, he made no mention of the relation between constructed expressions and the intuitively assumed field of complex numbers. Gauss' reticence, as in many other cases, did not enable him to pronounce a more specific hypothesis about his ideas of algebraic realms of numbers.

One encounters a similar situation in Gauss' study of binomial equations.[9] Here

---

[6] Cf. [167]. I. Bashmakova [20] presented a modern interpretation of this proof giving some of the historical connections.

[7] More accurately, for each equation

$$x^n + ax^{n-1} \ldots + a_n = 0 \tag{1}$$

he determined the equation $y^2 + b_1 y + b_2 = 0$ such that both roots y_1 and y_2 are roots of Eq. (1), and coefficients b_1 and b_2 are (rational) functions of the coefficients a_1 and the indeterminate x, i .e., $b_j = f_j(a_i, x)$.

[8] Bashmakova especially stressed this connection in the said paper ([20] 218 nn). However, I am of the opinion that the similarity is more in the effort to preserve a purely algebraic method. Only this provides the possibility of finding similar features in both approaches. The substantial thing is that Kronecker constructed fields in his formal way, whereas Gauss only constructed elements which made it possible to resolve a polynomial. Bashmakova did maintain: "Gauss construisit le corps dans lequel le polynôme donné a le facteur de second degré. La méthode de Gauss pour la détermination d'un tel corps fut plus tard développée par Kronecker, qui l'appliqua à sa construction du corps de décomposition des polynômes" ([20] 221). However, Bashmakova did not provide sufficient proof for these final statemets of her work. It will be shown later, and this is also the opinion of historical literature, that Cauchy had considerable influence on Kronecker's procedure of "symbolical" adjunction, and this Bashmakova did not mention. As regards Gauss' second proof of the fundamental theorem of algebra and the relation of the method used in it to Kronecker's opinions, see E. Netto's comment in the German edition of Gauss' four proofs from the end of the century ([167a], 81).

[9] I was thinking mainly of the corresponding section of the *Disquisitiones arithmeticae* (1801), which affected the development of algebra in various ways. Its significance for the further treatment of the theory of algebraic equations and for the Galois theory, as well as for the theory of numbers, is mentioned elsewhere.

Gauss always assumed the existence of a field of complex numbers, and his deliberations were conducted within it. The equation

$$x^n - 1 = 0 \qquad (4.5.1)$$

(it is sufficient to consider the case where n is an odd prime number) has complex roots with the exception of the root $x = 1$. They are the roots of the equation

$$x^{n-1} + x^{n-2} + x^{n-3} + \ldots + x + 1 = 0. \qquad (4.5.2)$$

this equation which is frequently called the "Kreisteilungsgleichung" is, as Gauss proved in a complicated way, irreducible in the field of rational numbers. Nevertheless, Gauss arrived at a resolution of this equation for every n. He built on the properties of the roots. He defined the concept of a primitive root r, and divided its powers, arranged conveniently into groups whose members were again powers of one of them. For example if $n - 1 = e \cdot f$, the powers of the primitive root can be divided into e groups of f elements. Gauss called the sum of the f elements the period. He pronounced several important theorems concerning them, and in particular used them to construct auxiliary equations. The resolution of the polynomial on the L. H. S. of Eq. (4.5.2) led him to polynomials whose cofficients are rational functions of the roots of the auxiliary equations.

It is considered advantageous to mention here some of the features of Gauss' approach which are discussed elsewhere in greater detail and from various aspects. They do have certain consequences for the problems investigated here. Clearly, the auxiliary equations do not have all coefficients rational (it is not important at this point that they are real). Morcover, this fact had been currently acknowledged in algebra for centuries, although one seldom encounters an intentional mention of the fact that the coefficients of the equation are not rational numbers. The commonness of the calculation used is one of the reasons for not interpreting Gauss' procedure as an intentional adjunction of the root of the equation to the original field. However, there are also other reasons. Gauss based his deliberation on the well-known properties of the roots of Eq. (4.5.2). Drawing on these properties he also carried out all the intermediate calculations which lead to the determination of the coefficients and eventually also to that of the roots of the auxiliary equations. In the course of these intermediate calculations it is not necessary to create a concept of special and unusual numerical or intermediate fields. That which Gauss had carried out could have been carried out within the scope of earlier current concepts and procedures. One may say that Gauss' approach was mostly arithmetical, and within it the algorithms were in themselves sufficient to master the problem without creating new concepts.

Perhaps this feature of Gauss' work has been stressed too much, but it appears that the great impression he created on his contemporaries was not only due to the achieved results, but also to a large extent to the means he used, which were, of course, available. In spite of the fact that he apparently did not exceed the scope of traditional thinking with regard to the investigated problem, his work represents the beginning of a decisive turn. It would seem that he approached the very limits which could be reached with traditional means and concepts. Therefore, the mathematicians who wanted to go one step further in the theory of equations had to try and create a conceptual and theoretical revolution.[10]

First of all, Abel's algebraic work was a step forward and also the inextensive studies of Galois. We shall not analyse their mutual relation and multilateral influence again, but let us concentrate on a narrow problem: how the related concepts of a field and of adjunction are defined and applied and what role they play in the works of these mathematicians. Both approach the problem along parallel lines; the algebraic aspect of their work is mostly based on Lagrange and on the traditions due to him while the arithmetical elements were considerably influenced by Gauss. Both at first consider the insolvability of a general equation of the 5th and higher degrees, and then go on to look for types of solvable equations.[11]

First of all, consider the form of Abel's idea of a field. Various formulations, which are evidence of how he had attempted to express the required concept for the theory of equations, can be found in his study.[12] In the first proof of the insolvability of equations of the 5th degree [2] Abel was satisfied to use traditional concepts. He in fact only used the rational functions of the coefficients a, b, c, d, and e of the equation (in this case of the fifth degree). Its roots, in the case of an algebraic solution, should be

---

[10] This is the reason why one can speak of two connections with Gauss' work. The one, which we shall call the traditional approach, was discussed in the last chapter; it did not bring anything new. Apart from this one could also go further, as was substantiated by the works of Abel and Galois in the first place. This was also the case in the question of introducing and applying the concept of a field. However, we shall be able to verify below whether Bourbaki's statement ([46] 75) is not exaggerated (or too general): "Dès les premiers successeurs de Gauss, l'idée de corps (de nombres algébriques) est à la base de tous les travaux sur la question (comme aussi des recherches d'Abel et de Galois sur les équations algébriques); son champs d'application s'agrandit lorsque Dedekind et Weber ... calquent la théorie de fonctions algébriques d'une variable sur celle des nombres algébriques."

[11] Historical literature mostly acknowledges the merits of Abel and Galois. Nevertheless, there are certain differences in expressing these merits. In the German edition of the *Encyklopaedie* ([136] I, p. 286) 286) one reads: Die frühesten Spuren des Körperbegriffs finden sich in N. H. Abel's Untersuchungen über algebraische Gleichungen, die Erkenntnis der Bedeutung der Adjunktion für die Theorie der Gleichungen gebührt E. Galois. The French version of this encyclopedia ([135] 236—7) is even vaguer in its statements; it only says that for quantities, which Abel called constant and Galois known, different names are used today (1910), also field. Since Gauss' arithmetical approach which introduced numbers of the type $a + bi$ (where a and b are integers) was also mentioned, the statements are even more obscure. However, at the same time it is said of Galois ([135] 239): La notion d'adjonction et de la découverte de son importance pour la théorie algébrique sont dues à E. Galois.

One should also mention the summary of Bell ([26] 214—5): The earliest recognitions of fields, but without explicit definition, appear to be in the researches of Abel (1828) and Galois (1830—1) on the solution of equations by radicals. The first formal lectures on the Galois Theory were those of Dedekind to two students in the early 1850's. Kronecker also at that time (1853) began his studies on Abelian equations. It appears that the concept of a field passed into mathematics through the arithmetical works of Dedekind and Kronecker.

[12] That is why the literature, some of the opinions of which were mentioned in the previous comment, quotes various passages from Abel's papers.

expressible "par une fonction des quantités *a*, *b*, *c*, *d*, et *e*, formée par des radicaux". He also mentioned ([2] 31) irrational functions which express the root of the original equation rationally. Abel took a certain step forward in reconsidering his proof for the first volume of Crelle's *Journal* [3], where he also put the problem in more general terms, and where he did not restrict himself only to equations of the 5th degree. He defined the algebraic solution of an equation here as expressing its roots "par des fonctions algébriques des coefficients", which then forced him to explain the concept of an algebraic function in detail ([3] § 1, p. 66). He used the following words to define it: Soient x^{I}, x^{II}, x^{III}, ... un nombre fini de quantités quelconque. On dit que *v* est une fonction algébrique de ces quantités, s'il est possible d'exprimer *v* en x^{I}, x^{II}, x^{III}, ... à l'aide des opérations suivantes: 1. par l'addition; 2. par la multiplication, soit de quantités dépendant de x^{I}, x^{II}, x^{III}, ..., soit de quantités qui n'en dépendent pas; 3. par la division; 4. par l'extraction de racines d'indices premiers.[13] These words also led him to the definition of a rational function which was formed by the first three operations, and of an entire rational function which was only defined by the former two. A rational function may also be understood to be the ratio of two entire rational functions.

It is now necessary to investigate the relation of the functions defined in this manner to the nascent idea of a number field. Like all the mathematicians of the time, Abel also had in mind its Bernoulli-Eulerian concept when he spoke of a function, i.e., a certain expression.[14] Abel also gave the form of this expression. He maintained, for example, that any entire function could be expressed by the sum of a finite number of expressions of the form

$$A x'^{m_1} x''^{m_2} x'''^{m_3} \dots \qquad (4.5.3)$$

The introduced concepts in themselves do not yet imply any arithmetical and algebraic consequences. The latter only become outstanding with Abel's classification of algebraic functions, and with looking for a "general form of algebraic functions". This is how Abel proceeded. Consider $f(x', x'', x''', \dots)$ an arbitrary rational function; the algebraic function will then be

$$p_1 = f(x', x'', \dots, \sqrt[n']{p'}, \sqrt[n'']{p''}, \dots), \qquad (4.5.4)$$

where p', p'', ... are rational functions of x', x'', ... He called this algebraic function an algebraic function of the quantities x', x'', ... Consider p_1', p_1'', ... to be more quantities of the form of p_1; the expression

$$p_2 = f(x', x'', \dots, \sqrt[n']{p'}, \sqrt[n'']{p''}, \dots, \sqrt[n_1']{p_1'}, \sqrt[n_1'']{p_1''}, \dots) \qquad (4.5.5)$$

---

[13] According to Abel this contains subtraction, powers with an integral exponent and *n*-th roots for arbitrary natural *n*. The second item of the quotation given in the text is interesting; as regards multiplication he apparently had in mind the multiplication of rational numbers (which are "independent" of the quantities x', x'', x''', ..., of which there is a finite number); on the other hand he did not give the same explanation for division and taking the root.

[14] Nothing is changed by the fact that a new concept of a function was formed exactly at the time, in the elaboration of which Abel took part.

represents the general form of the algebraic function of the quantities x', x'', ..., and of algebraic functions of the first order. He called the functions of the form of p_2 algebraic functions of the second order. In this way it is possible to proceed in general to the n-th order. Besides this, Abel also considered the degree of the functions, i.e., algebraic functions of order m may further be differentiated according to the number of expressions of the type $\sqrt[n']{p'}$ they contain (where p' is an algebraic function of order $m - 1$), and this number determines the degree of the function.[15] Abel's definitions are sufficiently clear. Their purpose is to prepare the ground for deciding whether there exists for each equation an algebraic function which would satisfy the equation. That is why in the spirit of the tradition the function itself is a quantity. There is the question of the extent to which it is a quantity dependent on the variables. This is apparently so in the same sense as the coefficients of the equations are "variable", or, considering "variable" coefficients, as "variable roots". That is why Abel himself rather spoke of the permutations of these "variables"; the function then also acquires certain values. However, it rather remains an "expression" for quantities which can be algebraically derived from the given ones. It can be said that a given algebraic function, or its given type (i.e., with a given order and degree), represents a kind of set (summary) of elements which can be expressed in a given way in terms of the given quantities. Perhaps Abel was close to understanding the problem in this way in his first algebraic studies. In this way he would have arrived at a definition of number realms into which the realm of complex numbers then divided. Nevertheless, these realms of Abel's were not fields. He spoke of the ratio of algebraic functions of the same type and he formed other functions from them (especially rational ones), but the operations he applied to them did not lead him to realms of numbers closed with respect to operations. The types of algebraic functions are in fact disjunctive, the higher order not including the lower. Abel expressly needed this in his proof of insolvability. Therefore, if the field theory is known, it is possible to interpret Abel's words as if his sets were fields, but the direct use of the idea, not to say the concept of a number field, cannot be proved in Abel's words. His means are different and represent only a slightly modified traditional terminology.

Of course, Abel's concepts developed further.[16] Considering the known connection of Abel's studies on elliptical functions with his algebraic studies, one can hardly be

---

[15] Abel's statement ([3] 69) that an algebraic function of order $m > 0$ and degree $n = 0$ is a function of order $m - 1$ and function of order $m = 0$ is a rational function, is then clear.

[16] Abel also repeated his opinions given in the mentioned treatise in an abbreviated form in *Férussac's Bulletin* ([3b] 87—88). He expressed the purpose of the treatise in the following words: "L'auteur démontre, dans ce mémoire, qu'il est impossible de résoudre algébriquement l'équation générale du cinquième degré; car toute fonction algébrique des coefficients de la proposée, étant substituée à la place de l'inconnue, conduit à une absurdité." This is where the understanding of coefficients becomes important. He wrote ([3b] 89) "...une équation quelconque dont les coefficients sont des fonctions rationelles des quantités x_1, x_2, ...", nothing in particular being assumed of the quantities x_1, x_2, ... In the treatise itself ([3a] 72) he proceeded in a similar way, only requiring that these quantities be "des quantités indépendantes quelconques".

surprised that the ideas mentioned above appear in them again. However, the other aspect of Abel's concepts comes to the fore here: he rather considered functions to be "dependent" variables. That is why he spoke of an arbitrary algebraic equation of degree n in the well-known treatise generalizing the addition theorem of elliptical integrals (e.g., [1], T. 1, 146−147), the coefficients of which are entire rational functions of the variable x. The concepts of a quantity and function merge here (e.g., [1] T. 1, 170) while "quantity" and "value" are differentiated ([1], 204−205). A little later, when he considered the relation between elliptical functions and the solability of algebraic equations ([5], p. 263 nn), the terms "expression" or "algebraic expression" (expression algébrique) became outstanding. Also the following statement appeared (p. 265): Il est à remarquer que les expressions de racines contiennent des quantités constantes qui, en général, ne sont pas exprimables par des quantités algébriques. Ces quantités constantes dépendent d'une équation du degré $m^2 − 1$.

Disregard the context in which the said statement appeared. Here again, under the term of constant quantities dependent on a certain equation one may anticipate the concept of a field, although not even here is there mention of a set of constant quantities. The dependence of constant quantities varies; algebraic expressions consisting of the coefficients of an equation are probably meant here. In other cases ([1] T. 1, 352) Abel used "algebraic expressions" consisting of non-algebraic quantities.

After several years Abel again returned to algebraic problems in an article published in 1829 [6]; he determined the classes of equations which could be solved algebraically by generalizing Gauss' ideas. The system of terms did not change markedly here. The initial property of the roots is that they represent a "rational function" of one of them. Nevertheless, even before paper [6] was published, the treatise[17] *Sur la résolution algébrique des équations* [8] was outlined in 1828, which showed how Abel gradually arrived at Galois' theory. He attempted here to determine the classes of equations which could and could not be solved algebraically. The initial term is "algebraic expression". The program of the whole treatise, which should be quoted in full, is founded on it: On sait que toute expression algébrique peut satisfaire à une équation d'un degré plus ou moins élevé, selon la nature particulière de cette expression. Il y a de cette manière une infinité d'équations particulières qui sont résolubles algébriquement. De là dérivent naturellement les deux problèmes suivants, dont la solution complète comprend toute la théorie de la résolution algébrique des équations, savoir:

1. Trouver toutes les équations d'un degré déterminé quelconque qui soient résolubles algébriquement.

2. Juger si une équation donnée est résoluble algébriquement, ou non. ([8], 218−219)

Abel's formulation of the program of the treatise quoted cannot be considered quite clear. The existence of an algebraic expression satisfying a given equation is identified with an answer to the problem of the algebraic solvability of the equation. However, this way of putting the problem corresponds to Abel's approach and terminology. What

---

[17] Cf. Abel, [1] T II, p. 329, where the said manuscript is dated in the second half of 1828.

should one understand by algebraic expression? This is why Abel tried to make its meaning more accurate.[18] He saw two possible ways; either one considered the coefficients of the equation to be independent variables, or constant quantities. In the former case the roots were algebraic functions of the coefficients, the functions containing arbitrary algebraic and non-algebraic constant quantities. In the latter the coefficients were formed of other constant quantities with the help of rational operations. If the constant quantities are α, β, γ, ... nous dirons qu'on peut satisfaire algébriquement à l'équation proposée, s'il est possible d'exprimer une ou plusieurs racines en α, β, γ, ... à l'aide d'opérations algébriques. ([8] 220) The main motive for both specifications (of which only the part in which we are now interested has been mentioned) probably was the expression of the generality of the studied equation and of the general form of the root.[19] However, the expression of a field is clearest with Abel in the second interpretation. He arrived at new quantities from constant quantities by means of "rational operations". He made no mention of the nature of the constant quantities; however, they need not be rational numbers. He formed quantities of a different nature by means of algebraic operations. But a set of elements or relations between sets of elements no longer appear here.

On the subsequent pages of the treatise mentioned, Abel presented a further clarification of the terms. He still spoke of an algebraic expression of a quantity by means of other quantities, of using a finite number of operations. He also arrived at the following differentiation of (known) quantities: If a quantity can be expressed algebraically by means of unity, it is called an algebraic number; if it can be expressed rationally by unity, it is a rational number, but if it can be formed from unity by adding, subtracting and multiplying, it is called an integer. On the other hand, if a known quantity contains one or more variables, it is in a similar way an algebraic, rational or entire function. In this case he said that each constant quantity was to be considered as being a known quantity ([8], 224). He then went on to differentiate between the various types of algebraic expressions with a view to the number of roots they must contain. Although Abel arrived at the concept of algebraic numbers and expressions, the concept of a set with an internal structure was still missing. Operations with expressions were contained explicitly here ([8], 224 nn), but all concepts were connected with the problems of solving equations.

---

[18] Abel [8]: Avant tout il faut fixer le sens de cette expression.

[19] Abel also differentiates between two kinds of equations in this connection, "celles qui sont résolubles algébriquement, et celles auxquelles on peut satisfaire algébriquement. En effet, on sait qu'il y a des équations dont une ou plusieurs racines sont algébriques, sans qu'on puisse affirmer la même chose pour toutes les racines". This statement indicates that Abel also allowed for non-algebraic solutions of algebraic equations, for which later authors criticized him. It is not quite clear, however, whether this criticism was justified. With Abel (cf. above) an equation either has or has not an algebraic solution, which means that the equation is algebraically solvable (i.e., in terms of radicals) or not; Abel said nothing of what the roots were in the latter case. I am afraid that this problem, which is particularly important for comparing Abel's and Galois' results, escaped the attention of historical literature which was quite taken up by the more modern interpretation and did not understand Abel's groping.

Nearly simultaneously with Abel's work on the articles last mentioned, Galois' first treatises were published in France. However, this does not in any way indicate a connection of Galois' ideas with the work of a mathematician so similar in fate and nature of scientific work.[20] Galois' approach to the considered problem is slightly different from Abel's. One can say that it is less dependent on traditional algebraic thinking and that it arrived at more clearly defined concepts. He did not use the terms algebraic functions or algebraic expression; neither did he look for a general expression of the root of the equation which with Abel is evidence of the traditional trend which originated in the time of Euler. He rather spoke of solvability of equations "par radicaux". He explained the concepts he introduced systematically in his paper *Mémoire sur les conditions de résolubilité des équations par radicaux* ([156]). The introductory definition to this paper may be considered to be an attempt at explaining the concept of a field. Understandably, Galois was more interested in acquiring means of presenting his own theory; whereas the following definitions of a group and several of its properties (this concerns groups of permutations) are clear, one can only say that the definition of a field was really just an attempt.

Galois first defined a reducible equation as allowing for rational divisors, which led him to explain the term "rational". His approach is basically such that he considered the rationality of rational numbers to be known. If all the coefficients of a reducible equation are not rational numbers, the resolution is rational provided the coefficients of the divisor can be expressed as rational functions of the coefficients of the given equation.[21] He even maintained that," ... on pourra convenir de regarder comme rationnelle toute

---

[20] A fragment of Galois' notes ([153] 24) was preserved, in which he retaliated and maintained that he did not even know Abel's name when he presented his first papers to the Institute. However, with his self-consciousness he kept his distance from Abel's too special results; he was sorry that death prevented Abel from concluding his algebraic work, which he had told Legendre about in a letter, and which treated the problem of solvability of algebraic equations.

[21] Galois, [156] pp. 34—5: Définitions. — Une équation est dite réductible quand elle admet des diviseurs rationnels; irréductible dans le cas contraire.

Il faut ici expliquer ce qu'on doit entendre par le mot rationnel, car il se représentera souvent.

Quand l'équation a tous ses coefficients numériques et rationnels, cela veut dire simplement que l'équation peut se décomposer en facteurs qui aient leurs coefficients numériques et rationnels.

Mais quand les coefficients d'une équation ne seront pas tous numériques et rationnels, alors, il faudra entendre par diviseur rationnel dont les coefficients s'exprimeraient en fonction rationnelle des coefficients de la proposée.

Il y a plus: on pourra convenir de regarder comme rationnelle toute fonction rationnelle d'un certain nombre de quantités déterminées, supposées connues a priori. Par exemple, on pourra choisir une certaine racine d'un nombre entier, et regarder comme rationnelle toute fonction rationnelle de ce radical.

Lorsque nous conviendrons de regarder ainsi comme connues de certaines quantités, nous dirons que nous les adjoignons à l'équation qu'il s'agit de résoudre. Nous dirons que ces quantités sont adjointes à l'équation.

Cela posé, nous appellerons rationnelle toute quantité qui s'exprimera en fonction rationnelle des coefficients de l'équation et d'un certain nombre de quantités adjointes à l'équation et convenues arbitrairement.

fonction rationnelle d'un certain nombre de quantités déterminées, supposées connues a priori" ([156] 34).

The latter formulation should perhaps be understood as an attempt at defining a certain set of elements (similar to rational numbers) which are determined by rational functions in terms of "known" quantities. The new elements are also known.[22] However, no actual mention of a set of elements was made; only their origin and that they were rational was mentioned. Nevertheless, he made progress in another direction. He began to differentiate the various spheres of rationality, to use a term introduced later. Other known elements may be added to given elements (or to a given set of "known" elements), they may be adjoined.[23] One then calls "rationnelle toute quantité qui s'exprimera en fonction rationnelle des coefficients de l'équation et d'un certain nombre de quantité adjointes à l'équation et convenues arbitrairement". This in fact then gives rise to a new rationality. He was aware that the adjunction of a quantity could change an irreducible equation to a reducible one; he chose an example for this from Gauss' study of the irreducibility of equations.

Galois did not go back to these concepts but used them quite currently, in particular as regards adjunction, where the adjunction of the roots of auxiliary equations is mostly concerned. The theorems he presented can be easily translated into modern terminology by using the term field, but the terms Galois used are only similar to it in the section discussed above.

The study mentioned was only published as part of the inheritance by Liouville in 1846. Since Galois' treatise published around 1830, did not contain an interpretation of a field and of adjunction, he could not have influenced development until the year of Liouville's edition.[24] Not even Abel's studies were known in this interim period. The last

---

[22] Cf. Comment 11. However, even here the way a function is understood is important; with Galois it is more an expression. He did not speak of variables; all the elements "used" in the function (expression) were known, but the expression is not a number. It seems that he only considered a rational number to be a number.

[23] Since he introduced his concepts solely for purposes related to his theory, he defined them quite specially; that is why he spoke of quantities adjoined to an equation (i.e., to the coefficients of the equation). It is worth mentioning that one can also find the "adjunction" of a quantity (algebraic) with Abel, and this changes the irreducibility of the equation. This is the case in the paper on elliptical functions ([1], T. I, p. 547), where it could have easily escaped one's attention. The way he expressed himself helped this. Nevertheless, he mentioned it in the comment on the paper *Sur la théorie des nombres* ([155] 17). To quote Galois: La proposition générale dont il s'agit ici peut s'énoncer ainsi: Étant donnée une équation algébrique, on pourra trouver une fonction rationnelle de toutes ses racines, de telle sorte que, réciproquement, chacune des racines s'exprime rationnellement en. Ce théorème était connu d'Abel, ainsi qu'on peut le voir par la première Partie du Mémoire que ce célèbre géomètre a laissé sur les fonctions elliptiques. Galois expressed himself in a similar way in his algebraic study published posthumously (idem p. 37). As regards this problem, as well as the relation between Abel's and Galois' studies, cf. L. Sylow, *Les études d'Abel et ses découvertes* ([9] 23—6).

[24] In his paper *Sur la théorie des nombres* [155] published in 1830, Galois treated the solution of algebraic equations

$$f(x) \equiv 0 (\bmod p),$$

and most stimulating treatise was also only in manuscript. It was then included in the first edition of Abel's collected works.

It has been mentioned elsewhere that considerable attention was drawn to the problem of algebraic numbers as they occurred, in particular due to the theory of numbers. Evidence of this can be found in the studies of Dirichlet and later of Kummer and Eisenstein, as well as others. Although this trend had, since the twenties, yielded a number of results which contributed to the understanding of the arithmetic of algebraic numbers and also of the structure of algebraic fields, it did not lead to the concept of a field. This trend gradually matured and led to the study of units in algebraic fields and multiplied the cases of studying the elements derived from algebraic equations, etc. It reached maximum intensity in the forties. The direct connection with Abel's and Galois' studies in algebra, however, still remained obscured. The dependence on Gauss' results was rather more outstanding. The study of the irreducibility of the equation of the n-th roots of unity, or the solution of the equation $x^n - 1 = 0$ in general, which indicated algebraic problems, did not lead to changes in concepts. The decisive role in the determination of new, concepts was played by Kronecker.

It is interesting to note that in Kronecker's work studies which developed the problem of algebraic equations further encountered studies dealing with the problems of the theory of numbers with a markedly arithmetical character. Kronecker carried on a study which connected up with Gauss', and which Dirichlet and Kronecker's older friend, Kummer, had undertaken. One may call this the arithmetic in fields of roots of unity. However, this is not the manner in which the problem was presented. Apart from the irreducibility of the equations of roots of unity for any arbitrary natural n,[25] i.e. the equations

$$x^{n-1} + x^{n-2} + \ldots + 1 = 0, \tag{4.5.2}$$

and also the equations

$$x^{n-1} + ax^{n-2} + bx^{n-3} + \ldots + p = 0, \tag{4.5.6}$$

for integers $a, b, \ldots, p$, Kronecker concentrated on the study of the arithmetic of formations like

$$a_0 + a_1\varepsilon + a_2\varepsilon^2 + \ldots + a_{n-1}\varepsilon^{n-1} \tag{4.5.7}$$

where ε is the n-th root of unity. He also devoted his thesis, which founded his scientific career, to related problems. However, at first there was no hint of intentional consideration of fields in his papers.

---

where p is a given prime number. If the equation is of degree n and i is one of its roots, he studied the properties of expressions of the type

$$a_0 + a_1 i + a_2 i^2 + \ldots + a_{n-1} i^{n-1},$$

where a_j are integers of mod p. He himself said that he was generalizing Gauss' procedure from the *Disquisitiones arithmeticae*. This part also contains the embryo of far-reaching ideas, whether due to introducing congruence into the theory of algebraic equations, or to indications of finite fields. The latter idea, which is stressed in literature, could not have been understood by his contemporaries.

[25] As is known, Gauss only dealt in detail with equations $x^n - 1 = 0$ where n was a prime number.

Kronecker's studies in algebra developed concurrently. He was influenced considerably by Abel's work, and later, after the beginning of the fifties, also by Galois'. The
arithmetical problems mentioned also appeared in this part of Kronecker's work, or, to
be more accurate, his capability of exploiting the transition from one region to another
could also be seen in his work. As a point of contact he considered Abel's equations, the
special case of which is the equation (4.5.2) which represents equations with a cyclic
group. In connection with this set of problems, which is also treated elsewhere,
Kronecker eventually began exploiting the concept of a realm in his first treatise on
this subject, *Über die algebraisch auflösbaren Gleichungen* [247], in 1853. He in fact began where Abel had left off.[26] He uses his terminology, which follows from the fact that
he was stimulated by Abel's work. In accordance with Abel, he sought the form of the
most function of the quantities $A, B, C, ...$, which satisfies an equation of a given degree
with coefficients that are rational functions of the said quantities ([247] 4). A modern
interpretation of these words would call for the use of the term fields. However, like Abel,
Kronecker only used descriptions which were nearly the same as Abel's. They therefore
differ just as much from the concept he might have anticipated. His deliberations on
equations however, are more general. He did not consider the coefficients of an equation
to be rational numbers, but rational functions of the quantities $A, B, C, ...$, of the
nature of which he made no further mention. Kronecker considered an equation (the
coefficients of which were the said rational functions of the quantities $A, B, C, ...$) to be
irreducible, provided it could not be resolved (non-trivially) into factors of lower degrees
with coefficients of the same form. Without actually saying so, the set of these rational
functions, also containing rational numbers, thus came to the fore.[27] It would not be
correct to seek the disparities between Abel's and Kronecker's concepts; Kronecker's
concepts are, after all, slightly clearer. Of course, Kronecker, who knew Galois' studies,[28]
went on immediately to formulate the above problem in a different way: For a given
number n (natural) he wanted to find the most general algebraic function of $A, B, C, ...$
which would yield various expressions on varying the roots it contains, among which
there would be n, such that their symmetric functions would represent all rational functions of the quantities $A, B, C, ...$

The terminology he thus chose enabled him to study the relations between the
various fields without explicitly using the term.[29] The next step forward was Kro-

---

[26] The first edition of Abel's collected papers was also published in 1839, and their 2nd volume also
contained Abel's unfinished manuscripts. The algebraic manuscript [8] of which mention was made
above, was quoted by Kronecker in 1853 ([247] 4).

[27] However. Kronecker was also aware that he would arrive at a special case, provided $A, B, C, ...$
were inters, and in some cases he excluded this (p. 10). In this connection he also expressed a beautiful
theorem which joined in a certain sense the problems of arithmetic and algebra, i.e.".... dass die Wurzel
jeder Abelschen Gleichung mit ganzzahligen Coefficienten als rationale Function von Wurzeln der
Einheit dargestellt werden kann..." ([247] p. 10); also cf. Comment 6 on p. 497 in [243] Bd. 4.

[28] He quoted him a few pages further on ([247] 7) without using his terminology for the time being
(especially adjunction and adjunction to the equation), which he adopted a little later.

[29] We have omitted other authors who worked their way along similar lines to Kronecker, of whom

necker's paper [246] of 1854, written in French, with which one can see more clearly the influence of French mathematicians including Galois.[30] To the terms he used earlier he added an important term used by Galois, i.e., adjunction. In the paper of 1854 he studied which complex numbers should be adjoined to preserve the ir-reducibility of the function $F_m(x) = 1 + x + x^2 + \ldots + x^{m-1}$. Kronecker also gradually began to speak of realms, the realms of complex numbers and the realm of rational numbers, also acknowledging the existence of other rational realms.[31]

For these reasons one may assume that Kronecker arrived at a concept which was equivalent in fundamental features to the concept of a field in the middle of the fifties, largely as a continuation of the tradition of Gauss and especially Abel and Galois. In the studies on algebra which he used to develop Galois' theory, Kronecker used this concept frequently. Nevertheless there is a certain difference in the way he used the concept at this time and in later treatises. Some of the features of this difference follow from the other concepts used at the time. Integers and rational numbers are real numbers; in order to describe the others he used the terms "function", and "expression". That is why the sets of elements still remained in the background; he did not used the term "element". It is also important that algebraic equations and their solutions still remained the centre of interest. The properties of coefficients and of roots only appeared from this point of view. The spheres to which they belonged, like their origin, were of secondary importance; he only devoted attention to them as means to an end. He did not consider them on their own. The formation of these spheres and their internal structure, as well as the endeavour to define them, did not come to the fore independently, which did not help to clarify the concept. Nevertheless, the situation was ripe to take this last step. However, one had to wait some time for this.

In the interim, the development of new algebraic theories continued and presented new results. Galois' theory, like other new results, became a part of university text-books (Serret). It is said that in 1857 Galois' theory was specially taught for the first time by Dedekind at the university, although only two students came to the lectures. Other components of the development of algebra, terms and results changed as well. In this atmosphere accumulated experience with the emerging concepts helped them to crystallize and generalize. As regards the concept of the field two components were chan-ging: the arithmetic of algebraic numbers based on the theory of numbers, on the one hand, and the theory of algebraic equations, on the other. They still formed a unified arithmetico-algebraic sphere in the minds of the mathematicians, but the unifying

---

we might mention Jacobi and Malmsten; however, none of these expressed the concept of a field so clearly.

[30] Cf., e.g., H. Weber ([395] 6). Of course, Weber had no direct proof that Kronecker acquainted himself with Galois' work during his stay in Paris in 1853. Kronecker did not publish any mathematical papers betweem 1845 and 1852, although he certainly studied mathematics apart from the work he did to earn a living.

[31] Therefore, he used the following words: "das Gebiet ... des (im gewöhnlichen Sinne des Wortes) Rationalen", or "das Gebiet der complexen Zahlen", etc., Cf. ([248] 31).

interpretation of the various results was published quite late. However, when it was published, it necessarily contained the interpretation of the concept of a field as one of the initial concepts of the whole sphere. This interpretation appeared in Dedekind's work for the first time.

It is not quite clear when Dedekind expressed his interpretation of the concept of a field for the first time. One might regard as very probable the connection with his university lectures, mentioned above. However, Dedekind belonged to those mathematicians who are reluctant to publish their results and who seek patiently for the most suitable form of expression. That is why few of his papers were published when he was young; he only published his earlier results later. Nevertheless he did boast two publications in 1857, which linked up directly with the topic treated by mathematicians of the time, of whom mention has already been made. As regards the paper *Beweis für die Irreduktibilität der Kreisteilung-Gleichung* [114] the title speaks for itself; this is based on the studies of Gauss, Kronecker, Eisenstein and Schönemann. As regards influence, his paper *Abriss einer Theorie der höheren Kongruenzen in bezug auf einen reellen PrimzahlModulus* [113], which he connected with Galois' study *Sur la théorie des nombres* [155], is more important. Of the latter it was said that it contained the study of finite realms. Whereas Galois used his finite systems (in which we identify finite realms in the present sense) of "imaginary" quantities for his studies, Dedekind tried to master similar problems by other means. Of course, with Galois and Dedekind alike, the interpretation of the congruence of entire rational functions with respect to the prime-number modulus is without any trace of connection with the system of conceptions prevelant at the time, which gave rise to the concept of a realm. Realms of numbers, as they were then understood, were always infinite, to use modern terms, necessarily of zero characteristic. Finite realms of numbers were then unknown in mathematics. There is another important approach which Dedekind also employed, though he was not the first. Gauss introduced residual classes according to a given modulus, inclusive of the operations with them, and Cauchy then employed congruence with respect to the modulus $(x^2 - 1)$ in order to formally introduce complex numbers in 1847. Especially the latter was stimulating for algebraists. Dedekind introduced congruence with respect to an entire rational function m as well as a complete system of functions representing residual classes of mod. m ([113] 46 nn) which also led him to study the roots of congruence. Although the interpretation was general and systematic, the predominating connection with the theory of numbers could be seen. In spite of the fact that ideas of a finite field could be identified in this study of Dedekind, there are no accurate data on the connection with the contemporary problems leading up to considerations of fields.

A clear interpretation of the concept of fields together with the term "Körper", can only be found in Dedekind's 10th supplement to the second edition of Dirichlet's "*Vorlesungen über Zahlentheorie*" of 1871 [122]. The author considered the problems of the theory of numbers, in particular the theory of binary quadratic forms. He referred to the work of Lagrange which was continued by Dirichlet, whose treatise

"...die Transformationen solcher Formen in sich selbst... oder, was dasselbe ist, die Theorie der Einheiten für die entsprechenden algebraischen Zahlen behandelt". It was said that Kummer's ideal numbers provided a more profound insight into the real nature of algebraic numbers. "Indem wir versuchen, den Leser in diese neuen Ideen einzuführen, stellen wir uns auf einen etwas höheren Standpunkt und beginnen damit, einen Begriff einzuführen, welcher wohl geeignet scheint als Grundlage für die höhere Algebra und die mit ihr zusammenhängenden Theile der Zahlentheorie zu dienen".[32]

After this introduction, which indicates Dedekind's awareness of the novelty and generality of the approach, the definition of a realm follows: By a field one understands any infinite system of real and complex numbers which is so self-contained and complete that addition, subtraction, multiplication and division of two arbitrary numbers always yield a number within the same system.[33] To this Dedekind added that the simplest field is formed by all rational numbers and that all numbers form the largest field. These statements were followed by a number of theorems, most of which translated the results known in other connections into language of the field theory.

The concept of a field was explained sufficiently clearly and was assigned the most important status. It was a number field, i.e., a field the elements of which were numbers; it was of zero characteristic. The definition in itself was not quite accurate; the terms used for special fields of numbers, however, were in current use. One can also see a change in the whole system of terms. Numbers, including complex numbers, are mutually equivalent. They are all numbers; no circumlocutions with the help of expressions or functions are necessary. A field is composed of elements and this leads to an expressly set-like approach in the theorems on divisors and multiples, and also permutations, etc.[34] A similar approach cannot be found with any of Dedekind's predecessors. The conceptual basis, from which modern mathematical deliberations on algebraic structures could develop, had already been formulated here. It was not the only one, nor was it perfect. But it was there; the concept of a field was no longer anticipated, its form and meaning were no longer sought, nor was the relation to other concepts. Positive development could begin, and this was in fact the beginning of the next stage in the development of algebra.

However, the fact that Dedekind's interpretation was published could not quite move another of Kronecker's studies into the background. We have seen what

---

[32] Dedekind presented this explanation as well as the subsequent definition of the concept of a field in § 159 of the quoted reference [122b], p. 423 nn.

[33] Unter einem Körper wollen wir jedes System von unendlich vielen reellen oder complexen Zahlen verstehen, welches in sich so abgeschlossen und vollständig ist, dass die Addition, Subtraction, Multiplication und Division von je zwei dieser Zahlen immer wieder eine Zahl desselben Systems hervorbringt. ([122b] 424).

[34] These theorems also appear in Dedekind's work in connection with his theory of ideals, which he considered to be a tool in the theory of numbers, although an ideal represented a set with a different structure to that of a number field.

means he employed in working his way through to the concept of a field and its meaning in the fifties. He presented his comprehensive and general interpretation very late, not until 1882, but there is no doubt that it was born much earlier and that it was in direct connection with his earlier results and attempts.[35] Kronecker himself pointed out the connection between this interpretation and the studies of the beginning of the fifties in the introduction to his treatise published in 1882. He also restated his earlier results using his new terms.[36] One can also see the genetic connection with the statements and terminology of the studies which appeared in the interval 1853 – 1882,[37] although they are not identical. It is even possible that at the moment Dedekind published his interpretation of the concept of a field Kronecker, like other mathematicians, thought that he was actually thinking of something which was already known, of a term which they were already using.

Kronecker used the followisng words to define the new concept ([244] 4): "Der Rationalitäts-Bereich $(R', R'', R''', \ldots)$ enthält, wie schon die Bezeichnung deutlich erkennen lässt, alle diejenigen Grössen, welche rationale Functionen der Grössen $R', R'', R''', \ldots$ mit ganzzahligen Coefficienten sind". He then went on to explain in what sense he spoke of quantities. The concept, according to his opinion, should make it possible to include "quantities" which have certain identical properties into closed circuits, and he used the term "quantity" in the widest possible sense of arithmetic and algebra. He reacted to the traditional question and expressly said that a quantity need not be connected with the "to be larger or smaller" property. He was aware that the defined object was identical with Dedekind's "field". In accordance with Dedekind he repated some of the fundamental properties of the "realm of

---

[35] Unfortunately a more reliable analysis of Kronecker's life work which would make it possible to determine more accurately Kronecker's ideas is lacking. Much could probably be explained by Kronecker's lectures and letters. But instead of historical facts only reminiscences remain. An interesting piece of information came from Klein who probably knew the relations and facts of German mathematics very reliably, and who said: Es ist nun schwer, einen historisch zutreffenden Bericht zu machen, weil Kronecker seine Ideen oder doch das Vorhandensein seiner Resultate von 1858 an geschprächweise verbreitete, aber seine Abhandlung darüber: *"Grundzüge einer arithmetischen Theorie der algebraischen Grössen,"* erst 1881/82 in *Crelle*, Bd. 92, der Festschrift zu Kummers goldenem Doktorjubiläum, veröffentlichte, während Dedekind die zweite Auflage der von ihm herausgegebenen Dirichletschen Zahlentheorie (1871) benutzte, um in einem Supplemente seine Theorie zu entwickeln. [234], p. 323.

[36] When he mentioned the results of the study [247] of 1853, which has been mentioned above, he also used the term "Rationalitäts-Bereich", which is not in the study. He admitted in the preface that he had been thinking of summarizing the results in a more extensive study for a long time, but that he had constantly been dealing with gaps in the theory.

[37] Ich fixire, wie in meinen früheren Aufsätzen, z. B. in denjenigen, welche in den Monatsberichten der Berliner Akademie vom Juni 1853, vom Februar 1873 und vom März 1879 (sic!) abgedruckt sind, durch die Grössen $R', R'', R''', \ldots$ einen bestimmten Rationalitäts-Bereich $(R', R'', R''', \ldots) \ldots$ Die dort zuerst eingeführte Fixirung eines solchen Bereiches war, wie a. u. O. näher dargelegt ist, für die Klärung der Theorie der algebraischen Gleichungen durchaus nothwendig ([244] Erster Theil, § 1, p. 3). Kronecker also maintained that the requirement for the concept already followed from the work of Abel and Galois, whom he criticized for certain deficiences. Kronecker also affirms originally he used in lectures the term "Rationalitäts-Bereich".

rationality". For example, each realm contained rational numbers, various realms can be formed by "adjoining" algebraic or even non-algebraic quantities, etc.

In contrast to his earlier studies, one can clearly feel the awareness of a realm of elements (he sometimes used the term "die Elemente eines Rationalitäts-Bereiches" ([244], 8), and these realms have a certain structure given by operations and elements[38] which create it. These realms are different and there is a possibility of determining their mutual relations. Therefore, the aspect of a set is also present here.

However, there are certain insubstantial differences between the points of view and interpretations of Dedekind and Kronecker. First of all, with Kronecker one observes his general philosophical approach and the fact that he derived all numbers from natural numbers, which are the only ones really existing. He did not apply them as much to the concept of a field as to the manner of constructing them, where the symbolic element comes more to the fore.[39] The other difference is in the different approach: Dedekind sees the way out rather more in the theory of numbers (including the theory of divisibility), Kronecker in the solvability of algebraic equations and thus in building up Galois' theory. However, both of them build up theories of algebraic fields, including their arithmetic, for their own purposes.

One could certainly complement the facts mentioned about the origin of the concept of a realm in many ways.[40] They make it possible to determine at least the main stages of the generation of this concept, so important for modern algebra. One can speak of three stages in the period investigated. In the first, up to about the middle of the twenties, only the fundamental realms of numbers were recognized, and neither the theory of algebraic equations nor the originating facts about algebraic numbers required this concept. Basically, the existing, intuitively understood realms of numbers were sufficient in this respect. The nature of the second stage was determined by the studies of Abel and Galois. They first discovered the necessity of knowing the properties of the realm of coefficients and the realm of roots, as well as of the relation between them. In order to cope with these problems which formed the foundation of the development of the theory of algebraic equations, Abel and Galois looked for adequate means. They helped themselves by circumlocutions based on traditional concepts (function, later expression). The aids they found could easily be translated into the laguage of the field theory, but at the same time there are certain differences with both as regards the attempt to establish the concept of a field. An important

---

[38] Ein Rationalitäts-Bereich ist im Allgemeinen ein willkürlich abgegrenzter Grössenbereich, doch, nur, so weit als der Begriff gestattet. Da nämlich ein Rationalitäts-Bereich nur durch Hinzufügung beliebig gewählter Elemente R vergrössert werden kann, so erfordert jede willkürliche Ausdehnung seiner Begrenzung zugleich die Umschliessung aller durch neue Elemente rational ausdrückbaren Grössen. Es gibt aber auch natürlich abgegrenzte Rationalitäts-Bereiche, so das Reich der gewöhnlichen rationalen Zahlen, welches als das absolute in allen Rationalitäts-Bereichen enthalten ist ... ([244]9).

[39] Kronecker ([244], 42—4) connects with Gauss' second proof of the fundamental theorem of algebra which in fact led to symbolical adjunction, as pointed out earlier.

[40] Of course, the origin of the concept of a field would justify an independent monograph, but it is only possible to outline some of the features of the whole complicated process here.

concept was the adjunction of elements, introduced by Galois, which can be found in a different from with Abel. The efforts to establish a new concept were subject to objectives which existed at the time in connection with research into the solvability of equations. They could at most become an auxiliary means which could be replaced by intuitive knowledge.

Along with the studies of Abel and Galois, an increasing influence was exerted in this sphere by the results of the study of algebraic numbers in the theory of numbers. In this way requirements grew, as well as experience with elements of various fields, to use modern terminology (e.g., units). The algebraic and arithmetical approaches towards a new concept began to join materially in the fifties, especially in the work of Kronecker. In the middle of the fifties a further step had already been made in his work towards creating a concept which had the present day meaning, but which was introduced in connection with older ideas. Rational functions of certain elements are still in the fore, or certain expressions of a set of elements, were still in the mind, a set with certain properties. However, it took a relatively long time to take this small step forward. Not until 1871 did Dedekind express the concept of a field clearly in his publications, expressly as an infinite number field. However, it is a set of elements in which the operations of addition, subtraction, multiplication and division were defined (intuitively). Kronecker defined the concept in a similar way, but published it only in 1882, although one may assume that he had arrived at it earlier. However, from then on the concept of a field would become one of the fundamental concepts for further progress in algebra, a concept which was capable of further generalization and which would also become the initial concept for other algebraic theories. This development, however, belongs to the next stage in the development of algebra.

4.6 Conclusion

The properties of the fundamental algebraic structures were adopted by mathematics in numerical sets when operating with numbers. This process was not straightforward. The most important period of this process was between the end of the 18th century and approximately the seventies of the 19th century. In this chapter we investigated in what way the traditional topics of the sphere of algebra and arithmetic contributed to the crystallization of new concepts, which describe the fundamental algebraic structures in the said period, i.e., excluding the influence of new concepts and theories which were only being created in the 19th century. It was found that the traditional effort to present a reasonable explanation of the whole sphere, continued in arguments mentioned earlier. In attempting to overcome them, different properties of the fundamental numerical realms (natural numbers, integers, rational, real and complex numbers) were gradually discovered, as well as operations with their elements. In this way the theory of opposite quantities was established and propagated, and the fundamental properties of equality and operations were formulated. The authors introduced associativity, commutativity and distributivity without giving them special names. The climax of these efforts is in Ohm's papers, published in the twenties of the 19th century, which attempted to present a comprehensive, logically sound and, one may say, axiomatic interpretation of the whole of pure mathematics, and Bolzano's manuscript, which originated a little later and which had the same objective. With these, as well as with other authors, the objective is a description of an intuitively given numerical model. Therefore they tried to present an accurate description, as in Euclid's system, and efforts for further generalization were outside their reach, although outlines and possible methods for future exploration are provided implicitly in their work.

The traditional effort for a logically comprehensive interpretation did not in fact make any headway. However, in the old problems in the sphere of arithmetic and algebra were hidden problems whose solution could contribute substantially to the understanding of the structures of numerical realms. The old problem of the properties of irrational numbers, in particular under the influence of the theory of numbers, led to the differentiation of the realm of complex numbers, hitherto united, into various realms of algebraic numbers. However, the developments which led up to this presented the mathematicians with new inconsistencies which first had to be dealt with. Since the time of Euler there had been many indications that in algebraic fields, to use modern terminology, the definitions introduced for the rational field would not suffice. Gauss' demand for the extension of the realm of deliberations concerning the theory of numbers, and the fact that he introduced complex integers, opened the way to a more general definition of the divisibility of numbers, and to encounters with unexpected phenomena such as the invalidity of the uniqueness of the factorization of an integer into prime factors. The discussion of these problems which also led to the creation of new theories (Kummer) contributed substantially to realizing the existence of various kinds of algebraic numbers, peculiarities of their arithmetics, and the fact that traditionally accepted definitions and objects were not the only possible ones.[1]

From another aspect, even the old problem of expressing complex numbers and their arithmetic contributed to development. It is true that the strongest trend working to overcome the distrust in the real justification of complex numbers, suggested a geometrical interpretation and that this received the sanction of Gauss' authority. But, in reality, Gauss wanted to once again subject these numbers in their consequences to a geometrical model, which was quite against the spirit of arithmetic and algebra of the time. This trend certainly had a favourable effect on the origin of vector analysis and the theory of complex numbers, but it did not satisfy the more calculation oriented mathematicians. The various efforts displayed indications of a purely axiomatic approach, this could be seen most markedly in Hamilton's work, whatever its philosophical context.

---

[1] That is why the discovery of the arithmetic of algebraic numbers is sometimes compared with the origin of non-Euclidean geometry, Cf. [108].

The decisive stimulus for clarifying the common properties and differences in the various numerical realms was the theory of solving equations. It required the differentiation of numerical realms of coefficients and roots of the investigated equations, but the establishment of the concept of a field was delayed while the need for this concept increased, and while experience accumulated with deliberations which were not quite clear and in which the anticipated concept was used. This algebraic trend, in connection with other stimuli, especially from the theory of numbers, led to an intentional, clear and sufficiently accurate formulation in Dedekind's study of 1871, although the author had already arrived at the concept earlier. Kronecker could also have claimed that he had established this concept and in particular that he had introduced it into fundamentals of arithmetical and algebraic deliberations, although he published his accurate and clear formulation in 1882. However, in this way the discussion of the fundamental concepts of the sphere of arithmetic and algebra, or the system of these concepts, has obtained a different form. It in fact now belongs to a different period in the development of these disciplines. It does not follow that the whole change in the method, the approach to and concept of algebra resulted from the traditional problems of algebra and arithmetics; the exact opposite is true as will be seen in the subsequent chapters.

CHAPTER 5

STUDY OF THE STRUCTURES OF "UNTRADITIONAL" REALMS

5.1 Introduction

In the last chapter the way in which the knowledge of numerical realms was gradually accumulated was discussed. The subject, the knowledge of which did not exceed its scope in any way, was the realm of complex numbers which had been intuitively understood and used since the 16th century and quite currently in the 18th. Within this scope the need was to define more accurately its elements and operations during the period recorded in the previous chapter, also to arrive at a finer differentiation of numerical realms, i.e., realms of algebraic numbers in the first instance. The acquisition of knowledge of the structures of the realms was only possible if the subject studied, i.e. the numerical realms, helped this effort. This was mostly the case when the discovered structures could be identified with the description of the realm. This in no way means that new, apparently paradoxical properties (e.g., the uniqueness of the factorization into prime factors) did not appear; however, they rather led to realizing the differences between the individual realms than to a more general view. The generalization called for new comparative material.

Various representations of algebraic structures without the scope of numerical realms, described traditionally, appeared in and outside algebra a long time before the beginning of the 19th century, although their relation to the algebraic realms studied was only realized later.

It is understandable that these representations were made in various "coded" forms, and that is why their substance and algebraic meaning were missed. Their form was affected by the mathematical problems from which they originated. A remarkable example is combination calculus. The sources of its origin and development are different. An important role was played by probability calculus, the logical ideas of Leibniz and mere "playing around", only to make way later for special mathematical problems such as the determination of the powers of the binomial and polynomial in general. Combination calculus, dealing with combinations (permutations) of elements, was perhaps the only significant discipline in the 17th and mainly 18th century which was not compatible with the idea of mathematics as a science of quantities. The initial concept was the order of a given number of elements; combination calculus considered the various orders in which the elements were arranged to be different qualities, and according to the manner in which these various orders were formed, they were arranged in sequence. The numerical order[1] or the number of different orders of a given number of elements (perhaps with certain conditions given) was only the result of deliberations on the properties of these orders. However, the results were important with regard to the application of combination calculus in mathematics.

As regards the study of permutations, credit is due to the foremost mathematicians including Pascal, Leibniz, Wallis, Joh. Bernoulli, et al. The deliberations on permutations became quite usual and the terms used became known generally. Calculations of the number of permutations under various given conditions also required that adequate means be established. In the 17th century (Leibniz) a considerably vague idea of a cyclic permutation had appeared, this was then used in various forms by other mathematicians to study various problems of combination calculus in the 18th century.[2] This is perhaps where the idea of "composing" permutations was the clearest, i.e., the understanding of permutations not only as the order of elements, but also as operations prescribing a certain change of order. However, no more progress was made in the 18th century.

The idea of a finite realm emerged in the same connection: by repeating the same "permutation" one arrives at the original one. It is possible that the finiteness of the realm of permutations of a given number of elements, a realm in which an operation has been established, made it very difficult, if not impossible, to find the similarity between this "non-numerical" finite realm and infinite numerical realms.[3] This is the way that the ground was prepared for various applications of combination calculus, but this did not provide even the beginning of the study of the structure of sets of elements studied by means of combination calculus.

Towards the end of the 18th century a school of combination calculus, the principal representative of which was C. F. Hindenburg, flourished in Germany. The propagation of its results and opinions was helped by its own journal which was edited by Hindenburg, and which was one of the first specialized mathematical journals.[4] The papers of Hindenburg and of several of the other mathematicians of the time yielded some elements important for the development of mathematics, the symbolics used amongst others, but at the same time they only provided a means of solving problems of other branches of mathematics, e.g., of differential calculus, of

---

[1] In textbooks of the 18th century the following problems were given: Which permutations of the alphabetical order is the given word. Some of the letters might have occurred twice in the alphabetical order.

[2] The example Monge studied is well-known: composition and overlapping of maps. Cf., e.g., Cantor [55b] p. 221.

[3] There is no direct proof for this statement. It is a fact that similar features between these realms were not even sought in the 18th century, nor is there proof that they were anticipated. The separation of finite and infinite realms, including such concepts as groups and fields, lasted until well into the 2nd half the 19th century.

[4] The journal was first published in Leipzig in 1794 under the title *Archiv der reinen und angewandten Mathematik* and it tied up with an older journal the *Leipziger Magazin zur Naturkunde, Mathematik und Oekonomie*.

computing the powers of the polynomial, etc. On the one hand one could observe Leibniz's faith in the mission of combination calculus, on the other, the narrow scope and insufficient creative capabilities of the advocates of combination calculus. Hindenburg himself tries to classify permutations in different ways, and to form various classes of them, but his means were mostly too specialized and focused on concrete objectives. In general he only restricted himself to determining those groups of permutations of *n* elements in which a given number of elements did not change their order. An interesting idea, which was not however original (rather was it Hindenburg's complicated symbolics that were original) was the composing of a certain group (like sets) of permutations of disjunctive subgroups. However, since with Hindenburg the permutaions were exclusively a certain order of elements and not of operations, there is no indication of identical permutations.[5]

Although combination calculus did not play a direct role in the origin of the theory of groups, it has many features in common with it. Apart from the concepts of permutation, cyclic permutation, the number of permutations of *n* letters, etc., the fact is important that combination calculus as well as the theory of groups operated with finite sets, the elements of which were not numbers. Some mathematicians realized the special status of combination calculus, and this also led them to doubt whether mathematics really was a science of quantities. Of course, the same properties (finiteness and the study of "unquantified" properties) separated the originating theory of groups from the study of the structures of numerical realms. Although many mathematicians discovered some important properties of groups, possibly thanks to special cases, in the period considered, the abstract concept of a group, like the concept of infinite groups, only originated towards the end of the period, though both became widespread and adopted much later. The abstract concept of a group was preceded by an effort to find a more abstract approach to algebra in general, and more specifically to the development of some of the algebraic realms, which could be understood as the representations of groups. Therefore, the theory of groups will not be considered further in this chapter, although the study of groups in general, and in particular the study of groups of permutations, undoubtedly represented a new and untraditional element in mathematics at the time. Before the theory of groups found its place in the concept of algebra, various representations of algebraic structures gradually appeared, providing data for a more general view and for differentiating the substantial structures of characterizing properties from the insubstantial. These representations were nearer to numerical realms and one may say that they were derived from them in various ways, some as complex numerical expressions, others as concepts facilitating procedures of calculation. They did not lose their quantitative character, but were closely connected with the problems which were then being treated by algebra; and

---

5 Even Gauss used the concept of permutation in this current sense in his *Disquisitiones arithmeticae* in Art. 41, where he proved theorem: Consider *p* to be a prime number; the number of permutations of *p* elements which are not all the same is then divisible by *p*. In the proof Gauss referred to the "well-known theory of permutations".

therefore they were more comprehensible and necessary. The objective of this chapter is to investigate how these realms were formed, and how and when the elements which influenced the study of algebraic structures originated within them. The most important probably are the stimuli stemming from Gauss' work (5.2) from the beginning of the century. Apart from them an important role was also played by establishing the concepts of determinants and matrices, including the elaboration and exploitation of the so-called theory of determinants and matrices (5.3). Hamilton's creation of the concept of quaternions (5.4) displayed a different approach, and this was also due to a mere extension of the concept of complex numbers; Wessel had also worked quite close to it at the end of the 18th century.

5.2   Influence of Gauss' theory of numbers

"Untraditional" numerical expressions, forming realms with specially defined operations, appear in Gauss' theory of numbers more than in any study of his predecessors. The important concept of congruence according to an integral modulus which Gauss introduced by a symbol which later became currently used, is explained in the first section of his *Disquisitiones arithmeticae* which, compared with the other sections, is very inextensive and in its way forms an introduction to the others.

Gauss did not create a new or incomprehensible concept here. In the number-theoretical deliberations of his predecessors one frequently encountered divisions of integers into certain types,[1] e.g., numbers divisible by four and those indivisible by four, which then had the form

$$4m + 1, \ 4m + 2, \ 4m + 3.$$

With regard to its content, the division is equivalent to Gauss' congruence, and the congruence of two numbers only means that both numbers are of the same type. Gauss was aware of the similarity between his deliberations, which he used to introduce the new concept, and those of his predecessors. He also said (Art. 2) that all residues

---

[1] This is clearly the case with Euler, especially as regards his studies of primitive roots to which Gauss himself referred. However, Euler used the same division in the number-theoretic part of his introduction to algebra. Here, e.g., ([141] IInd part, 2nd section, § 65) he wrote: "... in Bezug auf den Theiler 3 sind die Zahlen von dreierlei Art. Die erste begreift diejenigen Zahlen in sich, welche sich durch 3 theilen lassen, und durch die Formel 3 *n* dargestellt werden. Zu der andern Art gehören diejenigen, welche durch 3 dividirt 1 übrig lassen, und in der Formel 3*n* + 1 enthalten sind. Die dritte Art aber begreift die Zahlen in sich, welche durch 3 dividirt 2 übrig lassen, und durch die Formel 3*n* + 2 dargestellt werden." In the continuation he went on to study what residue (Rest) the squares of integers had, and he proved that they were either divisible by three or produced a residue of 1. This is only proof that Euler not only divided numbers into classes, as we would say, with respect to a given modulus, but he also introduced operations with these classes, sometimes only considering the minimum residue to be the representative of a whole "kind" of numbers. This does not require special explanation, since it follows from the calculations.

of a given number with respect to the modulus *m* are contained in the formula $a + km$, where $a < m$ and k is an "undefined" integer. He found the same concept with Legendre, who was said (comment to Art. 2) to have used a simple sign of equality to denote that numbers were congruent. Thus the concept of congruence was used before Gauss. However, Gauss placed the definition of congruence at the head of his whole interpretation of the theory of numbers, being aware of its simillarity with equality,[2] and he established a suitable system of symbols. This new generalized equality defines the division of the integers into the disjunctive classes for which he chose various representatives, either the least residues, the absolute least residues, or powers of the primitive root, etc. The advantage of Gauss' interpretation is that it becomes easy to survey, the transitions between the various class representatives are clear, and the similarity with equality comes to the fore.[3] On the other hand, Gauss introduced the concept and the symbolics of congruence as a mere convenient abbreviation of the text, he did not deal with the properties of congruence in any way, he did not discuss the problems of the resolution of the realm of "integers into disjunctive classes", and therefore he did not even encounter the problem of operations with these classes. The fundamental of the concept is experience in calculating with integers from which the properties and operations follow. That is why the concept and its application were clear to his contemporaries, since everything could be reduced to divisibility or indivisibility of the difference of two integers by a third.

The advantages of Gauss' interpretation based on the concept of congruence in treating the problems of the theory of numbers, can be seen in comparing his work with the subsequent editions of Legendre's theory of numbers. Gauss' text is simpler and more elegant, even in the sections where the contents agree. Nevertheless Gauss' concept and symbolics did not become widespread immediately. Although the *Disqusitiones arithmeticae* drew very considerable interest of the French mathematicians, and, mainly due to Laplace, were soon translated into French, the concept of congruence was adopted earlier and more spontaneously by German mathematicians.[4] They were mastered in particular by mathematicians who dwelled on Gauss' stimuli, like Dirichlet, and later Kummer, Eisenstein, Kronecker et al. However, the understanding of congruence as a convenient means of recording, on which calculations can also be founded, prevailed. This was also the case in Dirichlet's first papers which treated the proof of the insolvability of certain indeterminate equations of the 5th degree.[5] It is

---

[2] The sign for congruence $\equiv$ was introduced by him "wegen der grossen Analogie, die zwischen der Gleichheit und der Congruenz stattfindet...". Cf. DA, note to Art. 2.

[3] Gauss, of course, not only introduced congruence between integers, but also between polynomials. Cf. DA, Art. 9, 10, 11. In the latter paragraph reference is made to the eighth, unpublished and unfinished section of the whole work.

[4] However, there also were German mathematicians who avoided the term congruence and its symbol. Cf., e.g., Crelle [107].

[5] This refers to the paper *Mémoire sur l'impossibilité de quelques équations indéterminées du cinquième degré*, which Dirichlet presented to the Paris Academy on 11th July 1825, and which he published in

interesting to note that Lebesgue was able to avoid the use of the concept of congruence as well as of Gauss' system of symbols when he returned to Dirichlet's treatise in Liouville's *Journal*[6] several years later, and either spoke of residues due to the division by a given number, or about numbers of a given form. Cauchy, when he published his attempt for an "algebraic" concept of complex numbers[7] in 1847, did not quite trust Gauss' symbolics either, and he proposed that one should speak of a divisor (diviseur) and also use the abbreviation "div", rather than of a modulus which has more than one meaning. He then derived several properties for this new equivalence, in particular for addition and multiplication of these equivalences.

Without making a more detailed analysis which would require devoting much more space to the way in which the concept of congruence was understood and exploited in the first half of the 19th century, one could make do, as a result of the cases discussed, with the statement that the concept was accepted in the first instance as a means of simplifying the form and record of deliberations, and that its general acceptance rather remained in the form of calculation, closer to what it was with Euler. The division of the realm of integers into disjunctive classes, inclusive of the fact that the operations were transferred here from the infinite realm of integers to a finite realm, can only be discerned in the background. Neither did anybody raise a more expressive objection that there existed divisible zeros in this realm (for a non-prime number modulus) in the first half of the 19th century.

Like with other concepts stressed by Gauss, the conept of congruence was also subject to a new approach in a disguised form, and it was only a question of time when and to what extent the concept of congruence would be deprived of its traditional meaning. It seems that there was only one mathematician in the first half of the 19th century who grasped that Gauss' concept of congruence established the study of a finite field of elements, and he was E. Galois. In a paper, published in 1830, under the title *Sur la théorie des nombres* [156], he spoke of the significance of the expression

$$a_0 + a_1 i + a_2 i^2 + \ldots + a_{v-1} i^{v-1}, \tag{A}$$

Crelle's *Journal* [124] three years later. That is where he wrote (p. 25): "...en se servant du signe employé par M. Gauss...", and used the relation

$$\vartheta \equiv a\varepsilon^2 \ (\text{mod } l).$$

[6] Lebesgue [270] referred to Dirichlet's paper mentioned in the previous comment several times, e.g., on pp. 49, 53, etc.

[7] Cauchy ([82] 88) said: Lorsque deux nombres entiers *l, m* étant divisés par un troisième *n*, fournissent le même reste, ils sont dits congrues ou équivalents, suivant le module ou diviseur *n*. Pour indiquer cette circonstance, on peut écrire, avec M. Gauss, $1 \equiv m$ (mod. *n*). He then referred to Kummer, and, as opposed to Gauss' "arithmetical equivalence", he introduced an "algebraic equivalence" which holds for the polynomials $\varphi(x)$, $\Psi(x)$, $\omega(x)$, but instead of Kummer's $\varphi(x) \equiv \Psi(x)$ [mod $\omega(x)$] he wrote $\varphi(x) = \Psi(x)$ [div. $\omega(x)$], which means nothing else (p. 89) than that there exists an entire function $u(x)$ such that $\varphi(x) = \Psi(x) + u(x) \, \omega(x)$.

where a_i are integers of mod p and p is a prime number.[8] En donnant à ces nombres toutes les valuers, l'expression (A) en acquiert p^v..., wrote Galois (p. 16). From this and from further context it followed that Galois really only distinguished p various elements when considering congruence with respect to a given modulus p, and that he did not connect any other concepts with them which would depend on the way these elements were "created" from the realm of integers. However, he did not consider the operations with these elements, which are quite clear as regards computing, in any way.

That was not until Dirichlet, who presented a more comprehensive interpretation of the concept of congruence in his lectures of the fifties, which Dedekind then summarized and published under the title *Vorlesungen über Zahlentheorie* [122].[9] The concept of congruence is one of the fundamental concepts here and the second chapter (pp. 32−73) was devoted to its interpretation. Following the usual definition (two numbers are congruent if their difference is divisible by a given number) and the quotation of the notations introduced by Gauss, he named some of the fundamental properties of congruence. He considered the reflexiveness of congruence to be a matter of course.[10] The concept of congruence then yielded theorems, e.g., that for two arbitrary numbers (integers) a and k it always held that $a \equiv a \pmod{k}$, and similarly $a \equiv b \pmod{k}$, $b \equiv c \pmod{k}$ led to $a \equiv c \pmod{k}$. He proved these properties by means of the definition of congruence. In a similar way he also proved the following theorems, among which there are statements like: if $a \equiv b \pmod{k}$ and $m \equiv n \pmod{k}$, then $a \pm m \equiv b \pm n \pmod{k}$ and $am \equiv bn \pmod{k}$. He pointed out that division was more complicated and that $am \equiv bm \pmod{k}$ need not necessarily mean that $a \equiv b \pmod{k}$; if $S = (m, k)$ it only held that $a \equiv b \left(\bmod \dfrac{k}{S} \right)$. This statement which

---

[8] The treatise was published with the note that it formed a part of Galois' investigations of the theory of permutation and algebraic equations. Mention was made of it in the previous chapter in the section on number fields, since the example given in the text is usually considered to represent anticipation of finite fields. The way Galois arrived at Eq. (A) has been omitted. As Galois himself said, it is based on Gauss' idea (mentioned in a comment above) to study the solution of algebraic equations with respect to a given modulus, i.e., $F(x) = 0 \pmod{p}$. Galois considered the roots of this congruence to be a certain kind of imaginary quantities. The object of the study were then the imaginary quantities.

[9] Dedekind visited Dirichlet's lectures in the years 1855—1858 (Cf. Preface to the 1st edition [122]). Dedekind's paper *Abriss einer Theorie der höheren Kongruenzen in bezug auf einen reelen Primzahl-Modulus* [113] also belongs to this period. Apart from referring to Gauss' manuscript on higher congruences he also referred to Galois' study mentioned above, and to the papers of Serret and Schönemann. In the editorial comment, Ore compared Dedekind's work with that of his predecessors and continuators, and he showed that the realm of residual classes, with which Dedekind worked, formed a finite field of characteristic p, and he summarized his point of view in the following words: "... die vorstehende Dedekindische Abhandlung gibt daher sogleich die arithmetische und zum Teil die algebraische Theorie der endlichen Körper" ([113] 67).

[10] He simply stated (p. 33): "Da die beiden Zahlen a und b in dem Begriffe der Congruenz dieselbe Rolle spielen, so darf man offenbar die zur Linken und Rechten des Zeichens $\equiv$ stehenden Zahlen mit einander vertauschen." Only then does he name the eight properties of congruence, the first of which are mentioned in the text.

can easily be proved by calculation, hides the fact that the created realm has zero divisors. However, this undoubtedly paradoxical statement was not expressed directly by Dirichlet, although using his own means one could have easily reached the conclusion that

$$(a - b) \cdot m \equiv 0 \pmod{k},$$

where neither $(a - b)$, nor $m \equiv 0 \pmod{k}$. It is interesting that he as if stopped before this conclusion and only presented it in a coded form. He then gave priority, like Dedekind, to prime-number moduli.[11]

In the next section Dirichlet maintained that congruence divided all numbers (integers) into classes of congruent numbers which have "common properties" with respect to the modulus, so that all the numbers of a class can nearly "play the role of a single number". It is therefore possible to take a representative of each class and form a complete system of incongruent numbers, i.e., a complete system of residues. As regards the non-prime number modulus he particularly stresses the classes which did not have a common divisor with the modulus, and thus he in fact approached problems of indeterminate analytics.

Apart from indeterminate equations, the properties of congruence were frequently applied to the study of the n-th roots of unity. An important stimulus was provided in this direction by Gauss in his *Disquisitiones arithmeticae* when, one may say, he established isomorphism between these roots and various systems of residues with respect to the modulus n. The finiteness of the number of n-th roots, however, was clear as regards calculation, and the operations too. In this case it was also only a question of time before calculations with the classes of residues with respect to a given modulus, or the realm of n-th roots of unity isomorphic to it (for any arbitrary n), would begin to be understood as a representation of more generally defined algebraic structures.

The examples we have spoken of were very close to the practical calculations currently used in the realms of complex numbers or even integers. The subject which Gauss studied in the fifth section of the *Disquisitiones arithmeticae*, i.e., binary quadratic forms, was considerably more remote for them. It was not a new topic, since it had earlier been treated by some mathematicians. Gauss himself admitted in a historical note[12] that much of what he had explained in Art. 153−221 was the

---

[11] In his textbook of Algebra ([352a] 23rd lesson, p. 298) in explaining congruence, Serret used the terminology and symbolics of Gauss. He gave the following reasons for doing so: L'avantage de la notation de M. Gauss, pour représenter les congruences, consiste surtout en ce qu'elle rapelle la grande analogie qui existe entre les congruences et les égalités, sans qu'il y ait pourtant de confusion à craindre. He then showed that congruences may be "adjusted" in the same way as equalities. Only as regards dividing is he carefull and says (p. 300) that congruence could be divided by any arbitrary number relatively prime with respect to the modulus, i.e., for $(m, p) = 1$, $ma \equiv mb \pmod{p}$ could be divided by the number m, yielding $a \equiv b \pmod{p}$. In the same way it is possible to divide the congruence $aa' \equiv bb' \pmod{p}$ by the congruence $a \equiv b \pmod{p}$ if $(a, p) = (b, p) = 1$.

[12] [161] Art 222. Fifth Section which is called "*On the Forms and Indeterminate Equations of the Second Degree*" takes up more than half the whole book, including paragraphs 153—307.

result of the work of other mathematicians. On this occasion he named Euler, Legendre and Lagrange. He gave credit to the latter[13] for the discovery of the finiteness of the number of classes of equivalent forms of a given discriminant. In this connection he gave himself credit for the expressly pronounced differentiation between proper and improper equivalence, besides making more accurate the proofs of some of the theorems which had hitherto only been explained by induction. Let us consider this concept, which may perhaps be used to characterize the difference between Gauss' approach to the problems and that of his predecessors.

Gauss gave the binary quadratic form as

$$ax^2 + 2bxy + cy^2 \tag{5.2.1}$$

where a, b, c are integers, and he denoted it symbolically by (a, b, c). The properties of the form (a, b, c) depend in the first place on the number $b^2 - ac$, as Gauss said (Art. 154) the determinant of the form, and which we are calling the discriminant. Forms are equivalent if they have the same discriminant and if they can be mutually transformed with the determinant of the transformation equal to $+1$. The following forms are equivalent:

$$f = ax^2 + 2bxy + cy^2, \tag{5.2.2}$$

$$f' = a'x'^2 + 2b'x'y' + c'y'^2 \tag{5.2.3}$$

for which the transformations

$$x = \alpha x' + \beta y', \qquad y = \gamma x' + \vartheta y' \tag{5.2.4}$$

exist, such that $\begin{vmatrix} \alpha & \beta \\ \gamma & \delta \end{vmatrix} + = \pm 1$, where $b'^2 - a'c' = b^2 - ac$. The forms f and f' are properly equivalent if it holds that

$$\begin{vmatrix} \alpha & \beta \\ \gamma & \delta \end{vmatrix} = +1; \tag{5.2.5}$$

and improperly equivalent if the determinant is equal to -1. Like his predecessors, Gauss considered the problems of quadratic forms to be connected with the solution of indeterminate quadratic equations,[14] but instead of an auxiliary means, which the investigation of the forms had hitherto represented, the knowledge of the properties of binary quadratic forms, their classification, and relations between their classes, including the study of substitutions, became the inherent, and one may say principal, mission of the whole extensive section.[15] That is why Gauss' interpretation differs from Lagrange's, even where both in fact presented the same result. Gauss studied the function of two indeterminates (Art. 153), while Lagrange's investigations ([268] 695) "... ont

---

[13] In the first place he quoted here Lagrange's study "*Recherches d'arithmétique*" of 1773—5 [268].

[14] In the introductory paragraph of the section he wrote that the purpose of the whole section was the solution of indeterminate quadratic equations.

[15] This is also indicated by the division of the contents of the whole section. This mission is also indicated, e.g., by Delone in his description of this section ([116], p. 41 nn).

pour objet les nombres qui peuvent être représentés par la formule $Bt^2 + Ctu + Du^2$ où B, C, D sont supposés des nombres entiers donnés et t, u des nombres aussi entiers, mais indéterminés". The formulation "the number expressed by a given formula" is not just a turn of speech with Lagrange, but it represents an approach to the problems as well as an indication of the role which the proper quadratic form is playing. The formulae, of course, have various properties, e.g., they may have the same discriminant (of course, Lagrange did not use this word, but he spoke of the value $4 BD - C^2$). What is substantial is that the given formulae can represent (given a suitable choice of the values of the indeterminates) the same set of numbers. Secondary is the fact that in the latter case, the substitution had a determinant (as it was later called) equal to $+1$ or -1; from Lagrange's point of view the differentiation of the two cases had no meaning.[16]

From the point of view being investigated, it is important that Gauss was aware in his interpretation how the introduction of equivalence (he sometimes spoke of congruence) and non-equivalence of forms introduces the division of the "sets" of forms of a given discriminant into classes. He said so quite clearly at the point where he departed from the topic treated by Lagrange and others, and where he summarizes, to a certain extent generally, the results achieved: There is an infinite number of forms of a given discriminant but there is a finite number of classes of properly equivalent forms. From each of these classes one can chose a form which will be representative of the whole class. Understandably, it is possible to choose this representative especially conveniently, and Gauss also gave criteria for this.

Gauss then went on to look for some special relations between the classes and defined their orders and types, only to go over soon to the problem of "composition of forms" (beginning with Art. 234), which he proudly claimed nobody had yet considered. In principle he maintained that the form

$$F = Ax^2 + 2BXY + CY^2 \tag{5.2.6}$$

was the product of two forms

$$f = ax^2 + 2bxy + cy^2, \tag{5.2.2}$$

$$f' = a'x'^2 + 2b'x'y' + c'y'^2, \tag{5.2.3}$$

i.e., $F = f \cdot f'$ or form F is composed of forms f and f' if the following substitution exists,

$$X = pxx' + p'xy' + p''yx'' + p'''yy', \tag{5.2.7}$$

$$Y = qxx' + q'xy + q''xy' + q'''yy', \tag{5.2.8}$$

---

[16] In comparison to Gauss' interpretation Lagrange's has its advantages. His approach to the proof of the finiteness of the number of classes of forms of a given discriminant is not perturbed by numerous subsidiary considerations which lead to the classification of classes and substitutions; nevertheless, it is not less tedious. Gauss, who moreover presented a large number of numerical examples, was probably interested in these classifications.

where the six numbers $pq' - pq''$, $pq'' - qp''$, $pq''' - qp'''$, $pq'' - q'p''$, $p'q'' - q'p'''$, $p''q''' - q''p'''$ are indivisible (Art. 235). Drawing on this definition, which led him to the proof of a larger number of theorems, he eventually arrived (Art. 249nn) at the composition of orders, types and classes of forms. In connection with this Gauss then introduces something rather clever: whereas he had as yet only used the concept of a product, apart from the tentative term for composition, he now considered it convenient to use the addition sign for class composition; similarly, he said, to denoting the identity of classes by an equation sign. He then immediately exploited it to introduce integral multiples of classes, i.e., for K he introduced $K + K = 2K$, $K + K + K = 3K$, etc.

We shall not continue to follow Gauss' interpertation even in an outline. Gauss' composition of binary quadratic forms and classes of these forms (a little later even an attempt at generalization for ternary forms) is in fact the first operation introduced by definition, moreover, for a non-numerical realm, so that its properties do not directly follow from the operations of the realm of complex numbers.[17] The operations introduced in this way for forms, are then applied by Gauss to classes of forms, multiplication being replaced by addition as regards notation. Moreover, he was dealing with (infinite and then finite) realms with one introduced operation.[18] Retrospectively it was clear that this realm formed an Abelean group in which the main class introduced by Gauss was a unit of the group, and that there are an opposite class to each class, etc.[19] However, these concepts remained hidden to the readers of Gauss' work for a long time.

As in other cases, Gauss' work was also important here in three different ways. In the first instance, it contributed to solving the problems of the theory of numbers and of the problems connected with the solving of indeterminate quadratic, equations, and to the capacity of quadratic forms for expressing numbers; here Gauss rounded off the results of his predecessors rather than bringing something new. By turning attention to the study of the properties of forms, to their classification, and to the study of classes of equivalent forms of a given discriminant, a wide range of new mathematical investigations was opened up. The indications which can be found in Gauss' studies helped the tendency to not restrict research just to binary quadratic forms, but to study quadratic forms of three indeterminates (which was begun by Gauss himself), but also in general of n indeterminates, and there was also the possibility of studying forms of higher degrees. Apart from equivalence, which produced the problem of the finiteness of the number of classes, a fruitful and comprehensible problem arose as to the number of classes of given forms of a given discriminant (first probably Dirichlet, later Kronecker and others).

---

[17] Nothing is changed by the fact that Gauss built the whole theory within and for the purposes of the theory of numbers.

[18] The idea of how to introduce at least the integral multiples of the classes of forms is too isolated.

[19] This interpretation can be found in a number of papers, e.g., with Dieudonné [120]; it was treated in greater detail and given more careful and historically more justified formulations by Delone ([116], 55 nn) who showed that the section of Gauss' study considered, contained indications of other results and of modern mathematical theories.

From the very beginning the study of forms was also connected with the problem of certain invariants with respect to certain groups of substitutions. Gauss also had this problem, but later he led Eisenstein to the first intentional study of invariants and covariants of cubic forms. Nearly simultaneously with Eisenstein, G. Boole introduced the demand for studying invariants in a general form in England; this was taken up by the English school represented by Cayley, Sylvester and Salmon from the fourties their developed work at the beginning of the second half of the century was followed by the studies of Hermite, but mainly Clebsch and Gordan in Germany .This brought the study of the invariant theory into a new stage as regards subject and the extent of research. The intensity of development was also helped by the awareness of geometrical connections.

It is probable that Gauss already had in mind the geometrical interpretation of the study of quadratic forms, however, probably in the form of geometrizing the theory of numbers.[20] This is probably where Gauss' effort for the elaboration of the theory of algebraic numbers began, in particular its arithmetic. Kummer also connected up with this aspect of Gauss' theory of quadratic forms in the forties[21] when he worked out the theory of complex numbers of the type $x + y\sqrt{D}$, and when he pointed out that his investigations of ideal complex numbers "had die grösste Analogie mit dem bei Gauss sehr schwierig behandelten Abschnitte: De compositione formarum, und die Haupt-resultate, welche Gauss für die quadratischen Formen ... bewiesen hat, finden auch für die Zusammensetzung der allgemeinen idealen complexen Zahlen statt".

One could find many such stimuli in Gauss' work; the *Disquisitiones arithmeticae*, like other of Gauss' studies, provided the initial impulse for many mathematicians in their creative work, also in the period immediately after Gauss' death the responses mentioned and also notes the elaboration of other parts of the DA, one will see that they do not appear immediately after the study was published. On the contrary, the examples which were given in connection with the theory of binary quadratic forms, provided they did bring really new results (excepting Dirichlet in the twenties) only appeared in the forties and were developed and further treated in the second half of the century.

The importance of Gauss' work also has a third aspect which the contemporaries practically missed until Gauss' death. This is a more abstract understanding of the subject of the deliberations in which Gauss anticipated the more general, and one might say structural, properties of the subjects he was studying. These he considered to be analogies at best when in fact it was a question of different representations of the same abstract algebraic structure. The historical significance of Gauss' work from this point of view is in the examples which were actually assembled in it for the first time, and which could provide material for establishing more general concepts and a more general

---

[20] Cf. Klein ([234] 35 nn), Delone [116]. Klein's opinion seems to be correct in that Gauss had already been aware of the connections between the theory of quadratic forms, problems of algebraic numbers and the theory of elliptical functions when he was preparing the DA.

[21] [258] p. 324—5, cf. also Wussing [410], p. 35; also cf. Chap. 4.5 in this book.

approach. For purposes of multiplying the data hitherto available, Gauss' theory of quadratic forms was probably important, in particular in its introductory part, beginning with the composition of the classes of these forms, where an operation which did not follow directly from the properties of numerical operations was introduced for the first time.

But this material could have been more effective for creating abstract concepts only to the extent to which the theories stimulated by Gauss' work began to be elaborated. That is why it is important that the creative responses to Gauss' work only began in the forties[22] (naturally with various deviations according to the problem considered), and gained in intensity in the second half of the 19th century. Only then do these untraditional spheres provide stronger impulses towards establishing more abstract concepts and conceptions, and also contribute to the changing understanding of the subject of algebra.

5.3   Determinants and the beginnings of the theory of matrices

The origin of the concepts of a determinant and a matrix as well as understanding their properties, are closely connected historically. Both these concepts stemmed from the study of the system of linear equations. The pattern of coefficients in this system already led Leibniz[1] to represent them with pairs of numbers which later was reflected in the introduction of two indices. As regards Maclaurin and later Cramer (1750), the study of geometrical problems led to the solving of the system of linear equations in a manner which bears Cramer's name to this day, and which presents the solution in the form of a ratio of certain determinants.[2] The connection between the determinant and its calculation, and the original pattern of coefficients in the system of equations was clear, although the various methods of elimination used could have obscured it. The idea of a square pattern of coefficients can also be seen in Laplace's development of a determi-

---

[22] As if this time factor did not exist in the descriptive studies stressing the logical conclusions. This is also true of the appropriate part of Wussing's study which was quoted above. In ([410] 40 nn) Gauss, Kummer and Kronecker have been bunched together and the reader can only tell from the context that they represent an interval of seventy years. If Wussing was only investigating one aspect of the development, i.e., the origin of the abstract theory of groups, Kronecker's approach and his explanation, which he quoted (p. 46), are incomprehensible. It is true, on the other hand, that here is no reliable literature which one might draw on yet. Especially a more accurate interpretation of the development of the theory of numbers following Gauss' DA is lacking, where there exists a void between the Gaussean researchers (also Klein and Delone mentioned above) and Dickson's history [118] This is reflected in the deficiences which will necessarily appear in any attempt to treat the development of the algebraic disciplines of the time, since the theory of numbers was then really algebraic.

[1] The case referred to in connection with Leibniz has been mentioned many times in literature, but one could also give older examples for the connection between the concept of a matrix and determinants. The most illustrative is the example given by Juskevic [220] in which, instead of a system of equations, only a system of coefficients was used in China; the system was subject to similar treatment as matrices.

[2] Cf. C. B. Boyer [48].

nant with respect to the elements of a line or column. The concept of sub-determinants can also be derived from the pattern of coefficients. In the second half of the 18th century the study of elimination concentrated attention on the calculation of determinants. Although a number of the foremost authors (Bézout, Euler, Lagrange, Laplace) worked in this field, the concepts remained very vague, and neither the terminology nor the symbolics made any progress. An exception which did not draw much attention[3] was Vandermonde's treatise [384] which was presented to the Paris Academy at the beginning of 1771. The author wanted to present formulae for elimination and when he had treated a system of n linear equations, he presented some fundamental theorems on determinants. The coefficients of a system of equations represent a couple of numbers (or, in general, of letters) written one above the other, the lower representing the column and the upper the line. In general, he used the Greek alphabet for the upper and the Latin alphabet for the lower. He gave three equations of three unknowns as an example:

$$
\begin{aligned}
{}^1_1\xi_1 + {}^1_2\cdot\xi_2 + {}^1_3\xi_3 + {}^1_4 &= 0 \\
{}^2_1\xi_1 + {}^2_2\cdot\xi_2 + {}^2_3\xi_3 + {}^2_4 &= 0 \\
{}^3_1\xi_1 + {}^3_2\cdot\xi_2 + {}^3_3\xi_3 + {}^3_4 &= 0
\end{aligned}
\tag{5.3.1}
$$

If there are two coefficients

$$
\frac{\alpha}{a} \quad \text{and} \quad \frac{\beta}{b}
$$

he wrote their product simply as $\frac{\alpha}{a}\cdot\frac{\beta}{b}$. He then went on to introduce what he called "a system of abbreviated annotation" in the following way:

$$
\frac{\alpha\,|\,\beta}{a\,|\,b} = \frac{\alpha}{a}\cdot\frac{\beta}{b} - \frac{\alpha}{b}\cdot\frac{\beta}{a}
\tag{5.3.2}
$$

$$
\frac{\alpha\,|\,\beta\,|\,\gamma}{a\,|\,b\,|\,c} = \frac{\alpha}{a}\cdot\frac{\beta\,|\,\gamma}{b\,|\,c} + \frac{\alpha}{b}\cdot\frac{\beta\,|\,\gamma}{c\,|\,a} + \frac{\alpha}{c}\cdot\frac{\beta\,|\,\gamma}{a\,|\,b}
\tag{5.3.3}
$$

$$
\frac{\alpha\,|\,\beta\,|\,\gamma\,|\,\vartheta}{a\,|\,b\,|\,c\,|\,d} = \frac{\alpha}{a}\cdot\frac{\beta\,|\,\gamma\,|\,\vartheta}{b\,|\,c\,|\,d} - \frac{\alpha}{b}\cdot\frac{\beta\,|\,\gamma\,|\,\vartheta}{c\,|\,d\,|\,a} + \frac{\alpha}{c}\cdot\frac{\beta\,|\,\gamma\,|\,\vartheta}{d\,|\,a\,|\,b} - \frac{\alpha}{d}\cdot\frac{\beta\,|\,\gamma\,|\,\vartheta}{a\,|\,b\,|\,c}.
\tag{5.3.4}
$$

He indicated how this method could also be used for $n > 4$. Although the upper symbol of $\frac{\alpha}{a}$ always means the ordinales member of the equation and the lower the ordinal number of the coefficient in it, he pointed out that the resultant value of the expressions did not change if the "lines" and "columns" were interchanged. He had some difficulties only with the signs to the individual terms. However, he knew that the number of addends on the r.h.s. (for n columns and n lines) would be $n!$, and that half of them would be positive and half negative. He pronounced several theorems on determinants, also the theorem

---

[3] This is also substantiated by the fact that the symbolics Vandermonde used did not become current, and not even Laplace exploited them in his papers which followed Vandermonde's publication.

which is in fact identical with the statement, that an even number of transpositions of columns or lines would not change the sign, but an odd would change the sign of the determinant. The theorem which says that a determinant is equal to zero if two lines (or columns) were identical he said was valid (as indicated) for $n = 2, 3, 4$. He did not carry on with induction which would not have been difficult.

Apart from pronouncing a number of theorems on determinants and introducing symbolics, Vandermonde presented relations between the calculation of determinants and the theory of permutations.[4] This was developed in an interesting way by A. – L. Cauchy in his paper of 1812 [58]. However, in the meantime several new and important ideas appeared in mathematics. And thus one again encounters in this connection Gauss' *Disquisitiones arithmeticae*. In the fifth section Gauss studied in detail linear substitution, an important characteristic of which was its determinant. If the linear substitution (of a binary quadratic form) is expressed by the equations

$$x = \alpha x' + \beta y', \quad y = \gamma x' + \vartheta y' \tag{5.3,5}$$

Gauss stressed the "number" $\alpha\vartheta - \beta\gamma$, and according to its value he differentiated between proper and improper substitutions, etc. With ternary quadratic forms, which he denoted briefly by

$$\begin{pmatrix} a & a' & a'' \\ b & b' & b'' \end{pmatrix} = ax^2 + a'x'^2 + a''x''^2 + 2bx'x'' + 2b'x'x''x + 2b''xx', \tag{5.3.6}$$

he only used the coefficients to express the substitution:

$$\begin{matrix} \alpha & \beta & \gamma \\ \alpha' & \beta' & \gamma' \\ \alpha'' & \beta'' & \gamma'' \end{matrix} \tag{5.3.7}$$

(Cf. Art. 268), and he spoke of the associated determinant as of a number k which playd the same role in the transformation of the form as the number $\alpha\vartheta - \beta\gamma$ did in the case of quadratic forms. If form F with discriminant E becomes form f with discriminant D due to substitution (5.3.7), it holds that

$$E = k^2 D, \tag{5.3.8}$$

which for $k = \pm 1$ again leads to the concept of equivalence and to the division of the forms of the same discriminant into a finite number of disjunctive classes. In this connection he pronounced a theorem which in fact introduced the multiplication of matrices. Given (Art. 270) three ternary forms f, f', f'' such that f becomes f' due to substitution (5.3.7) and f' becomes f'' due to substitution

$$\begin{matrix} \vartheta & \varepsilon & \xi \\ \vartheta' & \varepsilon' & \xi' \\ \vartheta'' & \varepsilon'' & \xi'' \end{matrix} \tag{5.3.9}$$

---

[4] In this case Vandermonde referred to his studies of permutations in the paper on solving algebraic equations of 1770 [383].

also form f will become f'' as a result of the transformation

$$\begin{array}{lll}
\alpha\,\vartheta + \beta\,\vartheta' + \gamma\,\vartheta'' & \alpha\,\varepsilon + \beta\,\varepsilon' + \gamma\,\varepsilon'' & \alpha\,\xi + \beta\,\xi' + \gamma\,\xi'' \\
\alpha'\vartheta + \beta'\vartheta' + \gamma'\vartheta'' & \alpha'\varepsilon + \beta'\varepsilon' + \gamma'\varepsilon'' & \alpha'\xi + \beta'\xi' + \gamma'\xi'' \\
\alpha''\vartheta + \beta''\vartheta' + \gamma''\vartheta'' & \alpha''\varepsilon + \beta''\varepsilon' + \gamma''\varepsilon'' & \alpha''\xi + \beta''\xi' + \gamma''\xi''.
\end{array} \qquad (5.3.10)$$

If f is equivalent to f', and f' to f'', f is also equivalent to f''. This statement, which he presented without proof like many other theorems, actually means that the determinant of the resultant matrix (5.3.10) is also equal to ± 1 if the determinants of substitutions (5.3.7) and (5.3.9) are equal to ± 1. At the end of this short explanation Gauss briefly mentioned that these theorems clearly held for more forms without further explanation.

However, the expression of the substitutions and the numbers associated with them represented a sufficient foundation for Gauss' objective, and he did not give them any other conceptual form, nor did he look for any of their other properties or relations.[5] This was done much later by his followers. However, Cauchy, who also reacted to the stimuli due, e.g., to Vandermonde, was the first to connect up with Gauss' ideas on determinants, regardless of the context in which they appeared in his book.

As in his other treatise, Cauchy was very consistent in obscuring the connection in which the investigated problem originated, and he tried to analyse it in greater detail as an isolated problem. In his study the connections between determinants and matrices[6] were pushed into the background. Cauchy considered determinants to be special cases of alternating functions.[7] In the first part of his study Cauchy divided permutations of n letters into even and odd, he spoke of cyclic permuations, etc. Cauchy introduced the determinant in the following manner: He formed a product of n elements a_i,

$$S(\pm a_1 a_2^2 a_3^3 \dots a_n^n) = a_1 a_2 \dots a_n(a_2 - a_1)(a_3 - a_1) \dots$$
$$\dots (a_n - a_1)(a_3 - a_2) \dots (a_n - a_{n-1}) \qquad (5.3.11)$$

---

[5] In spite of a certain abbreviated record of the ternary quadratic form itself, he never presented it in a form analogous to the arrangement of the coefficients of a substitution; one will therefore not find the arrangement

$$\begin{array}{lll}
a & b'' & b' \\
b'' & a' & b' \\
b' & b & a''.
\end{array}$$

The determinant appropriate to this matrix would be

$$ab^2 + a'b^2 + a''b''^2 - aa'a'' - 2bb'b''$$

which Gauss co-ordinated with this form without giving any further reasons for doing so.

[6] The terms determinant and matrix are used in the text although this is not historical. Laplace's term "resultant" is more usual for determinants, but in concept it is identical with determinant. Neither the term nor the concept of a matrix existed, but Gauss had worked with a square arrangement of coefficients of a linear substitution.

[7] This means functions which only display two "opposite" values if all the "letters" are permutated, i.e. $\pm k$. The connection of the concept of a determinant with the theory of permutations means that Cauchy connected up directly with Vandermonde, but is also indicative of the originating problems of groups of permutations. Moreover, paper [58] was also close in certain features to Cauchy's first study belonging topically to studies of the theory of permutations.

and put[8] $a_r^s = a_{rs}$. The resulting function he then called a determinant. He also arranged the elements of the determinant into a square table:

$$
\begin{matrix}
a_{11}a_{12} & \cdots\cdots & a_{1n} \\
a_{21}a_{22} & \cdots\cdots & a_{2n} \\
\vdots \ \vdots & & \vdots \\
a_{n1}a_{n2} & \cdots\cdots & a_{nn}
\end{matrix}
\tag{5.3.12}
$$

In this connection he also introduced the concept of conjugate elements, main elements, etc. The determinant is then in fact associated with this system. This enabled Cauchy to derive a larger number of theorems on the properties of the determinant. He also formed a system of terms, frequently based on the square arrangement of elements and under the influence of Gauss, which were suitable for describing the properties of determinants. The theorems (known in some cases even earlier like, e.g., the expression of a determinant by means of sub-determinants) were proved by means of permutations of coefficients, his proofs being satisfactorily accurate.[9] The method of proof, the concept of determinants and the comprehensiveness of the interpretation led some historians to the opinion that Cauchy was the formal founder of the theory of determinants.[10]

As regards the development of algebra, the creation of the concept of the determinant as well as the discovery of its fundamental properties did not only mean an acquisition of a technical means for solving certain problems. The determinant itself was only a number, but it represented a number co-ordinated in a certain way with n^2 elements. The multiplication of determinants as introduced by Gauss, and considering his results, was explained in a general way by Cauchy; it did not represent the multiplication of two numbers, but the multiplication of two systems of elements which created a third system. It is true that the elements of these systems were numbers, and the manner in which the elements of the products were obtained from the elements of the factors was defined by a simple algorithm based on the multiplication and addition of numbers. However, it was the result of the square arrangement of these elements, and it again led to the same. Therefore, a determinant was a number and it was not. It claimed an exceptional status and it had a part in creating a realm which was not purely numerical.

Although the determinant had been clearly defined, and its definition had been derived from the square arrangement, no progress was made towards understanding the concept of the matrix and the operations relevant to it in the first decades of the

---

[8] $S(\pm a, a_2^2, \ldots, a_n^n)$ represents the sum of products in the brackets for all permutations of the indices, the positive sign being appropriate to the even, and the negative to the odd permutations.

[9] Fundamentally, he used complete induction in this case, or, more accurately, a transition between $(n-1)$ elements and n elements, whereas his predecessors (including Vandermonde) were satisfied with incomplete induction based on examples for $n = 2, 3, 4$. In this he differed from his contemporary Binet.

[10] Cf. Studnička [362]; Muir was of a different opinion ([307] Vol. I, p. 131) in that he attributed more credit to Vandermonde.

19th century. The square arrangement of the elements then only had a single characteristic feature, which was the determinant. There was no reason really for differentiating between various square arrangements, the determinants of which had the same value.[11] Gauss worked very frequently with various substitutions, the determinants of which were equal to ± 1, but he never arrived at the concept of a matrix unit; this would not have been a "substitution" anyway. However, on the other hand, the substitution led him to the concept of an adjoined matrix. For similar reasons one cannot find with Gauss, Cauchy or Binet traces of a discovery which would have contained in itself the multiplication of matrices in the form found with Gauss, i.e., the non-commutativity of multiplication. A discovery had thus been within reach which would have multiplied abundantly the ideas of possible algebraic structures, but instead, the concept of non-commutativity did not appear here or elsewhere at the time. In the case considered it had been obscured by the multiplication of determinants, which represents the multiplication of two numbers and, therefore, is commutative.

It was necessary to wait for a very long time for the concept of the matrix, the relevant operations and the discovery of its properties to crystallize. In the interim the number of papers on determinants gradually grew.[12] Their study and the search for suitable applications with a good deal of incentive only began to flourish in the forties. The most important were three papers of C.G.J. Jacobi [209−211] of 1841, which connected up with the definition of the determinant introduced by Cauchy, and also with Cauchy's method of proof.[13] The determinants became a generally known tool in the forties; they were subject to exaggerated attention. The number of papers and textbooks published since the fifties is not proportional to the scientific contribution, but exceeds it considerably. Nevertheless, a wider circle of mathematicians became acquainted with the properties of the square arrangement, its multiplication, transpositions, etc., in a comprehensible form. This favoured the creation of the concept of a matrix. However, this was preceded by changes in the treated mathematical topics. Their result was that the square arrangement of elements began to appear in other connections than before, where it only represented the preparatory stage in calculating a determinant. Several tendencies could be observed here. The most important was probably the ever more general study of forms, where the trend led from binary quadratic forms, gradually to ternary forms and in general to n-ary forms; and also cubic forms and forms of the n-th degree were considered. The

---

[11] Gauss differentiated between the same matrices with the same determinant only with a view to the advantages or disadvantages in calculating the substitution represented by the matrix.

[12] There is reliable information, mostly in Muir's work, on the increase in the number of papers on determinants. It was treated some time ago by Price [335], and recently May [283] considered these data again.

[13] The author of one of the first textbooks on determinants, R. Baltzer, characterized the credit due to Jacobi's three treatises in that only a few mathematicians had used determinants up to the time they were published, but thanks to Jacobi they became universally used.

substitution of these forms acquired a considerably complicated form and the record of the transformed forms called for special attention and symbolics. An important impulse was born in England, in the forties[14], when G. Boole pointed out the existence of other invariants than the discriminants of forms. As a result of this Cayley and Sylvester immediately produced a series of papers.[15] Cayley, for example, used the term hyperdeterminant originally, instead of the term invariant, in order to stress the connection between determinants and invariants. The study of forms was no longer motivated only by the application of the results to the theory of numbers, but also problems of algebraic geometry, which was then modern, were being considered. The search for suitable symbols and terms for recording the coefficients of forms, for the classification of forms, for their substitutions and for invariants was a mighty stimulus for dividing the concept of the square arrangement of elements and the concept of the co-ordinated determinant, in particular as regards the work of Cayley and Sylvester. With these authors it can be seen that the matrix is being considered separately from the determinant in particular where problems leading to considerations of rectangular arragements of elements are being studied.[16]As regards Cayley this was already in evidence in his paper on n-dimensional geometry [88] of 1843, in which he considered, among other problems, the co-ordination with rectangularly arranged $m \cdot n$ elements ($m < n$ integers) of a system of determinants of the m-th degree. Similarly Sylvester (1850) treated a system of determinants co-ordinated with a rectangular matrix [367].

Although Cayley had a sufficiently clear idea of matrices in the middle of the forties, he first spoke of the theory of matrices in the middle of the fifties [98], and it seems that his idea of matrices and of their theory had become quite clear by then.

---

[14] These studies formed part of the effort of the whole English school of algebra, but as to its opinions, in spite of the fact that they stem from the thirties, they will be discussed separately in the next chapter.

[15] For lack of space it is impossible to go into the origin and development of the theory of invariants. The fundamental facts and especially the bibliography can be found in [284]; cf. also the historical notes of Weyl [400]. An interesting opinion is voiced in [142]. Cayley and Sylvester considered the problems of determinants in their first papers and thus joined the general flow. Sylvester even, without knowledge of his predecessors, again discovered the fundamental properties of determinants and expressed them in an original and complicated system of symbols. Credit is due to Cayley for introducing one of the ways of denoting a determinant, used to-day, i.e.,

$$a_{11}$$
$$\cdot$$
$$\cdot$$
$$a^{nn}$$

since the term "determinant" had been used from the beginning of the forties exclusively in the present sense.

[16] At my instigation J. Tvrdá [380] analysed the origin of the theory of matrices in the early papers of Cayley and Sylvester, and I have adopted her results.

He also introduced the multiplication of two matrices, and what is important, he also used matrices for recording bilinear and quadratic forms. Thus, the theory of matrices acquired a new and independent meaning for him, separated from the theory of determinants.[17] It was not until 1858 that Cayley explained his ideas of the theory of matrices in a paper called *A Memoir on the Theory of Matrix* [99]. Here Cayley had already been influenced by the results of other spheres to which he referred, in particular by Hamilton's quaternions. Therefore, it is not surprising that he was capable of discussing the characteristic features of operations, including associativity and commutativity.

Cayley's interpretation of the theory of matrices had the signs of a purposeful attempt at comprehensiveness. The concepts which had been materializing in various connections for a long time, were defined generally here, and included in the comprehensive conceptual system. Cayley defined the addition and subtraction of matrices, and showed the properties of these operations. He also defined the properties of multiplying a matrix by a number. He also introduced the multiplication of matrices, and in this connection he also determined the meaning of the concepts of a zero and matrix unit, introduced earlier.

In this part of this study Cayley treated square matrices in the first place (their elements could also be complex numbers); his theory of square matrices can be considered an example of linear algebra.[18] With a view to the period, Cayley built up a new system of elements which were not all numbers in the usual sense, but a system which did contain numbers. Operations were defined within this system, which had not been directly derived from operations known in numerical realms and which had considerably different properties. The realm was such that one could introduce "into it" a number of problems from "traditional" realms. For example in the paper referred to above, Cayley treated the solving of equations in this realm. The latter then represented one of the possible generalizations of numerical realms. It provided another example of the possibilities of various algebraic structures, in this way ground was being prepared for generalizing the subject of algebra.

However, Cayley's interpretation of the theory of matrices arrived at a time when other non-numerical fields, the structure of which was being studied, were already known. Therefore, there was nothing to hinder the rapid propagation of the concept of a matrix, which had already been prepared anyway by long years of study and

---

[17] Cayley already used (in French) the term "la matrice" in 1854, but in contrast to previous papers he only considered square matrices. He even spoke of the theory of matrices, and said: "Il y aurait bien des choses à dire sur cette théorie de matrices, laquelle doit, il me semble, précéder la théorie de Déterminants." For more detail see Tvrdá [380].

[18] In spite of the fact that Bourbaki interpreted the algebraic results of the mathematicians of the time modernly, it seems that in this respect he underestimated Cayley ([40] 121): "... Cayley... ne considère pas encore les matrices carrées comme formant une algèbre...". Tvrdá [380] tried to prove the opposite and spoke of the representation of linear algebra. Since linear algebra as a concept of a certain structure did not exist, I used a more careful word in the text, "example".

applications of determinants. However, Cayley and the British school were not the only ones who had reached the concept of a matrix. Some mathematicians in Germany soon began to contribute to the theory of matrices with their own original papers. It is not quite clear when Kronecker who, like Weierstrass, was known to delay the publication of his ideas and results, arrived at the concept of the rank of a matrix and in what connection this occurred.[19] It is probable that this happened during the period which is being investigated. Nevertheless, Sylvester touched on the concept of the rank of a matrix (its nullity) in a slightly obscured form prior to 1858. The meaning of the rank of a matrix became generally acknowledged in the seventies when the theory of matrices became domesticated and was well on its way to becoming established as the basis for interpreting the theory of determinants, on the one hand, and as a tool for solving problems in some spheres of algebra, algebraic geometry and the theory of numbers, on the other. The abstract interpretation of the theory of matrices as presented by Cayley in 1858, however, was not expanded upon in the subsequent years, being only of marginal interest. This fact is evidence that there existed possibilities of abstract treatment in algebra at the time, but that they were not being exploited. In any case, there existed a theory of matrices from the fifties, not only as a technical means, but also as an example of a set of elements, different from numbers; a set which had a different structure from that of the numerical realms.

5.4 Beginnings of the study of quaternions

With a view to operations in the real realm, complex numbers represent a quite natural extension of the realm of real numbers, and their acceptance was forced by the theory of algebraic equations. Their geometrical interpretation was only proof of their realness and justification, and in the form of a model, also a certain proof of their truthfulness. However, around the end of the 18th century, requirements for further extension began to appear, in this case of the realm of complex numbers. The initial impulse stemmed from geometry. At the time the sphere of arithmetic and algebra became the principal mathematical sphere, and geometry (and geometrical quantities) were considered to be special, the requirement for a multilateral algebraic calculus of geometric quantities grew stronger. In contrast to the general and in many ways vague demand of Leibniz in this respect, the effort to create an algebraic calculus which would be capable of handling calculations with directions and line segments in a three-dimensional space began to acquire a concrete form. It has been shown[1] that perhaps the first to attempt this with some measure of success was Wessel, who worked his way close to the discovery of quaternions. However, his work remained unnoticed for a long time.

---

[19] Kronecker only published the concept of the power of a matrix in 1878.
[1] Cf. Chap. 4.4

At the beginning of the 19th century several mathematicians tried to find algebraic expressions to describe formations and operations in three-dimensional space in connection with the search for a geometrical interpretation of complex numbers, or directly with establishing a geometrical calculus. One must admit that Argand, or the others, had not made as much progress in treating the problem as Wessel, whose work they did not know, at the time. However, apart from the impulses stemming from geometry for an expansion of the fields in algebra, the internal development in geometry was progressing. In the latter, frequently without any algebraic parallel whatsoever, mathematicians worked their way to a concept of a vector, and thus looked for the fundamentals of vector calculus.[2] The calculus began to develop as an important tool of geometrical and also mechanical considerations in the forties. Although the geometrical contact is important for the origin and further development of vector calculus, nevertheless it is possible to abstract the algebraic aspect from it. This is the more justified as these geometrical connections are very well known from literature,[3] but at the same time also because vector calculus also displayed certain non-geometrical stimuli in its development. It formed a part of the trend leading finally to linear algebra. The linear systems grew especially under the influence of the theory of numbers and within it, as a result of Gauss' influence.[4] The seventh section of his *Disquisitiones arithmeticae* drew attention to the algebraic connections of the n-th roots of unity. This started the fruitful studies of certain algebraic realms[5], which were then considered to be studies of rational functions (expressions) of the said n-th roots. The second part of the study on cubic residues was used by Gauss to draw attention to the possibilities provided by complex integers. Dirichlet then went on to elaborate on these ideas in an interesting manner at the beginning of the forties. He connected

---

[2] These problems have recently been treated by J. Folta who also showed [143] what great obstacles B. Bolzano encountered due to his non-algebraic approach in the process of reaching a more general understanding of vectors in his elementary geometrical deliberations. J. Folta also touched on the problem in his paper [144] in which he brought more extensive arguments from the world development of mathematics.

[3] In this respect one can refer to a recently published paper of M. J. Crowe [108]. Although he studied the development of vector calculus summarily, the set of data he provided on its algebra is more abundant than presented here on this relatively narrow problem. From our point of view the advantage of Crowe's study is in his accurate and detailed method with the data, which enabled him to investigate and correlate the various twists and turns in which the idea of the vectorial system was created. Thus is it possible to consider only those features which affected the development and understanding of algebra. It should be added that Crowe's evaluation of these features is basically considered as correct, although it is exaggerating slightly to compare the discovery of quaternions with the discovery of non-Euclidean geometry (pp. 30—1).

[4] In this respect cf. Boubaki's notes ([46], p. 84—7). However, I am afraid that a more detailed analysis would show that not only time continuity, but also demonstrable historical connection have been omitted here. But one must admit that, as in many parts of his historical comments, he tried to accumulate excessive historical material. More detailed historical studies of these important historical features are lacking.

[5] Cf. above Chap. 4.5.

up with much older studies of Lagrange. Dirichlet's studies in their historical significance led to Dedekind's theory of ideals, as well as to Kronecker's studies of the algebraic realm, mentioned above.[6] However, at the same time it also contained the study of linear expressions

$$a_1 x_1 + a_2 x_2 + \ldots + a_n x_n = \Sigma a_i x_i \qquad (5.4.1)$$

where a_i represented integers or rational numbers, and x_i were not variables but numbers of a certain type (i.e., n-th roots of unity, or roots of a given equation). A similar idea may be found with Galois [155] who spoke directly of some sort of imaginary elements, similar to $i = \sqrt{-1}$, instead of x_i. However, with a view to the spirit of contemporary opinions, the linear expression (5.4.1) also represented a set of elements of a given type, and therefore it hid the study of the structures of sets of these elements.

These facts have been pointed out without analysing how they originated from various stimuli, because they played an important role in that they existed, and by providing the possibility of forming algebraic (linear) systems which could be tested for describing the "subjects" of one kind or another.[7] This also stressed the possibility of investigating systems formally similar to complex numbers, and expanding the realm of real numbers, as well as the sphere of applications of complex numbers. Gauss studied these possibilities, although he never published positive results.[8] Beginning with the thirties several British mathematicians, among them Hamilton, the Graves brothers, Gregory, de Morgan and Cayley, made the liveliest bid for expanding the field of complex numbers.

As regards these British mathematicians, the search for and the study of new "numerical forms" also represented a part of an analysis of the nature of mathematical and in particular algebraic deliberations; in this way they also arrived at the first modern attempts at a new understanding of algebra as will be shown in the following chapter. These general connections, which will be omitted here, can be observed in the first important study which predetermined further efforts in many ways, i.e., Hamilton's concept of complex numbers as ordered couples of real numbers. His concept

---

[6] The nucleus of the idea, which the mathematicians of the time probably missed, was accurately formulated by Bourbaki [46], p. 87.

[7] The linear groups, contained in Galois' study, are an example of a "linear" system in algebra which we have not considered yet, and which is capable of expressing the internal relations of a larger number of algebraic realms then investigated.

[8] In the conclusion of the announcement of the second part of his study on cubic residues, where he presented his geometrical interpretation of complex numbers ([166], 178) he wrote: Der Verf. hat sich vorbehalten, den Gegenstand, welcher in der vorliegenden Abhandlung eigentlich nur gelegentlich berührt ist, künftig vollständiger zu bearbeiten, wo dann auch die Frage, warum die Relationen zwischen Dingen, die eine Mannigfaltigkeit von mehr als zwei Dimensionen darbieten, nicht noch andere in der allgemeinen Arithmetik zulässige Arten von Grössen liefern können, ihre Beantwortung finden wird. Bourbaki ([46], 84—5) interpreted this as Gauss being convinced of the impossibility of such an expansion, provided it would preserve the fundamental properties of complex numbers (i.e., probably commutativity). At this point Bourbaki also indicated the future development.

was connected with an effort to explain algebra in the spirit of the Kantean philosophy as a theory of pure time. However, Hamilton's concept of complex numbers became very fruitful independently of the philosophical context. The geometrical interpretation of these couples offered itself and was indicative of treating three-dimensional fields by the same means.

The way led from number-couples to quaternions and it has already been described many times; the fundamental data are sufficiently available.[9] Of course, the most important are published papers. It is important that during the period after Hamilton's work on complex numbers had been published (1833), at least four mathematicians of the British scientific world, independently of each other, tried to work out a system of more independent units. Later Hamilton wrote[10] that he had been trying to present a theory of couples, triples and multiples in his correspondence with his friend John T. Graves, and he went back to the problem from time to time. It seems that he calculated over and over again all the possibilities of chosing the rules for calculating the resultant coefficients in multiplying triples of a general form, $mx_1 + nx_2 + px_3$, which he also wrote ([192], Preface, 128−9) just as the triple (m, n, p). The failure of these efforts, for which he demanded a suitable geometrical interpretation, probably led him to return to the problem again and again. However, he was always convinced of the success and of the significance of this problem. The beginnings of his correspondence with A. de Morgan in 1841, indicate how interested he was in elaborating the triples. He even spoke of this theory as a branch of the "future algebra".[11] Nearly simultaneously with Hamilton's discovery of quaternions

---

[9] Hamilton himself tried to explain the trains of thought which led him to the discovery of quaternions in the prefaces to his later papers on quaternions [186, 192]. One cannot exclude a certain amount of subjectivism, in particular in relation to other authors of the time, but the fundamental facts may be considered to be reliable. The evolution of Hamilton's opinions is substantiated in many ways by his extensive biography, written by his friend and collaborator R. P. Graves [175], which contains excerpts from Hamilton's extensive correspondence. In the third volume, Hamilton's correspondence with A. de Morgan, which began at the time of the origin of the quaternions, was published. Unfortunately, as Graves himself wrote in the preface, he left out the specialized mathematical sections from his other correspondence. In spite of this drawback one can only be sorry that similar material has not been published about other mathematicians of the time considered. Many historical works have devoted justified attention to Hamilton's quaternions. One must consider Hankel's *Theorie der complexen Zahlensysteme* [196] to be one of the most important. Apart from the important historical notes, its importance is in that it summarizes the development of the last decades into a comprehensive and generalized form, and thus also indicates the relations of many discoveries with a view to contemporary understanding. What interests us is the fact that it was written towards the end of the period which is being investigated. Hankel only devoted the last two chapters to quaternions, and in a historical note he wrote: Wollte ich diese Theorie auf deutschen Boden verpflanzen, so war es nothwendig, die Darstellung total zu verändern ([196], 195).

[10] [175], II, 433; cf. preface to his *Lectures on Quaternions* (Dublin 1853) [192].

[11] "But, if my view of algebra be just, it must be possible, some way or another, to introduce not only triplets but polyplets, so as in some sense satisfy the symbolical equation $a = (a_1, a_2, \ldots, a_n)$; a being here one symbol, as indication of one (complex) thought; and $a_1, a_2, \ldots, a_n$ denoting n real numbers, positive or negative..." [175], Vol. III, 247—8.

de Morgan published a study on triples, the Graves brothers studied them and apparently also Gregory. It seems that they calculated similar cases as Hamilton, but with no result for a long time. Apart from Morgan's work, this was also substantiated by Cayley's independent discovery of "octaves", at which J. T. Graves had probably already arrived at Christmas of 1843, though he had not published anything about them at the time.[12] We shall go back to the problem of octaves presently.

The existence of the investigated systems for finding various alternatives is important because, perhaps for the first time in the history of mathematics, systems of elements were being formed intentionally with certain rules of composition, i.e., with a certain structure. Systems were sought which would not lead to a contradiction. The main difficulty was in finding satisfactory rules for introducing operations. As regards the multiplication of triples this meant, e.g., to find rules for mutual multiplication (in Hamilton's annotation) of the symbols x_1, x_2, x_3 such that the product

$$(mx_1 + nx_2 + px_3) \cdot (rx_1 + sx_2 + tx_3) \qquad (5.4.2)$$

was again a triple.[13] The objective was to form a system satisfying certain conditions, such as linearity and the number of symbols x_i, and also having certain rules of composition. The next objective was for this artificially formed system, which only expressed a certain abstract structure, to be tested for application. But these applications were not a necessary aspect; Morgan gave evidence of this in the conclusion to his study on triples; he said ([301] 254) that he had not considered the problem of applications, in particular the geometrical ones.

In this intellectual atmosphere, after long years of effort to create a linear system of triples or more general multiples, Hamilton came up with the ideas of quaternions. This occurred in October of 1843, and soon, at first in correspondence and then by printed word, other mathematicians became acquainted with the idea. The response was relatively fast; a large number of authors showed interest, and some of them even contributed to the development of the theory of quaternions in an original way.[14] The speed of the response is evidence that Hamilton's quaternions were comprehensible

---

[12] Cf. Hamilton [181], Appendix 3, pp. 648—656, where data on the independence of the said discovery are given apart from the actual discussion.

[13] Hamilton remembered ([192], Preface, p. 128) that this led him to calculate the multiplication which depended on 27 constants "...which might all be arbitrarily assumed, before proceeding to operate".

[14] In the Preface referred to (p. 155) Hamilton had named 12 mathematicians from the English world in 1853... "who have at moments turned aside from their own original researches, to notice, and in some instances to extend, results or speculations of mine".

One of the first (if not the first) mathematicians who connected up with Hamilton's discovery was A. Cayley in 1845 [91]. Crowe ([108], p. 110 nn) also presented proof of the rapid propagation of the theory of quaternions. He maintained, among other things, that a total of 594 papers were published on quaternions in the 19th century, of which 132 were published by the end of the fifties. It is true that during this initial period the author of most of them was Hamilton and only 33 (i.e., exactly a quarter) were by other authors.

to the mathematicians of the time, some of whom had made efforts at establishing similar systems.[15] One of the fundamentals of success was the geometrical (and therefore also mechanical) application of quaternions. Hamilton himself not only admitted that the idea of this applicability helped him in establishing quaternions,[16] but he soon produced geometrical and mechanical papers in which he exploited quaternions. He also introduced a convenient terminology, basically founded on geometry; he used the terms vector and scalar[17] in the present sense, and spoke of scalar and vector parts of the quaternion.

The greatest difficulty which had to be overcome in forming the quaternions, and which Gauss probably had in mind as early as in 1831,[18] was the parting with the "rules" of operations in traditional arithmetic, with rules derived from operations with integers or rational numbers. Hamilton finally decided to desert commutativity in determining the rules for multiplication. In this way he also managed to make the multiplication of quaternions compatible with the composition of the rotations of solid bodies. At the time, Hamilton, but also Morgan and J. Graves, were already thinking of deserting associativity. Morgan[19] really tried to establish a system of triples with non-associative multiplication.

It was not easy to desert these "rules". It was also made difficult by the fact that with Hamilton the multiplication of various square roots of -1 were being considered, i.e., something similar to complex numbers. He therefore constantly spoke of "imaginaries". Moreover, as regards the contemporary concept of operations "rules" like associativity, commutativity and distributivity were expressed in forms of theorems, which were also derived. Now these properties had become independent and they took

---

[15] Cauchy's statement is interesting; he claimed in the *Comptes Rendus* (10. I. 1853, p. 75) that his idea of "clefs algébriques" also included the theory of quaternions. Cauchy himself in his paper [85] in which he published his idea, proceeded from a totally different realm of thought than the English school; one must admit that he also arrived at linear systems under the influence of his study of determinants, although in a less marked form.

[16] This is where the effort to find an algebraic description or calculus for the three-dimensional space became effective, and that is why Hamilton ([192], Preface) stressed the geometrical representations of complex numbers.

[17] The term radius-vector was used before Hamilton's time, also in papers on the geometrical interpretation of complex numbers.

[18] Cf. above Comment 8. One must also admit that Wessel encountered these difficulties in trying to extend the calculus with complex numbers to render it applicable to three-dimensional space.

[19] Graves, [175], Vol. III, 251—2. In October of 1844 he compared Morgan's theory of triples with Hamilton's quaternions, and wrote: "Sir W. Hamilton's quaternion algebra is not entirely the same in its symbolical rules as the ordinary algebra; differing in that the equation $AB = BA$ is discarded, and $AB = -BA$ supplies its place. Those of Mr. De Morgan's system... all give $AB = BA$, but none of them... give $A(BC) = (AB)C$, except in particular cases."

In his letter of 9. 12. 1844 Hamilton replied that he was not in principal "against triplets"; he also gave arguments in favour of quaternions, first mentioning their algebraic simplicity, by which he meant that they were in "analogy to ordinary algebra, as to the rules of addition and multiplication (the commutative property excepted)".

part in defining operations. The nature of the elements entering the operations is of secondary importance,[20] given only by the interpretation of the system.

As soon as mathematicians became aware that they could define operations by other than current rules, attempts at new systems immediately followed. As soon as Hamilton had told J. Graves of his discovery by letter (17th and 24th October, 1843), Graves began to consider from a different point of view the possibilities of the rules of composition, and in this way (still in 1843) he arrived at octaves, i.e., at a system of eight units. He built it up analogously to quaternions[21] when his own earlier deliberations did not yield a result. When Hamilton received the news of Graves octaves (letters of 26. 12. 1843 and 18. 1. 1844) he developed his ideas and showed that their composition was not only not commutative, as Graves himself had said, in which respect the octaves were analogous to quaternions, but not even associative.

However, Hamilton forgot to publish Graves' ideas, and so Cayley who arrived at them independently published them first in a brief note "On Quaternions"[22] in 1845. The elements of the system have the form

$$\Lambda = 1 + \lambda_1 i_1 + \lambda_2 i_2 + \ldots + \lambda_7 i_7, \tag{5.4.3}$$

where λ_l are quantities (probably real numbers), and $i_l = \sqrt{-1}$.

Nevertheless, this did not mean the end of forming "artificial" systems of elements by English mathematicians. Further efforts were made. Hamilton's later idea (beginning with October 1855) is also interesting; as he said it brought him to a whole "family of systems" of which the best known he called "Icosian Calculus". It is a system of elements defined by the fundamental realations $1 = i^2 = k^3 = \lambda^5$, $\lambda = ik$. This new system of Hamilton's was uncommutative just like quaternions, but it differs from them by its "non-linearity" and in that it is only partly distributive, as the author himself said ([181], 609). He also praised the geometrical interpretation if the new system which he also connected with graphical representation. Discussion of these problems would lead us far astray from the objective set in this book.

The origin of quaternions, the long years of experiments which led up to them, as well as the efforts to form further systems, brought considerable changes in the understanding of operations, and showed that it was possible to form new systems of "calculation" with a new structure, unusual at the time. A possibility also appeared of not only searching for the structure in describing "known" mathematical objects, but also of forming systems first and only then the appropriate model.

---

[20] Cauchy, who approached the problem from a different angle, also spoke of symbols which are later eliminated from the relations.

[21] Data concerning these questions were presented recently in the 3rd volume of Hamilton's work ([181], p. 609 nn), and in particular also Appendix 3 (pp. 648—656) also published there, where a list of references is given towards the end. Graves also tried to treat systems with a different number of independent units.

[22] Comment [91] begins with the words: It is possible to form an analogous theory with seven imaginary roots of (—1) (? with $v = 2^n - 1$ when v is a prime number).

These features, brought by the effort leading up to quaternions and to Cayley's octaves, the importance of which for the development of algebra must be stressed appropriately, can also be found in various forms in other papers of the time.[23] The fact that authors began to arrive at them from various approaches beginning with the forties is outstanding. Although a number of studies from this period have to be omitted and not even the further development can be discussed, the beginning of which is formed by the systems mentioned above, one cannot omit the author Hermann Grassmann who arrived at similar results to those of Hamilton simultaneously, independently, and for different reasons, but whose work, unnoticed at the time, contained stimulating ideas indicating various possibilities.

Grassmann in his *Ausdehnungslehre* (1844) [174] was affected by other intellectual impulses than Hamilton and the British mathematicians.[24] The algebraic aspect withdrew into the background and the idea of an algebraic extension of complex numbers did not appear. Possibly a more significant role was played in this case by the German philosophical tradition and by the efforts of the contemporary German mathematicians, in particular those who dealt with geometry. Grassmann wanted to form a new, and one might even say fundamental, mathematical discipline in his *Ausdehnungslehre.*.He wanted to interpret again and anew the fundamental concepts (and "truths") which were applicable to all mathematical disciplines. His method is based on explaining concepts, developing concepts from previous ones, and determining their relations.[25] He first explained the concepts of equality and difference and carried on with the concept of conjunction (Verknüpfung – § 2) of two quantities or forms.[26] He tried to assume as little as possible of this "conjunction". He only considered associativity and commutativity to be two of the possible properties of conjunction.[27]

In the course of his interpretation he finally arrives at differentiating between the various types, or even better, degrees of joining elements; in other words, also to the second degree which is multiplication. As an indivisible property of multiplication he

---

[23] Hankel pointed out that H. Scheffer in two papers (1846, 1851) formed "eine Art complexen Zahlen zur Behandlung räumlicher Verhältnisse", whose multiplication was commutative but for which the distributive law did not hold in general. Cf. Hankel [196], p. 105.

[24] The part Crowe devoted to Grassmann ([108], 54—96, also certain data on p. 110 nn) is very instructive. It presented several excerpts from Grassmann's work, on the one hand, and some data on the conditions of their origin and response, on the other. It also mentioned the influence of Grassmann's father, with whom one can already observe the tendency to clarify the fundamental problems and the "proper" substance of concepts.

[25] In some respects Grassmann's method reminds one of Bolzano's speculations. As regards the sphere of algebra and arithmetic this approach had its dangerous points which even Grassmann did not avoid.

[26] Grassmann also spoke of "Formenlehre". This has been brought up because mathematics were not a mere science of quantities for him, but a formal science (Formenlehre), not dealing with forms of thought but with forms created by thinking, if one might interpret Grassmann's speculations in this manner (cf. pp. 22—3).

[27] He did not use the terms commutativity and associativity. He only spoke of the possibilities of eliminating brackets or of interchanging the terms of the conjunction.

mentioned (§ 10) its relation to a lower degree of conjunction, i.e. the distributive law. This fact (and one could mention others) indicates that Grassmann was influenced by the models he knew, in spite of his effort for a general and abundant diversity of ideas.

The effort for generality is the reason why he had already rejected the strict adherence to a three-dimensional space in the preface, and that is why he also acknowledged systems of higher degrees with different types of operations, etc. These abstract deliberations apparently in themselves and their logics (in reality probably different geometrical ideas) led him (§ 28 mm) to external multiplication which is of course distributive, but which can be applied to quantities of different types, mutually inconvertible, the multiplication of quantities (forms) of the same kind leading to a quantity of a higher type. However, if one introduces the current symbolics,[28] for any arbitrary elements (forms) $a \neq 0$, $b \neq 0$, $c \neq 0$ the following holds according to Grassmann:

$$a \cdot b = -b \cdot a, \qquad a \cdot a = 0, \qquad a(b + c) = ab + ac. \qquad (5.4.4)$$

Basically, therefore, Grassmann's external multiplication is a vector product which supplements the multiplication, which may be called the multiplication by a scalar (with Grassman, i.e., a number).

Let us now omit the form in which Grassmann expressed his ideas, and which rendered his study difficult to understand.[29] Grassmann undoubtedly presented a certain system, equivalent with vector calculus, and he formed some fundamental concepts which could also have affected geometrical deliberations.[30] Although his work did not directly influence algebraic thinking (as opposed to the studies of Hamilton and others), it played an important role in the history of algebra. He showed how the analysis of fundamental concepts of contemporary algebra, i.e., in particular of operations, could indicate various alternatives on comparing arithmetic, geometry and perhaps also mechanics, in which these concepts could be attributed with different properties and mutual relations.

---

[28] A similar approach was used by Crowe [108], p. 72; Crowe also rightly pointed out that with Grassmann the external multiplication changed the "type" of quantity.

[29] In this respect the well-known letter in which Gauss thanked Grassmann for sending him his Ausdehnungslehre (cf. Gauss [159], Bd. X/I, pp. 436—7; 447) is characteristic. In connection with the contents Gauss spoke of "die concentrirte Metaphysik der complexen Grössen", however, in order to be able to assess the study he would have to have more time, because one must first "mit Ihren eigenthümlichen Terminologien zu familiarisieren".

[30] Cf. note Study and Engel in [174], p. 405—6.

CHAPTER 6

EFFORTS TO FIND A NEW APPROACH TO ALGEBRA

6.1 Beginnings of formalism

In the 18th century and far into the 19th the opinion prevailed that algebra was a science of algebraic equations, or more generally, a "universal" arithmetic. This approach also led to efforts to find an involved interpretation of the sphere of arithmetic and algebra which would finally yield an axiomatic system similar to the axiomatic systems of geometry, the origins of which went back to Euclid's Elements. The logical structures of the sphere of arithmetic and algebra, and of geometry, had basically the same characteristic features. These systems were meant to describe, in accurate and clear terms with theorems which were logically interconnected, a single model considered to be known. This was the source of the realism so typical of contemporary mathematics. This realism also contained a simple, and one may even say, naive criterion of truthfulness; in geometry referring to the observable, while in the sphere of arithmetic and algebra referring to a calculation.[1]

Realism was explained in different ways, according to the philosophical point of view advocated by the author. On the one hand there are the opinions most pregnantly formulated by the encyclopedists in the 18th century, namely d'Alembert, according to whom the subject of mathematics, roughly speaking, is the external world deprived of all qualities, so that it is only represented by spatial and numerical quantities. However, Bolzano's contradictory opinion, which admits the independent existence of truthful theorems in its consequences, requires the description of a single existing model (even ideally existing as in Plato's world of ideas). In its way Kant's conception was the same; it also considered spatial observation and quantitative relations in mathematics to be given, and thus representing a single model, which could be described accurately in an axiomatic manner.

However, the prevailing realism had its difficulties. Even in the sphere of arithmetic and algebra the well-known problem as to how compatibility was ensured between the results of abstract speculations and the "behaviour" of the model to be described,

---

[1] If this opinion was voiced in the 18th century (e.g., in some form even by mathematicians such as d'Alembert), it appeared with the knowledge that it was not sufficient in all cases. One may therefore consider admitting allegiance to such an opinion to be a personal solution to a dilemma where reference to experience was not sufficient to prove a result. On the other hand, the axiomatic system was so imperfect that it gave rise to justified doubt in too many problems, even those of fundamental significance.

appeared in between lines. It is true that this question interested philosopheres in the first place, but it was also reflected in the philosophy of mathematics. The mathematicians themselves avoided direct and general answers to such philosophical questions, although they were undoubtedly aware of the importance of these problems. The difficulties of the realistic point of view grew within the sphere of arithmetic and algebra, e.g., where results were outside the scope of the traditional realistic approach. As is known (cf. 4.5) the embarrassment was greatest where the meaning of negative and complex numbers was concerned. These concepts also showed up the deficiencies of the contemporary axiomatic system which was based on and whose indivisible part was "verbal" definitions of the subjects of mathematical speculations which (as with Euclid) actually described objects whose existence and properties were defined intuitively (whatever the meaning various philosophical systems attributed to the word).

Various attempts were made to overcome the difficulties mentioned above, but they were not fully successful. These efforts were mainly based on traditional positions, traditional approaches to the subjects of arithmetic and algebra, as well as their axiomatic systems. They had their climax in the efforts of Ohm and Bolzano in the twenties and thirties.[2] Apart from the latter, however, there began to appear, although scattered and obscure in form, certain opinions of a quite different concept which were mainly based on the contradictory character of algebra.

Algebra was frequently considered a general theory of quantities, as mentioned several times before. This Newtonian tradition was represented by Lagrange, among others, at the turn of the 18th and 19th centuries. Lagrange spoke in favour of this concept, e.g., in his lecture at the École Normale.[3] In the preface to the 2nd edition of the *Traité* (1808) ([269d] 14) he again referred to Newton's approach which he complemented and modified. He considered the study of the operations of algebra to be its substantial feature. These operations must be understood generally; they are valid in arithmetic and geometry alike[4], because he considered arithmetic and geometry, without further reasoning, to be isomorphic, to use modern terms.

An operation which is to be applied to some more specifically undetermined quantities, whether geometrical or arithmetical, is expressed by a "formula". The dependence of the "formulae" Lagrange then called a function ([269d] 15). He then

---

[2] Neither Ohm nor Bolzano were algebraists; the sections on algebra in their work were among the weakest and least original. One may even say that neither of them grasped the tendencies of contemporary algebra. This could also perhaps contribute to explaining their whole approach to the sphere of arithmetic and algebra.

[3] Cf. *Séance de l'École Normale*, [351] Tome I, p. 16 nn; Tome 5, p. 276 nn.

[4] Le caractère essentiel ... (de l'Algèbre) ... consiste en ce que les résultats de ses opérations ne donnent pas les valeurs individuelles des quantités qu'on cherche..., mais représentent seulement les opérations, soit arithmétiques ou géométriques.... L'Algèbre plane pour ainsi dire également sur l'Arithmétique et la Géométrie, son objet n'est pas de trouver les valeurs mêmes des quantités cherchées, mais le système d'opérations à faire sur les quantités données pour en déduire les valeurs des quantités qu'on cherche d'après les conditions du problème.

also considered algebra to be, in the wider sense of the word, "l'art de déterminer les inconnues par des fonctions des quantités connues". In this way Lagrange (like Laplace, his co-lecturer at the École Normale) came to stress the "algebraic" language, using simple symbols, which is "very general and of peak accuracy, and this is why it is simple to understand it and to speculate on the most complicated relations between objects" ([269d] 17). Traduire en langage algébraique, c'est former des équations.

However, Lagrange's approach also has its weak points. The most important is that, in spite of stressing operations, he says no more of them. Thus, one is forced to accept them in the current sense, in the way they have been empirically introduced in arithmetic, and to transform them into arbitrary expressions (formulae). This brings out some of the other features of Lagrange's algebra. The operations which can be applied to arbitrary formulae without further discussion, obscure the differences between numerical realms; this is connected with the non-arithmetical character of Lagrange's algebraic speculations, which has been mentioned above. The stress on operations[5] and formulae brings out an element in algebra which was not unknown in the development of mathematics. In algebra itself calculation with complex numbers was in fact carrying out an operation with a given formula. In the same way one can find a number of formal procedures in calculus where the operation is carried out with a "form"; an expression was the initial and principal feature without any accurate explanation of the meaning of the form. These approaches were common especially in the theory of series,[6] but they can also be found elsewhere. Lagrange's calculus itself is penetrated by formal elements although the main feature is the formula and the algorithm, the meaning and justification of the algorithm being of secondary importance.

---

[5] The stressing of operations could also be seen in Ohm's approach to the whole sphere of arithmetic and algebra. Cf. Chap. 4.2.

[6] In this connection it should be interesting to recall Gauss' opinions. Gauss himself was very sober when he had to speak of the nature and method of mathematics. In his manuscript notes ([159] X_1/396) he wrote: "Die Mathematik ist so im allgemeinsten Sinn die Wissenschaft der Verhältnisse, in dem man von allem Inhalt der Verhältnisse abstrahirt." One may relate Gauss' words to the formal approach to mathematics from his letter of 1. 9. 1850 to Schumacher, written as a response to a treatise on the theory of series which he did not consider to be accurate enough ([159] X_1, pp. 434—5): Es ist der Character der Mathematik der neueren Zeit (im Gegensatz gegen das Alterthum), dass durch unsere Zeichensprache und Namengebungen wir einen Hebel besitzen, wodurch die verwickeltsten Argumentationen auf einen gewissen Mechanismus reducirt werden. An Reichthum hat dadurch die Wissenschaft unendlich gewonnen, an Schönheit und Solidität aber, wie das Geschäft gewöhnlich betrieben wird, eben so sehr verloren. Wie oft wird jener Hebel eben nur mechanisch angewandt, obgleich die Befügnis dazu in den meisten Fällen gewisse stillschweigende Voraussetzungen implicirt. Ich fordere, man soll bei allem Gebrauch des Calculs, bei allen Begriffsverwendungen sich immer der ursprünglichen Bedingungen bewusst bleiben, und alle Producte des Mechanismus niemals über die klare Befügnis hinaus als Eigenthum betrachten. Der gewöhnliche Gang ist aber der, dass man für die Analysis einen Character der Allgemeinheit in Anspruch nimmt, und dem Andern, der so herausgebracht Resultate noch nicht für bewiesen anerkennt, zumuthet, er solle das Gegentheil nachweisen.

However, Lagrange was not the only one in his time who stressed the significance of formal relations in mathematics. Some mathematicians stressed the formal character of the whole of mathematics. Therefore, they spoke of algebra as the language of mathematics which accumulated within itself various meanings. For example, Buée differentiated between algebra as a universal arithmetic and as the language of mathematics when he tried to show the analogy between geometrical and arithmetical operations in 1806. In this second sense algebra combines within itself the expression of arithmetical and algebraic quantities. In this respect the unfinished book of Condillac, *La Langue des Calcus* [103], is interesting, in particular its preface. His opinions, which were not quite clearly voiced, may perhaps be interpreted as follows: The accuracy of our thinking improves with the accuracy of our language. The only language which has been "well formed" is algebra, the language of mathematics; this is also the source of accuracy in mathematics. The role of analogy is decisive; it leads from one expression to another. Algebra is a language formed by analogy in which there is nothing random, no conventions or customs: l'analogie, qui fait la Langue fait les méthodes; ou plutôt la méthode d'invention n'est que l'analogie même (p. 7). With this approach one need no longer speak of abstraction; the relations expressed algebraically can be modified by other relations. This language is closed in a way, although one and the same algebraic expression may express different cases, determined by analogy.

It is not our intention to draw conclusions from the author's own words, which were briefly paraphrased and altered. However, I am of the opinion that there is an anticipation here of some kind of new approach to algebra (and to mathematics in general) in which the relations between symbols determine the possibilities of different interpretations. Further, his text indicates that the author was till incapable of understanding arithmetic other than traditionally. In the spirit of his ideas, he stressed many times the difference between "word", which is a symbol, and number which is its representation. Although Condillac's study was read because of the philosophical and social attitude of the author, its direct effect on the development of algebra was in no way remarkable.[7] In the development of algebra he only illustrates how the stress on form, on formal relations, and on the possibility of interpretation of these forms could be connected, even in different ways, with the empirical, or outright sensory, philosophical point of view.

However, let us go back to the influence of formalism in mathematics at the beginning of the 19th century. Its propagation in algebra at the time, where Lagrange's influence could have acted, was considerable. The formulae, the relations between them, various equations recorded symbolically, or equalities occurred frequently. Understandably, the use of symbolical records formed formalism at the time. The latter should be viewed as such where formal, symbolical relations become decisive, where

---

[7] A certain negative reponse may be observed in Gergonne's paper [168] which contains an axiomatic structure based on axioms, i.e. on theorems "dont il suffit de connaître l'énoncer pour en apercevoir la vérité" (345—6).

they dominate over the content from which they originally stem, and vice versa where the form itself begins to determine the content of the concept. Calculus was a region where it was possible to speak of formalism very early. The efforts toward its algorithmation and algebration, of which Lagrange was one of the foremost advocates, led several authors to approach differential and integral calculus as an algebraic calculus. Among these authors the foremost was the Frenchman Servois who published his *Essai sur un nouveau mode d'exposition des principes du calcul différentiel* [359], the title voicing the author's ambitions, in the 5th volume of Gergonne's *Annales* (1814—1815).

In his study Servois wanted to introduce symbols and also an algorithm for mathematical analysis. He only considered the properties which were valid for a function in general; he devoted practically no attention to the function as a dependence of variables. For an inverse function he introduced the symbol $f^{-1}z$, for a derivative of a function Δ, for an integral of a function Σ, so that $\Sigma^n z = \Delta^{-n}z$, $\Delta^n \cdot \Delta^{-n}z = z$. Similarly for a logarithmic function $L \cdot L^{-1}r = r$, $\sin^{-1} z = \arc \sin z$, etc. As part of this approach, which we call formal, he also defined distributive and commutative functions. As an example of distributive functions he gave $a(x + y + \ldots) = ax + ay + \ldots$ (a being an arbitrary constant, $x, y, \ldots$ variables); an undistributive function is, e.g., $\sin x$, since $\sin(x + y + \ldots) \neq \sin x + \sin y + \ldots$. Functions F and f are commutative if (using his symbolics) $Ffz = fFz$. But $\sin x$ is not commutative since $\sin az \neq a \sin z$.

Understandably Servois' attempt to derive Taylor's development with these concepts, and from it the foundations of calculus, could have had no success. The significance of the study lies elsewhere; he introduced distributivity and commutativity as important concepts leading to the differentiation of functions. Servois not only presented the fundamentals of the present terms, but also expressed the possibility for the first time that the "operations" of certain subjects of mathematical speculations could be uncommutative and undistributive.[8]

Servois' study was followed by others, some of which tried to classify functions on the basis of whether the function satisfied certain formal equations. In this respect, the paper of F. Moth [305] is important; he was a native of Prague and later became a fellow of the Viennese Academy of Sciences. It is also interesting because Moth later referred to his contacts with Bolzano whom he considered his teacher and patron. His marked example, though of little originality, is being mentioned in order to show

---

[8] Without a suitable term, the concepts of commutativity and distributivity were known earlier. In the 1st volume of the *Annales* the editor discussed commutativity ([168] 52—8). He also came up with an interesting idea (p. 53): the multiplication of a line segment by an integer is understandable, but the reverse, i.e. the multiplication of a number by a line segment is impossible. He pointed out, e.g., Legendre's "proof" of commutativity and tried to present his own. An example of commutativity, as far as he was concerned, was the interchangeability of partial derivatives, i.e.,

$$\frac{\partial^2 z(t, u)}{\partial t \, \partial u} = \frac{\partial^2 z(t, u)}{\partial u \, \partial t}.$$

that the ideas of a formal approach appeared in different media and in various forms at the time.

Ideas stressing symbolic relations and the formal approach were adopted by the English mathematician Charles Babbage at the end of the second decade of the 19th century. English mathematicians were then looking for a way out of an extended period of crisis. Especially young mathematicians saw the causes of the crisis in the spasmodic adherence to Newtonian tradition in calculus; they were convinced that their terminology and symbolics, which rendered speculations obscure and did not lead to simple and comprehensible algorithms even in differential calculus, prevented the development of mathematical calculus in England, whereas Leibniz' concepts and symbols supported development. While the argument finally led to the adoption of Leibniz's symbols in England, some English mathematicians tried to generalize their opinions on the importance of symbolics. This was probably the way Babbage's paper, *On the Influence of Signs in Mathematical Reasoning* (1821, [16]) was initiated. According to him, suitable symbols unified the accumulated material, the accumulated facts. These symbols formed a language which made it possible to understand future research as well. In this way a convenient symbolical language facilitates the generalization of future knowledge too.[9] The mathematical language has advantages as compared to a "general" language. The meaning of the symbols must be controlled by definitions;[10] these are clear in mathematics, unique and more simple. Because of its advantages, the power of the mathematical language is also higher than that of the "general" language in generalizing.[11] Nevertheless, the mathematical language is not uniform. The role of the algebraic language is special; it has a different connection with the quantities which its symbols represent than in geometry. The connection of the symbols with the quantities is quite arbitrary (p. 338).

In the spirit of the algebraic tradition Babbage praises Lagrange, as well as his theory of functions, highly. He also referred to other authors, especially the French. He expressed his opinion on the three degrees of exploiting the mathematical language, or "the language of analysis". It is first necessary to translate the problem into this language, secondly, carry out the necessary system of operations which leads up to the solution, and thirdly, translate the results into ordinary language again.

These are the fundamental ideas of Babbage's study. Its extent (more than 50 pages) enabled the author to discuss the individual problems in detail, and also to quote

---

[9] Only as a marginal note I would like to point out that traditions of the English empirical philosophy can be observed in his article.

[10] ...in Geometry it has been well remarked... that its foundations rest on definition, and if this does not altogether hold in algebraical enquiries, at least the meaning of the symbols employed must be regulated by definition... ([16] 326—7). There is a certain amount of doubt in the quotation given, as to the identity of definition of the meaning of symbols in geometry and algebra.

[11] The power which language gives us of generalizing our reasonings concerning individuals by the aid of general terms, is no where more eminent than in the mathematical sciences, nor is it carried to so great an extent in any other part of human knowledge. ([16] 335).

authors, on whose ideas he drew. Examples showing the significance of working with
suitable symbols also appear in the paper to a lesser extent. This is quite understandable,
since he could refer to his older papers, especially his *Essay toward the calculus of
functions* [17] of 1816. In this paper he characterized (like Servois and others) functions
by means of certain functional equations, and he tried to derive some of their
properties by applying formal modifications. In the conclusion of the long study
he stressed (p. 256) the general significance of his symbolical approach, not only with
respect to differential and integral calculus, but also for mathematics as a whole. He
advocated the same opinion in his paper [15] of 1820. But here he also voiced other
more general ideas. For example, ([15] 64) he pointed out the profound connections
between mathematical disciplines, connections such that the introduction of a new
symbol or definition engendered many concequences. In his way he also indicated the
direction of future research.[12] The idea of the unity of mathematics which was
represented by mathematical language, i.e., by introduced symbols and definitions,
thus appeared.

He stressed the problem of the relations between the various branches of mathematics
even more in a special theoretical treatise of 1817 which dealt with the importance of
analogies in mathematics. He was aware that he could introduce analogies, but in his
treatise he wanted to show the usefulness of analogies. Without saying so explicitly,
he considered the carrier of analogy to be the similarity of symbolical expression
and the work with these symbols.

It is true that Babbage's ideas are very vague, and that Babbage made very few
comments regarding algebra directly.[13] Nevertheless, Babbage tried to stress new ideas
in the system of mathematics and, therefore, also of algebra. It is true that he mostly
remained within the realm of contemporary ideas about mathematics, and that his
ideas only stressed concepts advocated by his predecessors. But by summarizing them
and expressing them, and in some cases even applying them generally, he gave them
a different meaning. They not only represented the expression of a single element used
in speculations, which is more or less the case with Lagrange, but they touch on the
substance and, in particular, the method of mathematics. In this way they become
a methodical requirement to a certain extent.

Babbage expressed the fact that formulae, symbolical relations and their modifications also represented an analogy among the various spheres of mathematical research,
more accurately than the other mathematicians at the time. It is true that he was not
able to find convincing examples, not to mention examples pertaining to the most
modern developments of algebra, although he also spoke in this connection about

---

[12] So great however is the connection that subsists between all branches of pure analysis, that we
cannot employ a new symbol or make a new definition, without at once introducing a whole train of
consequences, and in defiance of ourselves the very thing we have created, and on which we have
bestowed a meaning, itself almost prescribes the path our future investigations are to follow. ([15] 64).

[13] Of course, in the English mathematics of the time "analysis" also included algebra; some of
Babbage's examples moreover, are expressly algebraic.

the set of roots of the equation $x^n - 1 = 0$. On the other hand, beginning with his *Disquisitiones arithmeticae*, Gauss worked with representations of different algebraic structures. He was probably aware of their similarity and the modern reader of his study will get the impression of purposefully constructed isomorphic systems. However, in Gauss' studies one would look in vain for proof that Gauss himself had expressly proclaimed isomorphism, not to say that he had explicitly pronounced the term isomorphism. Therefore, it is possible to consider the situation contradictory in this respect up to the twenties of the last century. On the one hand there are the efforts to formulate similarities and analogies between systems (if one does not want to use the term structure outright) of various branches and theories of mathematics, but without sufficient accuracy and suitable mathematical argumentation and examples. On the other hand, mathematical research at its peaks studied the structural similarity of various systems of elements, but this study was not connected with the corresponding general approach.

6.2 The English School's approach to algebra

The English, or British,[1] school of mathematics originated in a scientifically specific medium. At the end of the 18th century it was already clear that English mathematics were becoming stagnant with respect to the great expansion of continental mathematics in which mathematicians from an increasing number of countries were taking part. At the beginning of the 19th century a group of young Cambridge and Oxford mathematicians,[2] who were dissatisfied with the state of affairs, blamed this stagnation on blind adherence to Newtonian tradition and tried to find a way out of this situation by quickly establishing contact with contemporary continental mathematics.

The young English mathematicians not only tried to expand calculus again, but their effort also extended to other branches of pure and applied mathematics. They took a special fancy, which was perhaps connected with the features they stressed in calculus, to the algebraic nature of speculations, which could also later be observed in papers on logics and geometry. They certainly devoted considerable attention to algebra, in particular after becoming familiar with European calculus.

The English mathematicians also had to explore the state of algebra on the Continent.[3] Here, however, the situation was considerably different, since there was a problem as to what they should and should not adopt. The English mathematicians became widely acquainted with, and accepted with understanding, the latest results

---

[1] For example Bell ([26] 180—2); Hankel ([196] 15) had already called them "the Cambridge school" in the last century. However, both spoke in particular of the school of algebra.

[2] The nature and the role of this group was evaluated lapidarily, e.g., by Struik [361].

[3] As regards the origin of the English school of algebra and its further development and results, cf. [318, 319].

in algebra.[4] However, this did not prove satisfactory in one respect. They found the approach to and the interpretation of the sphere of arithmetic and algebra on the Continent in just as unsatisfactory state as in the books written in English. Possibly this was one of the main impulses which led them to try and work out a new approach to algebra. But, as is usually the case, however new their approach may have been, it was not quite detached from the previous development. It related to elements stressing formalism in algebra itself and to the whole trend of formal treatment in calculus, represented by Babbages work in England in the second and third decades of the 19th century. At the end of the third decade, immediately following Babbage's papers, the ideas of the new approach to algebra began to crystallize.

Three mathematicians tried to create the new approach to algebra in close co-operation: George Peacock, Auguste de Morgan and D. F. Gregory. The group was led by the ideas of Peacock to whom credit is due for introducing and publishing his book in 1830 [327a]. Three years later, at the meeting of the British Association for the Advancement of Science, Peacock presented his ideas in an extensive paper and connected them with "*the recent progress and present state of certain branches of analysis*" [328]. He advocated his point of view for the third time in the second edition of his algebra in the forties [327b, c]. Morgan also presented his views on the approach to algebra several times. The first brief outline appeared in his textbook *The Elements of Algebra* [296] in 1835, and then in several subsequent treatises of 1839 to 1844. He then went on to interpret algebra in the new spirit in his book of 1849 [302]. Gregory did not write an extensive and comprehensive textbook of algebra, but presented a series of papers towards the end of the thirties and the beginning of the forties.

Peacock built on the traditional difficulties of the logical structure of algebra. He was aware of the tendency to generalize which acquired two forms in contemporary mathematics (also in the mathematics of the previous century). Firstly, expressions formed to solve certain problems and therefore only derived for a certain field, were given a much more general meaning. This incomplete induction proved very fruitful, e.g., in establishing integral calculus in the 17th century. Peacock ([328] 199) pointed out that, e.g., the powers of the binomial (e.g., $(1 + n)^n$) were originally only introduced for natural powers. The second, similar form of the generalizing tendency was the aditional definition of the meaning of expressions like zero power, or, in combination calculus, the meaning of the expression $\binom{m}{n}$ for $n = 0$. In order to justify these procedures Peacock introduced into algebra the so-called principle of permanence. He of course had difficulty expressing it accurately and clearly,[5] which was due to the fact that it is impossible to extend a field in which the expression is valid arbitrarily, and

---

[4] We shall go back to this question in connection with Peacock's *Report* [328].

[5] It is not clear to what extent Peacock was influenced by geometrical speculations, and the discussions of French mathematicians in particular in using the principle of permanence (Poncelet—Cauchy). For further details of the principle of permanence see Hankel [196], § 3, p. 10 nn.

because the additional definition is not arbitrary either. Although he must have been aware of this, he presented the principle of permanence as follows: "Whatever form is algebraically equivalent to another form expressed in general symbols, must continue to be equivalent, whatever those symbols denote" ([328] 198). The difficulties with this general and discussible formulation are toned down in Peacock's text by the context, in which he understands the principle exclusively as an auxiliary means for explaining and giving reasons for his approach to algebra.

In the sphere of arithmetic and algebra he arrived at the differentiation of three mutually related units: arithmetic, arithmetical algebra, and symbolical algebra. Arithmetic dealt with "concrete" (i.e., special) numbers which were discrete quantities, and carried out operations which led from one or more numbers to a number again, provided the realm considered allowed for it. Peacock did not specify how he understood this field, he only mentioned that different realms might be applicable to different speculations. Natural numbers were the initial feature. Operations and number were introduced by definitions, operations could not be carried out without restriction, etc. Thus, he understood arithmetic as a science of numbers (i.e., of quantities) in the traditional way current at the time. The fact that Peacock stressed definitions meant nothing else but that he considered them to be explicit (and verbal) descriptions of a single known model; the truthfulness of conclusions (the results of calculations) was proved by a check. Peacock considered arithmetical algebra to be a generalization of arithmetic. This is the traditional "arithmetica litteralis" with all its advantages and disadvantages: it only differed from arithmetic in that it used letters instead of numbers. Its generalization was only in the symbolics, i.e. in using letters. The same definitions of numbers (numerical realms) and operations were valid, as Peacock himself indicated, and the same difficulties of logical construction were present. The same old problem was encountered again. The operations could not be carried out without restrictions; an example of this is subtraction where only the subtraction of a smaller number from a larger has been introduced.[6]

The third of the units mentioned is symbolical algebra. To a certain extent it is a generalization of arithmetical algebra. The transition from one to the other Peacock secured by the principle of permanence. The latter eliminates all restrictions with regard to the validity of the expressions in arithmetical algebra (i.e., inclusive of the meaning of $(a - b)$, $\sqrt{-a}$.) This very vague definition might be indicative of the train of Peacock's thoughts, but it gave no direct answer to the question as to what symbolical

---

[6] Peacock explained these opinions at various points in the three papers referred to. His concept of arithmetical algebra he summarized briefly in the preface to the second edition of his algebra ([327b] IV): In arithmetical algebra, we consider symbols as representing numbers, and the operations to which they are submitted as included in the same definitions... as in common arithmetic: the signs + and — denote the operations of addition and subtraction in their ordinary meaning only, ...all results whatsoever, including negative quantities, which are not strictly deducible as legitimate conclusions from the definitions of the several operations, must be rejected as impossible, or as foreign to the science....

algebra was, what its logical structure was, and what meanings its statements had. In answering these questions Peacock was aware that he first had to explain his concept of the logical structure of mathematics and of its axiomatic system. That is why he dealt with this problem on the first pages of his *Report* ([328] 185–187). He required that algebra, like geometry, be founded on assumed principles and definitions; he considered these principles and definitions to be the final and ultimate facts from which conclusions were logically derived. The truthfulness of these conclusions only resulted from their necessary connection with the initial facts.[7] Up to this point Peacock only reproduced the ideas on the structure of geometry which were then current. However, as soon as he began to apply these ideas to algebra,[8] he arrived at his own opinions, affected by a whole series of elements of actual mathematical thinking particularly stressed in England. The logical structure of symbolic algebra, as regards Peacock, stemmed from principles (rules) we have suitably chosen. These principles (rules) are only equalities between expressions which are similar or identical with the expressions of arithmetical algebra with one fundamental difference. In arithmetical algebra letters represent number's, in symbolical algebra the meaning of the symbols contained in the expressions is only restricted by the fact that they satisfy the chosen principles. Otherwise their meaning is "quite" arbitrary. That is why symbolical algebra for Peacock may "become essentially a science of symbols and their combination, constructed upon its own rules, which may be applied to arithmetic and to all other sciences by interpretation" ([328] 194–195). Of the symbols one must then assume that they are quite general and unrestricted as regards their value and expression; in a similar way the operations can be carried out in all cases, i.e., also for all the values of the symbols satisfying the principles.[9] As the principles are initial, the interpretation of the whole science, of its operations and results will follow, not precede. In this way Peacock arrived at the definition of the important term of interpretation which characterizes symbolical algebra. Peacock maintained that in arithmetic and arithmetical algebra the operations were defined, but in symbolical algebra they were interpreted according to the symbolical conditions which were valid for them.[10]

---

[7] Algebra and geometry are... "founded upon assumed principles and definitions"... "we consider those principles and definitions as ultimate facts, from which our investigations proceed in one direction only, giving rise to a series of conclusions which have reference to those facts alone, and whose correctness or truth involves no other condition than the existence of a necessary connexion between them, in whatever manner the evidence of that existence may be made manifest...". ([328] 186).

[8] At the same time he pointed out that in algebra the laying of the foundations was little developed, and that it made little progress during the last hundred years ([328] 185). He also added that only mathematics could have such a structure (with reference to the physical world).

[9] Peacock spoke of... "supposition or assumption that the symbols in the symbolical algebra are perfectly general and unlimited both in value and representation, and that the operations to which they are subject are equally general likewise..." ([328] 195).

[10] ...we define operations in arithmetic and arithmetical algebra conformably to their popular meaning, and we interpret them in symbolical algebra conformably to the symbolical conditions to which they are subject, ([328] 198 note).

Moreover, in symbolical algebra the meaning of the operations is determined by the accepted rules, which also provided the means and restrictions of the interpretations. "We call those rules, or their equivalent symbolical consequences, assumptions, in as much as they are not deducible as conclusions from any previous knowledge of those assumptions, in as much as they are arbitrarily imposed upon a science of symbols and their combinations, which might be adapted to any other assumed system of consistent rules". ([328] 201). Before this, on p. 198, he wrote that it need not necessarily follow, if one had the symbols and the rules for "combining" them, that there exists an interpretation which could be attributed to the operations, in order to satisfy the required symbolical conditions.[11] Therefore, the search for such interpretations, provided they could be found at all, is, according to Peacock, one of the most important and most substantial tasks of the deductive process which one might require of algebra and its applications.

In stating these views, Peacock's own formulations were adhered to as far as possible.[12] They indicate that Peacock considered symbolical algebra to be a science of "combining" symbols, in spite of many obscurities, the rules for these combinations being similar to the expressions of arithmetical algebra. Other statements are deduced from the fundamental principles. In this way, one may perhaps say that a kind of axiomatic system was created in which those statements were true which could be derived from the initial rules. One of the main objectives of algebra is to find suitable interpretations, their existence being in no way guaranteed; at the same time there may exist more than one interpretation.

This general form of the interpretation which mentions the arbitrary choice of the initial rules is intermingled with statements about the mutual connection between symbolical algebra and arithmetic. They at least partly clarified Peacock's ideas of "the rules". At several points he drew attention to the fact that these rules were initiated by the corresponding rules of arithmetical algebra, but that they did not follow from them.[13] This then considerably restricted the arbitrariness of the rules in

---

[11] ...neither does it necessarily follow that in such case there exists any interpretation which can be given of the operations, which is competent to satisfy the required symbolical conditions... ([328] 198).

Likewise he wrote in the preface referred to ([327b] VII n): "...the meaning of the operation or of the result obtained, whenever such a meaning can be assigned, must be determined in conformity with the conditions which it must satisfy, and consequently must vary with every variation of those conditions: upon this principle we shall be enabled to give a consistent interpretation to symbolical expressions or results; ...but in innumerable other instances, it will be found that the results obtained will admit of no interpretation whatever, or have hitherto failed to receive it."

[12] This has been based most widely on formulations of 1833, i.e., of the *Report* [328], but only because Peacock explained his opinions here most extensively and systematically. Between this paper and both editions of his algebra [327a; b, c] there are no differences in opinion, but only slight deviations in formulation, which are not the result of the change of opinion but stem from Peacock's effort to achieve the most accurate expression of his ideas.

[13] The rules of symbolical combination which are thus assumed, have been suggested only by the corresponding rules in arithmetical algebra. They cannot be said to be founded upon them, for they

symbolical algebra. Peacock even maintained ([328] 195, 201) that a general symbolical expression of a result in arithmetical algebra was also true in symbolical algebra; to this he expressly added that symbolical algebra did not contain arithmetic. It is then understandable that Peacock should express the rules of symbolical algebra by algebraic expressions.

The said restrictions undoubtedly stemmed from the contemporary state of mathematics. There was no reason for a complete arbitrariness in rules, which Peacock perhaps only anticipated. However, there was a reason for a difference in "interpretations". Peacock presented an abundance of geometrical interpretations of complex numbers ([328] 266nn), and referred to authors like Buée, Argand, Servois, Français, Warren,[14] but pointed out that these interpretations only represented one of the possibilities. Elsewhere he also spoke of other geometrical interpretations (e.g., [328], 189nn).

The points of view of Gregory and Morgan were very close to Peacock's concept of algebra. Therefore, we shall take special notice of the points in which they differed. Gregory, who was the youngest of the three, adopted Peacock's opinions and his formulations. In defining symbolical algebra and in expanding on this idea, however, he proceeded in the opposite way from Peacock. The latter built on the differentiation of arithmetic, arithmetical and symbolical algebras; these three degrees of generalization being conditionally related. Gregory did not consider the genesis of symbolical algebra, but he considered it to be the initial concept. This enabled him to free symbolical algebra of the difficulties which frequently obscured Peacock's interpretation; neither did he study the relation between symbolical and arithmetical algebra, nor use the problematic "principle of permanence". Gregory's principal idea, which determined his speculations on the nature of symbolical algebra, was "the separation of symbols of operations from those of quantities".[15]

Gregory defined symbolical algebra as a science, "which treats the combination of operations defined not by their nature, that is, by what they are or what they do, but by the laws of combination to which they are subject". He followed this up with the following idea:

"And as many different kinds of operations may be included in a class defined in the manner I have mentioned, whatever can be proved of the class generally, is necessarily true of all the operations included under it. This, it may be remarked, does not arise

---

are not deducible from them... ([328] 198). Similarly in § 2 of [327b] or ([328] 195) where he spoke of the fact that arithmetical algebra "suggests its (i.e., of symbolical algebra) principles, or rather its laws of combination".

[14] As regards the relation between the ideas of the English school of algebra and the geometrical interpretation of complex numbers, valuable data and new approaches can be found in the paper of S. Bachelard [19].

[15] At the very beginning of the article *"On the Real Nature of Symbolical Algebra"* he wrote: The following attempt to investigate the real nature of Symbolical Algebra, as distinguished from the various branches of analysis which come under its dominion (!), took its rise from certain general considerations, to which I was led in following out the principle of the separation of symbols from those of quantity ([176] 208).

from any analogy existing in the nature of the operations, which may be totally dissimilar, but merely, from the fact that they are all subject to the same laws of combination. It is true that these laws have been in many cases suggested (as Mr. Peacock has aptly termed it) by the laws of the known operations of number; but the step which is taken from arithmetical to symbolical algebra is, that, leaving out of view the nature of the operations which the symbols we use represent, we suppose the existence of classes of unknown operations subject to the same laws. We are thus able to prove certain relations between the different classes of operations, which when expressed between the symbols, are called algebraical theorems. And if we can show that any operations in any science are subject to the same laws of combination as these classes, the theorems are true of these as included in the general case ..." ([176] 208n).

This quotation, which records Gregory's point of view and its connection with Peacock's opinions most accurately, indicates in what direction he developed the idea of symbolical algebra. A new important concept has been recorded here by the words "class of operation", in which one might perhaps justifiably anticipate the originating concept of isomorphism; all sets of elements into which operations "subject" or the same rules have been introduced, have the same structure,[16] and the same theorems hold for them.

Gregory went back to the theory of algebra several years later in a paper which touched on the relation between arithmetical and symbolical algebra [179]. Here he considered one of the most important problems, "what is to be understood by an Algebraical symbol, and in what way it represents an operation". ([179] 153). He then presented an explanation which maintained that a symbol was defined algebraically if its laws of composition were given, and that the symbol represented the given operation if the laws of composition were identical to the laws valid for the given symbol. With the help of this he showed that arithmetical operations were one of the representations of symbols of symbolic algebra, but he also named other representations, in particular geometrical ones.

In both papers Gregory devoted considerable space to his attempts to find at least some of the laws of composition and classes of operations which they defined. However, the results were not too fruitful. He studied, as an example, two classes of operations, F and f, for which the following "laws" held:

$$F F (a) = F (a) \qquad (6.2.1)$$

$$f f (a) = F (a) \qquad (6.2.2)$$

$$F f (a) = f (a) \qquad (6.2.3)$$

$$f F (a) = f (a) \qquad (6.2.4)$$

---

[16] There were considerable terminological difficulties here, which make the direct translation of Gregory's speculations into present-day terminology impossible. With Gregory the term "set of elements", which he could have perhaps used instead of the term "number" or "quantity", which he wanted to avoid, does not appear. That is why he only spoke of "operations" without sets in which they were introduced, and he thought that this was a sufficient way of guaranteeing the generality of his speculations.

He then showed that addition and subtraction, represented by the special symbols $+$ and $-$, belonged to these classes of operations. Another class of operations are those which satisfy the laws

$$f(a) + f(b) = f(a + b) \qquad (6.2.5)$$

$$f'f(a) = ff'(a) \qquad (6.2.6)$$

In connection with this example he recalled Servois and his terms distributive and commutative functions ([176] 211). In a later paper he also studied the associative property as one of the laws of combination. These examples, which he complemented with their various "representations", indicate the narrow base of Gregory's speculations, which did not in fact reach very far beyond the examples given by Servois or Babbage.

Morgan as well as Peacock and Gregory,[17] were influenced by the "formal" theory of functions. There is evidence of this in Morgan's paper of 1836, which pointed out the uniformity of the trend stressing the study of "formal" and "symbolical" aspects in calculus and in Calgebra.[18] However, a year later Morgan presented the fundamental idea of symbolical algebra in his book *The Elements of Algebra*, which is intended as a pre-university course of higher mathematics. Here, as an explanation of the meaning of algebraic formulae,[19] he used the law of permanence, interpreted very originally, and the concept of symbolical algebra became obscured by a concept which Peacock called arithmetical algebra.

Only beginning with 1839 did Morgan begin to explain his concept of algebra comprehensively in four papers ([298 – 301]). He had the same train of thoughts as the two preceding authors. In their spirit he maintained that in algebra the problem is the separations of the symbols of operations from the meaning of these operations, and that "the symbols represent something more than simple magnitude". However, he made no mention of arithmetical algebra or arithmetic, and of their relations to algebra; he did not even use the term symbolical algebra (each algebra is symbolical);

---

[17] For example, Gregory wrote his own paper on the solution of certain functional equations [178]. His interpretations indicate that the words mentioned in Comment 15 express his ideas. The rules of operations at which symbolical algebra arrived, can be applied to the whole of calculus. He was of the opinion that differentiation was, e.g., distributive, since

$$\frac{dy}{dx} - ay = 0$$

could, according to Gregory, be written as

$$\left(\frac{d}{dx} - a\right) \Phi(x) = 0 \qquad \text{and then}$$

$$\Phi(x) = \left(\frac{d}{dx} - a\right)^{-1} 0 = Ce^{ax}.$$

[18] As representatives of the formal trend in calculus he named, apart from Babbage and Herschel, also Carnot and Cauchy, as representative of algebra he gave Peacock, cf. [296] p. 1.

[19] [296] Chapter IV. On exponents, and on the continuity of algebraic expressions, especially pp. 94—104; of this part he said that it exceeded the requirements of pre-university education.

he only spoke of algebra, which according to his opinion was composed of two parts: of technical and logical algebra. The technical algebra dealt with the rules of operations. The symbols used in it were defined by rules which made it possible "to accept or reject any proposed transformation of it, or by means of it". ([298] 173 – 174). The logical algebra dealt with the interpretation of symbols, i.e. with finding a representation. Moreover Morgan specially stressed that the possibilities in finding these interpretations were very considerable, since they were only restricted by the rules for working with the given symbols, i.e., with the prescribed structure of relations among them.[20]

Not until the second paper [229] did he make an effort – all on his own in fact – to find the laws of the operations "separated" from the meaning of the symbols with which the operation was carried out. He named eight rules (laws) of varied and quite complicated nature. The first, e.g., required that the symbols $a, b, c, \ldots$ should have the same meaning throughout the whole "interpretation". The second introduced identity and its symbol $=$; if $a = b$, one may be arbitrarily substituted for the other; also if A has been derived from a, and B from b, from $a = b$ follows $A = B$. The third begins with the words: "The signs $+$ and $-$ are opposite in effect; what one does the other undoes: and 0 is the symbol of a pair of such oppositions having been performed, thus $+a - a = 0$". These operations are "convertible", i.e., $+a - b + c = +c - b + a = -b + c = a$, etc. In a similar way he introduced the symbols $\times$ and $\div$, where of course $xa \div a = 1$. The fifth rule introduces distributivity: $(+a) \times (+b - c) = (+a) \times (+b) + (+a) \times (-c)$, etc. The sixth contains an obscure statement that an even number of symbols $+$ and $-$ yields $+$ (similarly for x and $\div$), and an odd number $-$. The seventh runs as follows: The signs 0 and 1 may themselves be considered as subject of operation, and $1 + 1$ is abbreviated into 2, $1 + 1 + 1$ into 3, ... The eighth and last gives the rules for using the symbols a^b which are as follows: $a^b x a^c = a^{b+c}$, $(a^b)^c = a^{bc}$. To this Morgan added that he did not consider these rules to be either sufficient or excessive "though I should be noways surprised to see them proved both the one and the other ..." ([299] 288).

In this treatise and in the subsequent continuation [300] he only deduced some of the consequences of his rules. The fourth and last continuation [301] has a considerably different nature; it was written in the autumn of 1844 and contained one of the first responses to Hamilton's paper on quaternions [186].

Morgan tried to establish a system of three fundamental units (whence the name Triple Algebra) which would be capable of meeting the same requirements as Hamilton's quaternions, the rules of operations of which would be more similar to the laws of the usual algebra. In particular he wanted to avoid non-commutativity. He first considered n units $\xi_1, \xi_2, \ldots, \xi_n$ and elements of the type $a_1\xi_1 + a_2\xi_2 + \ldots + a_n\xi_n$ generally,

---

[20] In Morgan's work a lot of space is devoted to general speculations. Amongst these is also the discussion on Hamilton's interpretation of algebra in the spirit of Kant's philosophy as a science of "pure time". Of course, apart from this, Morgan's study also contains some purely logical elements, probably an indication of his future studies in logic.

where a_i are numbers, and he wanted to determine the relations between the units. He said that their multiplication must conform to the following two rules:

$$AB = BA \tag{6.2.7}$$
$$A(BC) = (AB)\,C. \tag{6.2.8}$$

After general speculations he turned to a system of two units, which he then investigated for three units. He tested various systems of laws and in his way investigated the properties of the axiomatic systems.

Morgan was aware that he had not reached his objective, he acknowledged the success ([307] 254) of Hamilton's quaternions which had geometrical punctuation guaranteed. He added: "My object has been to detect systems in which the symbolic forms of common Algebra are true, without making sacrifice to interpretation"; these are words which we have already recalled in connection with Hamilton's quaternions. Of course, the significance of Morgan's systems was elsewhere: they established the principles of the new approach to algebra, hitherto expressed in a more or less abstract form, by discussing the various possible (one may say linear) systems in the spirit of this algebra. These systems (the same actually holds for Hamilton's quaternions) are only partial; they do not include the whole of the contemporary arithmetic, but they determined the rules of combination only for new artificial objects (distinct symbols as Morgan put it), whereas the rules for calculating with numerical systems were considered known.

Morgan tried again to explain his ideas "of a symbolic calculus", as he sometimes called algebra, in 1849 in a book called *Trigonometry and double algebra* [302]. The whole concept of the book is interesting in that it considered trigonometry as part of algebra, and in particular the structure of "double algebra" as a more simple parallel of "triple algebra".

These features which indicate how Morgan tried — essentially unsuccessfully — to apply general statements on the nature of algebra to actual though perhaps partial systems, will not be dealt with here. It should only be pointed out that Morgan came even closer to Peacock's opinions in the concept of algebra itself. He differentiated between "arithmetic and ordinary algebra", where "we use symbols of previously assigned meaning from which meanings ... are derived by rules of operation". As opposed to them he put the science of symbols itself which he developed (p. 89nn), in an even more general form than Peacock and Gregory. A parallel with ordinary speech can be seen here, which is also a science of symbols (words) having, as he said, its meaning prescribed by a dictionary and the rules of combination by grammar. Abstract speech, however, is constructed perfectly. This was as if he were preparing a way to some kind of logical theory of speech. This feature of Morgan's work has also been pointed out in order to recall that one might encounter an older background here which can already be found in the works of Condillac and other philosophers.

The ideas of the English mathematicians about the new approach to algebra as a study of certain abstract systems, in which the meaning of an element is defined algebraically, as they themselves maintained, i.e. by rules which were assumed for combinations within these systems (provided one does not want to use the modern words "for elements of these systems"), marked the modern approach to algebra and perhaps even of the whole of modern mathematics. However, they were not developed further after the end of the forties, and they were apparently forgotten for a longer period. It is possible that they did not admit of further treatment in their time, or that they were not necessary in view to the state of algebraic research at the time. They were created under special conditions in British mathematics which until then was subject to a different development than Continental mathematics. Outside England the whole vision of the new approach to algebra had no response. However, in England, although this approach is not even recalled, it did affect the nature of further mathematical work. The English mathematical logics which were coming into existence in the work of Morgan and Boole, like the development of some of the algebraic theories of Cayley, Sylvester and perhaps of their pupils, bear distinct traces of these ideas. However, these are problems which not only exceed the scope of this section, but also of the investigated topic and the period pertinent to this study.

CHAPTER 7

ALGEBRA IN THE MIDDLE OF THE 19TH CENTURY

7.1 Introduction

In the middle of the 19th century the topics of algebraic papers were acquiring a new appearance. The problems of the study of equations as well as their relation with respect to the other problems of algebra had also changed. Galois' principal papers were published in 1846 in one of the few mathematical journals, and the papers and manuscripts of Abel, published slightly earlier, propagated the facts about Abel's ideas and results. New proofs of the algebraic insolvability of a general equation of a higher degree than the 4th appeared (e.g., Hermite [199], 1842), and the older ones were being complemented and simplified,[1] but on the whole the conviction prevailed of this insolvability. In particular, the younger generation of algebraists left this problem for the first treatments of Galois' theory in the fifties and sixties, especially as regards the study of the component which was so substantial for Galois' inheritance, i.e., the study of the theory of groups. Although this trend was important, and people like Kronecker, Schönemann, and others in Germany, Liouville, Serret and later Jordan, in France, and Betti in Italy took part in it, it was not the only trend in algebra in the second half of the century. Beginning with the forties, considerable attention was being devoted to the theory of invariants whose rapid quantitative growth, and especially their surprising fruitfulness for establishing connections among the various mathematical disciplines, caused it to become the leading algebraic discipline in the eyes of some contemporaries.[2] The theory of invariants still bore the relations which were responsible for its origin, and which provided it with its importance for the subsequent decades: it connected algebra with the theory of numbers (as followed from Gauss' inheritance in the theory of quadratic forms), and with the geometry that was new at the time (especially projective geometry), but it also reflected the permanent effects of the algebraic problems from which it had evolved, i.e., the theory of elimination and substitution, which developed in a similar way as in the 18th century in the traditional algebraic form.[3] It was only a question of

---

[1] For example Wantzel, Schönemann, etc. Also the response to Abel's studies with Hamilton [184] and his correspondence, in which a certain requirement for modification of the proof, was voiced, is interesting.

[2] Cf. Matthiessen, [282] Preface.

[3] I have in mind especially Sylvester's early studies in which the investigation of eliminations is a way out (like in the case of Cayley and others).

time before the general nucleus of the theory of invariants would be found, and before it would begin to be studied as an example of one of the general algebraic structures.

The development of the theory of invariants, in spite of its great importance in mathematics, was only short-lived.[4] Besides the theory of invariants however, no less attention was being devoted to the theory of determinants. which was considered to be an independent field of algebra, although with time it became clear that it was not very stimulating scientifically. Its great quantitative growth is dated in the forties, fifties and sixties of the 19th century. However, at the time one could already have spoken of other, new and relatively independent algebraic disciplines. The matrix theory originated at the time, as well as the theory of quaternions and hypercomplex systems in general; concepts of linear algebra began to be brought to the fore. To these algebraic disciplines one may also add numerical solutions. But this complex of algebraic disciplines, the internal division of which is not so much the result of the increasing number of papers as of the relative independence and theoretical specificity of these disciplines, joined together very loosely, had in fact escaped from the originally unified sphere of arithmetic and algebra. Theoretical arithmetic[5] and the theory of numbers[6] belong to the components which, together with the algebraic disciplines, inherited the original field of science.

It is not possible to discuss in detail here the development of algebraic disciplines, not even to the extent their results belong to the fifties and sixties. This is the task of specialized literature; the elements which were considered substantially useful for creating new algebraic ideas were mentioned in the previous chapters. However, so far we have omitted one algebraic discipline which was then new and which has a special significance in the development of algebra in that it was the first in which one of the fundamental algebraic structures, which was one of the first conscious connections among the contemporary algebraic disciplines, began to be studied. This was the theory of groups, originally studied in a special form as the theory of substitutions (or the theory of groups of substitutions). It is one of the facets which characterize algebraic results towards the end of the period investigated. However, like the theory of groups, the view of algebra from the standpoint of modern textbooks, written in the middle of the century and used frequently, deserves attention. The second and third sections of this chapter are devoted to these two topics, which are intended to show how algebra was understood towards the end of the period investigated.

---

[4] Compare with Fisher [142].

[5] Discriminating between algebraic and transcendental numbers, like the theory of real numbers, the new definitions of natural numbers, etc., all form the topic and methods which separated algebra from arithmetic.

[6] Discriminating algebraic topics from the theory of numbers went so far that, e.g., the so-called theory of forms was divided into a part attributed to the theory of numbers and another to algebraic disciplines in the journal *Jahrbuch über die Fortschritte der Mathematik*.

7.2   The origins of the theory of groups

The idea of a group, like that of all abstract mathematical concepts, was formed gradually from the study of special material and under the influence of actual require-ments pertinent to the solving of certain mathematical problems. The individual facts which would now be included in the theory of groups were originally formulated in a special form. Only slowly and with difficulties was their common basis distinguished, thus leading up to the more abstract concept of a group and of the whole theory.[1]

The decisive role in the initial stage of development of the theory of groups was played by the study of the solving of algebraic equations. As has been indicated, Lagrange and Vandermonde alike pointed out the importance of the study of permu-tations in determining the degree of the resolvent; Lagrange considered this to be the inherent "metaphysics" of the theory of equations. The subsequent development, however, was considerably affected by the form Lagrange used to introduce the study of permutations into the theory of equations. The fundamental concept was the function of (all) roots of a given equation. The function, which he understood in the spirit of Euler and Bernoulli, however, was practically restricted to very simple rational expressions, and considering the various orders of the roots it could acquire different values, in general up to $n!$ values for an n-th degree equation with n roots, $x_1, \ldots, x_n$. The main task which confronted Lagrange as a result of his analysis of the solutions of equations up to the time, was to find a function of such form which would acquire a suitable number of values for n permutations of the roots.

Lagrange considered the summary of all permutations of the n roots as if given, he never discussed this point. He rather considered a permutation to be the order of elements, not an operation. He could make do with the current idea of permutating, i.e., the changing of the order of given elements. He never encountered explicitly the idea of combination (multiplication) of permutations. He did not require it for his speculations. Nor are permutations the object of his study; perhaps one could say that they were only a means borrowed from a different sphere of mathematics, combination calculus. In combination calculus the facts about permutations were discussed to a much wider extent than Lagrange required for his own purposes.[2]

It has been said that the fundamental concept was a function of n roots. In studying it Lagrange arrived at several concepts. Later considerable response was drawn to the concept of "les fonctions semblables des racines

$$x', x'', x''', \ldots \text{''},$$

---

[1] As has been shown in the review of literature (Chap. 1.2) the origin and development of the theory of groups in the 19th century has been treated relatively in great detail. This makes it possible just to give an indication of the main trends, their form and the most important changes (nodal points) which formed the principal stages in the development of the theory of groups. The arguments in favour of the presented image are only given in fragments, since it has been discussed in detail, e.g., in [314—317], where other references are also quoted.

[2] Cf. Chap. 6.1.

which represented functions whose values either changed together or remained unchanged when the variables were permutated ([267] § 88, pp. 358−9). However, in this case the permutation, without any further explanation, was considered to be a rule for changing the order.[3] But, in connection with the study of similar functions, Lagrange pronounced a very important theorem, that the number of various values of a function of n roots (considering all permutations of the roots) is either equal to $N = n!$, or it is the divisor of N. With time and with knowledge acquired on the theory of groups of substitutions one may distinguish the statement about the order of a subgroup of the finite group (or about the index of this subgroup) in this theorem, but in reality Lagrange only stated what has been said. I am of the opinion that with Lagrange one cannot speak of a group of substitutions, and therefore neither of finding some of the simple properties of groups. Lagrange's merits as regards the origin of the theory of groups is perhaps only in that he turned the attention of some of his followers to these functions by stressing the study of the functions of roots, which in its consequences led to the study of permutations.

Many authors adopted Lagrange's method of determining the resolvent and especially its degree after the end of the 18th century. However, the first to expand on the stimuli contained in it and tending towards the development of the theory of groups, was Paolo Ruffini. The effort to study systems of permutations of given elements led him to look for the appropriate system of expressions, and to distinguish the various types of these systems. With Lagrange and Ruffini alike the initial concept is the rational function. But Ruffini began to study these functions more generally (he also chose irrational functions), and, at least to begin with, he not only spoke of functions of roots of a given equation, but of functions of letters. The number of orders of n letters, which Ruffini proved in effect by complete induction, is $N = n!$ He called these various orders permutations.[4] The functions of n letters enabled him to determine among these N permutations various subsets, i.e., functions of n letters define the sub-set of permutations which leave this value of the function unchanged. Ruffini also called this set "permutazione". He then distinguished among the various kinds of these sets. Although Ruffini's interpretation is not quite clear, he distinguished among "permutations"[5] of several kinds according to which permutations

---

[3] Lagrange ([267], p. 374), for example, expressed himself as follows: "...les deux fonctions t et y étant supposées semblables, il s'ensuit que le nombre des valeurs différentes dont elles seront susceptibles par toutes les permutations possibles entre x', x'', x''', ... sera le même pour l'une et pour l'autre, et que ces valeurs seront dues aux mêmes permutations dans les deux fonctions".

[4] Ruffini ([343] in Section 43 defined the permutation in these words: Data una funzione, per esempio la $f(\alpha)$ (β) (γ), suponiamo di cambiare in essa fra loro le lettere che la compogno, scrivendo per esempio la α in luogo della β, e la β invence della α, come mella $f(\beta)$ (α) (γ), o siccome nell'altra $f(\beta)$ (α) (γ), ponendo la α in luogo di γ, la β in luogo di α, e la γ in vece di β. Questi diversi collocamenti delle lettere, e quindi i varii risultati, che ottengonsi nella funzione, quelli sono, che dai Matematici si dicono PERMUTAZIONI.

[5] For the sake of accuracy I shall denote by inverted commas the two principal meanings which Ruffini gave to the word permutation. Permutation or **permutazione** is an order of elements,

they contained. Essentially a simple "permutation" is created by repeating a given permutation, no element being left without change; a composite "permutation" is created from a number of various permutations.

Before we go on to another of Ruffini's classifications of "permutations", it should be pointed out that Ruffini's interpretation fully bears the traces of trying to find a suitable conceptual system, which he also developed in his further papers. His own definitions of various kinds of "permutations" can easily be expressed in terms which were later created in the theory of groups of substitutions;[6] without them, however, it is difficult to understand and can be better understood through the examples he gives of practical calculations. Ruffini did not introduce the concept of "cyclic permutation" in his textbook [343] of 1799; this remained in between lines[7] causing a number of terminological difficulties. Therefore, he only referred to the form of functions. That is why he defined a simple "permutation" as follows: Chiamo permutazione semplice quella in cui le radici che la compongono devonsi muovere tutte simultamamente dal lozo luogo ([343] 161). As an example he gave a "permutation" which did not change its value:

$$\frac{x'^2}{x''} + \frac{x'''^2}{x'} + \frac{x''^2}{x'''} \qquad (7.2.1)$$

or

$$x'x''x'''^2 + x'''x^{IV}x^{IV} \qquad (7.2.2)$$

The "permutation" defined by function (7.2.1) is called a simple "permutation" of the first kind, that defined by function (7.2.2) of the second kind.[8]

With composite "permutations" Ruffini distinguished among three kinds: Le permutazioni composte distinguonsi in tre generi. Il gebere primo comprende quelle nelle quali niuna delle radici esistenti in una qualunque delle permutazioni può passare tra le radici, o nel luogo occupato dalle radici di unáltra.

Il secondo abbraccia quelle nelle quali le radici di una permutazione componente

---

"permutation" the set of these orders, which does not change the value of the given function. Wussing's interpretation ([410] 58) is not quite accurate, since Ruffini did not use the term "sostituzione" in the place quoted ([343] 161—3); it will only be found in the notes of Bortolotti. When Wussing, following Burkhardt ([51] 133), put a sign of equality between Ruffini's concept of "permutation" and Galois' "groupe", it only represents an explanatory modernization.

[6] Cf. Bortolotti's note, [342a] p. 413.

[7] He was of course speaking of the permutation which would transform function $f(x')(x'')(x''')$ $(x^{IV})(x^{V})$ into function $f(x'')(x''')(x^{IV})(x^{V})(x')$ ([343] 162).

[8] The cyclic permutation (x', x'', x''') is of the first kind, and $(x', x''')(x'', x^{IV})$ of the second. He defined the said kinds of simple permutations as follows ([343], 162): Diremo poi permutazioni del primo genere quelle nelle quali non possono alcune delle radici cambiarsi fra loro separatemente dalle altre.... Diremo del secondo genere quelle nelle quali, mentre alcune delle radici che le formano cambiansi fra di loro, altre si cambiano fra loro disgiuntamente dalle prime.... It is worth noticing that Ruffini no longer referred to the permutations of letters here, but that he went back to the permutations of roots, which led him to some undesireable restrictions. Cf. Burkhardt [51], p. 133.

possono passare tutte ad occupare il luogo già prima occupato dalle radici di unàltra, senza però che le radici della prima si framischino a guelle della seconda.

Il terzo funalmente comprende le permutazioni, in cui le radici di una delle componenti possono passare a mescolarsi tra le radici di unàltra ([343] 163).

Ruffini's interpretation contains the beginnings of the concepts of a transitive, primitive and imprimitive group, together with the following examples, but Ruffini's terms are still not identical with ours.[9] In his study Ruffini was also coming close to other concepts and facts of the theory of groups of substitutions. For example, he spoke of the number of permutations which did not change the value of the function, i.e., of the order of the group, and was aware that for any arbitrary a and N, a/N, a group of order a need not exist.

The last formulation modernizes Ruffini's results. Ruffini formed concepts and arrived at facts which presaged the future theory of groups of permutations, but which were closely connected with the objective of his algebraic speculations. He considered permutations to be an auxiliary means serving an actual purpose. The set of permutations which do not change the value of a given function had no internal structure as far as Ruffini was concerned (as we vould say today, it did not have sub-groups). No operation was introduced in this set and Ruffini himself did not create means for searching directly (to use modern terms) for the subgroups of a symmetric group. That is why his proofs of theorems on permutations, which he needed to prove insolvability, are very cumbersome. They rely on enumerating and "trying out" all the permutations of a given number of letters.[10] Nevertheless Ruffini made considerable progress in comparison to Lagrange (and Vandermonde), because he created stimuli for further study in the essence of groups of substitutions, and initiated the appropriate conceptual system.

Gauss' *Disquisitiones arithmeticae* of the same period did not directly influence the trend represented by Ruffini which led to the theory of groups of substitutions, nor did they add to its results. Gauss did use permutations exceptionally, but there is no connection with the theory of groups of permutations, where he essentially expanded upon his ideas of his own theory of groups. Gauss arrived at ideas concerning the theory of groups, especially in the theory of combining classes of binary quadratic forms and in the theory of n-th roots of unity, where he also exploited in a masterly way calculations with residual classes, even with respect to non-prime number models. In both cases Gauss studied a finite set in which a binary operation was defined. The second case is closer to traditional numerical knowledge, whereas the former appeared to be "artificial" in comparison to the mathematical speculations of the time (this was also stressed by the way the operation was introduced), in spite of the fact that the result was substantiated by numerical and theoretical relations. Among

---

[9] Cf. Bortolotti's comment, already referred to; it follows immediately from what he said. For example, the first kind is "always" intransitive.

[10] Burkhardt [51] pointed this out anyway several times.

Gauss' main results, which were the result of the study of the theory of numbers, there is the search for subgroups and determining the chain of subgroups. In the theory of *n*-th roots of unity, where commutative groups are concerned his results are clear as regards calculation and did not require new conceptual means. Another of Gauss' results were the indications which led to the determination of the base of the commutative group of classes of quadratic forms, which was expanded upon much later, perhaps still under the direct personal influence of Gauss.[11]

The first to develop the stimuli relevant to the theory of groups of permutations with an original contribution was Cauchy. He was interested both in the traditional theory of permutations and in the study of modern algebraic problems. Since 1812 he had exploited facts about the theory of permutations for the study of symmetrical functions, and not long afterwards he turned to a new and very influential interpretation of the theory of determinants which was being created at the time [58] by the same means. In the context of these papers an important contribution was born to the theory of groups of permutations,[12] i.e., the paper *Mémoire sur le nombre des valeurs qu'une fonction peut acquérir, lorsqu'on permute de toutes les manières possibles les quantités qu'elle renferme*, which was published in 1815 [59].

In this treatise, Cauchy, as he himself proclaimed in the preface ([59] p. 1, and pp. 8—9), expanded upon the problems which the Italian mathematicians dealt with,[13] and which were important for the theory of algebraic equations. However, Cauchy formulated the principal problem of the treatise exclusively as follows: How is the number of various values which a function of several variables acquires when all the permutations of these variables are carried out, restricted? In the paper he only dealt

---

[11] Mainly E. Schering [348] in 1869; also in Dedekind's comments to Dirichlet's work, cf. [122]. A detailed analysis of Gauss' indications and in particular the response up to the time Schering published his paper, is lacking. Wussing ([410] 44) did not consider this problem. However, there is the question whether there exist to them data to span the development in the years 1801 to 1869. The fact that a detailed analysis of Gauss' merits with respect to the beginning of the theory of groups, was lacking, has been pointed out above. That is why there is no answer to the important question of Gauss' intentional use of isomorphism in studying the problems of the theory of groups.

[12] The terminology fluctuated for a long time, and the terms "permutation" and "substitution" were used side by side. Essentially it is not my intention to disrupt the terminology used at the time in places where there is no danger of being misunderstood. The theory of permutation, however, is also called a summary of facts which were then developing under the name of "combination calculus" (cf. Chap. 5.1). The terms "the theory of groups of substitutions" and "the theory of groups of permutations" are used as equivalents, precedence being given to the term permutation for the older period, in which one cannot yet speak of an independent discipline. Later, the independent terms permutation and substitution will be understood in such a manner that the permutation is an order of elements, substitution a transition from one order to another, and it will not be noted specially which author used which and in what sense.

[13] He quoted Ruffini's book [343] of 1799 and his paper of 1805 [345a]. He did not refer to other papers expressly, although one cannot exclude the possibility that he knew them. He also mentioned Lagrange and Vandermonde who, as he said, were the first to discuss the problem as to how many values a function of several variables may have when the variables are permutated, in papers published in 1771.

with the proofs of a few theorems which were currently known from literature. He first proved the so-called Lagrange's theorem, which in Cauchy's interpretation maintained that the number of various values which a function of n variables [14] may acquire if the variables are permutated, is the divisor of the number

$$N = n! = 1 . 2 . \ldots . n.$$

He went on to generalize Ruffini's results, and as the principal point of the treatise he proved the following theorem: Le nombre des valuers différentes d'une fonction non symétrique de n quantités, ne peut s'abaisser au-dessous du plus grand nombre premier p contenu dans n, sans devenir égal à 2.[15]

There are three things which are interesting in Cauchy's work (apart from the theorem proved). First of all, it was probably the first study with this topic, in which the relation with the theory of equations was only mediated. Cauchy said that he was led to these problems by the study of algebraic equations, and the fact that he quoted Ruffini's papers indicates that he was aware of the principal importance of the proved theorems for the proof of the algebraic insolvability of equations of degree $n \geq 5$. This was the last mention made directly of the theory of equations, and the consequences resulting from the principal theorem of the treatise were not specified. The second important feature was the fact that functions of n letters were treated in the paper generally, and that the letters were not considered to be the roots of the equation. The third new feature, which one would seek in vain with authors prior to 1844, is the effort for a comprehensive and clear interpretation some of the contemporary facts on the groups of permutations. This feature, which also characterizes of Cauchy's papers pertaining to different spheres, is important for classifying facts, for ridding them of unnecessary and heterogeneous ideas, and for suitable annotation.

Cauchy called the order of the letters "permutation", and the transition from one permutation A_1 to another A_2 he called substitution and denoted it by

$$\begin{pmatrix} A_1 \\ A_2 \end{pmatrix}.$$

He also proved the independence of the substitution of the original order. He introduced the multiplication of substitutions, which led him to the concept of the power of a sub- stitution, on the one hand, and to the possible expression of a given substitution as a product of other substitutions, on the other. He also included unit substitution among the substitutions, and thus arrived at the reverse substitution, i.e.,

$$\begin{pmatrix} A_s \\ A_t \end{pmatrix}^{-1} = \begin{pmatrix} A_t \\ A_s \end{pmatrix}. \tag{7.2.3}$$

---

[14] He understood a function in the sense of Euler and Bernoulli. As an example he mentioned the function $f(a_1 a_2 a_3) = a_1 a_2 \cos a_4 + a_4 \sin a_3$.

[15] He also drew attention to special cases, e.g., that if n was an odd prime number, the theorem would hold for $p = n$. Similarly, for $n = 6$ the function cannot have fewer than six values, with the exception of the case of a symmetric function, or of a function acquiring two values ([59], 26).

The study of the powers of substitutions allowed him to make the statement that for any arbitrary substitution

$$\begin{pmatrix} A_s \\ A_t \end{pmatrix}$$

there exists the least natural $m > 1$, such that

$$\begin{pmatrix} A_s \\ A_t \end{pmatrix}^m = \begin{pmatrix} A_i \\ A_i \end{pmatrix}. \tag{7.2.4}$$

He called m the degree (degré) of the substitution, because there were just m various powers of the given substitution.

Cauchy also introduced the notation of "indices" for the number of values of a function;[16] for the number of various substitutions which do not change a given function (as we would say, for the order of the group), he introduced the term "diviseur indicatif", but in another connection he used the term "degré" for the same concept. This summary of Cauchy's concepts justifies making the statement that Cauchy, with the exception of the term, has an essentially clear idea of the group of substitutions as the basis for his speculations, i.e., finite sets of certain elements in which an operation has been introduced, which also contains the unit element, and in which there is an opposite element to each element. However, Cauchy went further and tried to find means for operating with substitutions without having to introduce a defining function for defining their "sets". In this connection he introduced the concept of cyclic substitution, for which he did not use the present symbols, of course. He did use the resolution of some substitutions into a product of transpositions,[17] which he denoted by (α, β), or directly by the indices of the letters, i.e., $(1, 2)$. Apart from resolving the substitution into a product of substitutions, he also arrived at the concept of resolving a group. Each substitution is of a finite degree, and the different products of their powers form a group.[18]

Cauchy used these concepts to prove the theorems mentioned as well as to solve special cases, in particular to determine various subgroups of the order of 6!, to use modern terms, i.e., subgroups[19] with indices 6, 15, 20 and 10. Nevertheless, Cauchy's approach to the whole topic and interpretation, which was an attempt at comprehensiveness, indicates how he detached himself from considering the study of substitutions to be a mere auxiliary means, and how at least implicitly he asked questions leading up to further generalization and to the analysis of further cases of substitutions and

---

[16] By the number of different values one understands the number of different values of a function of n letters which the function acquires when the letters are subject to all permutations.

[17] Of course, transposition was already encountered with Ruffini.

[18] This presages of Cauchy's future definition of the concept "système des substitutions conjuguées" which is identical with a group of substitutions.

[19] For these subgroups Cauchy also found functions of six letters defining them. In Cauchy's speculations one will also find the terms "le groupe" and "grouper", but in a sense current in French.

their properties. Although Cauchy's study was published in a widely read journal, it drew no response in this respect, and its consequences with regard to the problems of solvability of algebraic equations were only exploited later by Abel and Galois.[20]

Although Abel did not treat groups of permutations in particular as part of his algebraic studies, he contributed to their theory by presenting several theorems. For example, he proved ([3] 79) the theorem, to use modern terms, that each subgroup of index 2 is alternating, and without proof he presented the theorem that in a symmetric group of permutations of degree 5, subgroups of index 5 have one element which does not change. Galois devoted considerably more attention to the theory of groups of substitutions; he had also reached more remarkable results. His ideas and some statements are unfortunately only very briefly indicated so that data are lacking for evaluating certain questions.

Galois considered a group—this term, though not consistently applied, was used by him in the present sense of the word for the first time—to be a group of substitutions.[21] He did not expressly define the concept of a group, but he used it as a known concept and later presented as an important feature of a group the fact that it necessarily contained the substitution ST if it already contained the substitutions S and T.[22] This statement is usually considered to be Galois' definition of a group.[23] However, considering broader algebraic connections, Galois also spoke frequently of functions of letters which acquire a certain number of values when the letters are subject to all permutations. Sometimes this function serves the purpose of defining a group. However, by defining the so-called Galois' group generally, the group being pertinent to an algebraic equation, the said means gets pushed into the background, and Galois—probably under the influence of Gauss whom he frequently quoted in this connection—investigated the structure of the group of substitutions directly.[24] An important concept at which Galois arrived was the invariant subgroup, which led him

---

[20] Abel quoted Cauchy's paper as early as in 1824 ([2] 31) and he also referred to it in his proof of insolvability in 1826 ([3] 79); Galois was also sure to know it ([152] 46).

[21] Galois' turn of speech "grouper les substitutions" is interesting; it led him to the idea of "les groupes des substitutions". Cf. [152] 35.

[22] Donc, si dans un pareil groupe on a les substitutions S et T, on est sur d'avoir la substitution ST ([152] 36).

[23] Cf. [52], p. 211, note 8, where Galois' text, however, has been understood with a view to the supplements of E. Betti. In the same place Burkhardt maintained that a similar understanding of a group of substitutions could be found with Schönemann, Serret, Jordan, and others.

[24] Burkhardt ([52] 217) was even of the opinion that with Galois the significance of the individual operations was disregarded, and only speculations on the laws of combination were considered, so that he did not consider two isomorphic groups to be different. This interpretation of Galois' words I do not consider to be sufficiently substantiated, since Galois only said: Les substitutions sont indépendantes même du nombre des racines ([156] 40, Scolie II). However, there is the question whether one would find other Galois arguments for verifying the hypothesis that he had arrived at the concept of the isomorphism of groups. For example, Bourbaki presented an argument in favour of this [46] p. 74.

on to distinguish composite and simple groups. He also expressed[25] the theorem, of course without proving it, that the smallest simple group, the order of which was a composite number, was a group of order 60.

An extensive complex of problems, which Galois stimulatingly affected, were the various ways of expressing a group of substitutions. Essentially this was a problem of a rule of interchanging the indices. In this way Galois created the idea of a linear (homogeneous) group which had its effect as a tool of research in the further development.[26]

Galois' stimuli with respect to the theory of groups were very profound. Whereas Ruffini (and also Gauss) connected the facts about groups with special knowledge of the theory of equations, Galois connected them for a longer period of time with the so-called Galois theory. That is why the invariant subgroup came to the fore. One can also observe further progress with Galois as regards the development of terminology, determining general properties (cf., e.g., Comment 22), and a tendency towards a more general understanding of groups. But Galois' work had no effect in this respect up to the middle of the forties. It was in fact only discovered as a result of Liouville's edition of 1846, so that its individual aspects were gradually becoming understood and further expanded upon in the subsequent decades. However, just before Galois' studies were published,[27] Cauchy published more than 300 pages, devoted to the problems of the theory of groups, over an interval of two years, 1844—1846, i.e. more than had been written until then.[28]

Historical literature justifiably indicated the abundance of new results in these treatises[29] of Cauchy. Great credit is being given to Cauchy, e.g., because he proved

---

[25] Of course, the theorem was not presented so generally, but for a group of permutations ([152] 26): Le plus petit nombre de permutations que puisse avoir un groupe indécomposable, quand ce nombre n'est par premier, est 5.4.3. This theorem is frequently stressed in historical literature as Galois' great credit. Cf. Burns [53], p. 143; Burkhardt [52], p. 224.

[26] Cf. [52] 211, 214.

[27] Wussing [410] spoke of the development of "implizite gruppentheoretische Denkformen" in the sphere of geometry and the theory of numbers at this time. As regards the theory of numbers, with the exception of the *DA*, he only spoke of the studies beginning with the middle of the forties.

In geometry the traces of the theory of groups, according to his data, are quite obscure at this time. Wussing's arguments at their best are only indications of new ideas in geometry which later led to Klein's *Erlangen program*, and earlier to the study of invariants, drawing its stimuli also from algebraic geometry.

[28] First Cauchy published an extensive treatise [60] in the 3rd volume of the *Exercices d'analyse et de physique mathématique* in 1844, and in the autumn of 1845 he decided to publish the main results again in the *Comptes Rendus*; however, he soon became dissatisfied with repeating his earlier results and he developed new methods and results. In February of 1846 he suddenly left this sphere. The unsolved problem is whether the preparation of the edition of Galois' studies influenced Cauchy's feverish activity in one way or another.

[29] Most of the new theorems presented or proved by Cauchy have been listed with great care in paper [53]. pp. 144—147, to which the reader is referred.

Sylow's theorem[30] on a special case (groups of substitutions); in this same way it can be proved that he had approached the concept of the normalizer. Another important feature was the introduction of the concepts of imprimitive systems, concepts which are identical with the present concept of transitive and intransitive groups, etc., in the present form. However, from our point of view, the concept, interpretation and historical significance of Cauchy's treatise are important.[31]

As a whole, Cauchy's treatises of the years 1844 — 1846 have similar features to those of his paper of 1815 [59]. This was again an effort for a comprehensive interpretation in which the connection with the problems of algebraic equations remained in between lines, although the objective of the treatise required answers to two questions: 1° quels sont les nombres de valeurs que peut acquérir une fonction de *n* lettres; 2° comment on peut effectivement former des fonctions pour lesquelles les nombres de valeurs distinctes sont les nombres trouvés.[32]

Cauchy's interpretations are based on the study of substitutions. The main tool for their study was the cyclic substitution for which he introduced the present notation. In order to distinguish among the various kinds of substitutions he introduced a complicated conceptual system in which he defined regular, similar, and other substitutions apart from the concepts of transitive and intransitive, primitive and imprimitive substitutions already mentioned.[33] The concept of the similar substitution was required by Cauchy to introduce conjugated substitutions, as we would say today. Cauchy proved that for two arbitrary similar subsitutions P and R there always exist such Q and S that

$$P = Q^{-1}RQ \qquad (7.2.5)$$

and

$$R = S^{-1}PS \qquad (7.2.6)$$

He also proved the reverse, i.e., that if there exist such Q and S for the substitutions P and R that (7.2.5) and (7.2.6) hold, P and R are similar ([60] 190 — 192). This is also a way out for the stuff of the interchangibility of substitutions; Cauchy then frequently worked with formulae of the $PS = SR$ types, which he called formal linear equations, also mentioning an "unknown" substitution.

---

[30] Cf. Bell [26], p. 240.

[31] We shall not discuss the details or the differences between the interpretation in [60] and the treatise in the *Comptes Rendus*, or Cauchy's failures. For an analysis of these questions and more detailed arguments concerning the statement about the historical significance of Cauchy's scientific work, the reader is referred to [314, 315].

[32] Cauchy [61] p. 278. This is where he also pointed out the fact that hitherto attention had been concentrated on proving that functions, yielding a certain number of values, did not exist. It seems that Cauchy was alluding to—without expressly naming—the change brought about by the work of Abel and especially Galois.

[33] The regular substitution is composed of cycles of equal lengths (cycles without common letters, i.e., disjunctive); similar substitutions are those which are composed of the same number of cycles, the corresponding cycles being of equal lengths (cf. [61] 186).

Cauchy's method of systematic interpretation should be apparent from the indicated data. He constantly spoke of substitutions, only to introduce a concept equivalent to the concept of the group of substitutions later on. He did this in the following manner: Consider one or more substitutions as given, (all have n letters at the most). He called derived substitutions all which are created from the given substitutions by their multiplication or raising to a power. The derived and given substitutions are then called a system of conjugate substitutions (le système des substitutions conjuguées).

One of the important tasks which ensued from the definition presented, was the determination of the "order" of this group of substitutions.[34] As he had already proved earlier, each substitution has a finite order. Consider the group to be formed by the substitutions $P, Q, R, \ldots$; its elements are then words formed by the substitutions. Cauchy was thus led to a study of the interchangibility of these substitutions. Here he presented general formal rules, e.g., of the type

$$P^h Q^k = Q^k P^h,$$

on the one hand, and he referred directly to the properties of the corresponding substitutions which he expressed in terms of products of disjunctive cycles, on the other. In this connection he also frequently used tabular records, since a group formed by substitutions P and Q contains elements expressed in two multiplicative tables, provided P is a substitution of order a, and Q a substitution of order b.

$$
\begin{matrix}
1 & P & P^2 & \ldots & P^{a-1} \\
Q & QP & QP^2 & \ldots & QP^{a-1} \\
Q^2 & Q^2P & \ldots \ldots \ldots & & \\
\vdots & & & & \vdots \\
Q^{b-1} & \ldots \ldots \ldots \ldots & & & Q^{b-1}P^{a-1}
\end{matrix}
\tag{7.2.7}
$$

$$
\begin{matrix}
1 & P & P^2 & \ldots & P^{a-1} \\
Q & PQ & P^2Q & \ldots & P^{a-1}Q \\
Q^2 & PQ^2 & \ldots \ldots \ldots & & \\
\vdots & & & & \vdots \\
Q^{b-1} & \ldots \ldots \ldots \ldots & & & P^{a-1}Q^{b-1}
\end{matrix}
\tag{7.2.8}
$$

This last feature of Cauchy's interpreatation was treated in slightly greater detail because it played an important role in the future development of the theory of groups of substitutions, and in my opinion it contains an indication of Cauchy's effort to

---

[34] I do not think it necessary to avoid the present terminology completely here, provided Cauchy's terms are identical with the present ones.

understand the theory of groups more generally.[35] However, Cauchy went much further; he arrived at the study of commutative groups (substitutions permutables) and at numerous special cases, which he expressed by means of various substitutions or defined by functions which leave the substitutions unchanged. One may also say that Cauchy was not consistent in his interpretation as regards using a uniform conceptual system, but that he presented several different attempts while searching (in vain) for the most convenient method of answering the outstanding questions.

Cauchy's work had considerable influence on the further development of the theory of groups. This was due to the following circumstances. The situation was much more favourable for the acceptance of the study of groups in the middle of the century than at the time Cauchy wrote his first paper. The whole theory of equations had reached a different level. Not only did new proofs of insolvability appear[36] but the publication of Galois' studies and their gradual treatment, like the progress in algebraic research due to Gauss, Abel, and Kronecker, were responsible[37] for the continuation of Cauchy's work on a much wider scale. Cauchy's considerable authority certainly contributed to this in the middle of the century; his work was stimulating in many ways. Finally, a substantial influence was exerted by the fact that Cauchy presented his ideas clearly, and, in many places, lucidly indicated many other problems. Whatever the outcome, the fact remained that an increasing number of mathematicians studied groups of substitutions in the years following the interval 1844 – 1846, when Cauchy's papers were published. Bertrand [29], who expanded upon Cauchy's older paper [58], published his study at this time. In the forties, papers on the groups of substitutions were being published by Serret, who also introduced a section on substitutions into his university course. This was soon followed by Cayley, Betti, Kronecker and others. Since the time Cauchy's studies were published (i.e., since the time Galois' studies were also published), during an interval of not quite twenty-five years, but prior to the publication of Jordan's extensive *Traité des Substitutions* (1870), more than ten mathematicians began to study the groups of substitutions and published more than 700 pages of specialized text on these problems.[38] Most of these authors referred to Cauchy's work, including Cayley, Kirkman and Sylvester.[39] These

---

[35] For example Taton ([204] Tome III, Volume 1 p. 13) maintained that this tendency existed, whereas Wussing disputed its existence without further analysis ([410]).

[36] For example, Hermite in 1842 [199]; as far as he is concerned, cf. E. Picard's note ([198], T. 1, p. VIII).

[37] I am of the opinion that one of the jutsified efforts in Wussing's work [410], is expressing the points in which mathematical thinking had changed in so far as it suggested the subsequent origination of the theory of groups. It is certain that the theory of equations was only one part of these changes, although it was decisive.

[38] These data only contain the studies concerned with the groups of substitutions, which is a quite narrow problem. The studies which only exploited the results of this sphere, or which developed the ideas of the theory of groups in another form, have been omitted. As indicated by Wussing's book [410], the quantitative data corresponding to this period were considerably more numerous.

[39] Cf. [315] p. 157.

data then, indicate that Cauchy's work was located at the boundary between two different stages in the development of the theory of groups. Whereas earlier one could only have found single papers and results on these problems, which are interpreted in direct connection with problems required for solving equations, excepting only Cauchy's paper of 1815, Cauchy's papers of 1844 – 1846 established systematic study in the sphere of the theory of groups. However, the systematic nature of this study was not only manifest in its external displays, which were determined by the fact that these problems were being treated by a number of authors whose studies were related and published one after the other. There are also other features. These were formed under the influence of Cauchy's attempts to find a comprehensive inter-pretation of the problems of the theory of groups, which was developing independently and putting forward in itself new questions requiring further generalization. In treatises which connect with one another, the initial concepts and terminology were refined. The theory of groups, formed in this manner, became capable of penetrating other branches of mathematics. Provided one is allowed to speak of the theory of groups, studied systematically with the help of finite groups of substitutions, beginning with Cauchy's studies, this stage has its climax in Jordan's book [219], mentioned above, which presented the contemporary results in a summarized and comprehensive form, like any significant work of science, and indicated the way toward progress.[40]

In the course of the twenty-five years, representing the span between Cauchy's papers and Jordan's *Traité*, the sphere of problems investigated was gradually extended and the study of some of them became more thorough. Attention remained concentrated on the study of groups of a "low" degree, various subgroups of these groups being constantly looked for (as well as the subgroups of these subgroups), and the main tool remaining the recording of substitutions in disjunctive cycles. But there was also an effort to express groups and their subgroups by means of suitable functions of letters. The main feature[41] of this period was the filling in of the space opened up by Cauchy's studies, more involved results having been reached only due to the work of later authors (Sylow, Jordan, v. Dyck, Frobenius). Nevertheless, during this period the relation of the theory of groups of substitutions to the other parts of mathematics changed. Whereas it had nearly exclusively stemmed from the theory of algebraic equations up to this point, connections were now being sought between the theory studied and some spheres of mathematics such as the theory of numbers, but especially with the study of invariants and geometrical transformations.[42]

---

[40] As regards the significance and the heritage of Jordan's other studies on the theory of groups, cf. Dieudonné [119].

[41] The results of the authors from the years 1846—1870 have not been comprehensively described and evaluated yet. The analysis of their studies, to prove the statements made in the text has to be omitted here for lack of space.

[42] Wussing [410] described these relations for the 2nd half of the 19th century well. However, when read carefully it will be found that the connections mentioned in the text were applicable to a larger extent towards the end of the sixties and especially later.

These connections came to the fore very clearly around the year 1870. With regard to the theory of groups, Kronecker's paper [255], in which literature [43] essentially found the postulates defining a group in a general form, although Kronecker did not point out the connection with the theory of groups, should be mentioned. Likewise, Dedekind's comments on Dirichlet's theory of numbers, in which ([122] 160) he explicitly spoke of groups in connection with combining classes of quadratic forms, and moreover exploited the theory of groups for his further study of the theory of forms, deserve to be recalled. For example, in connection with geometrical problems Jordan studied groups of motion, continuous groups were mentioned, and around 1870 ideas appeared which were the result of the theory of groups. These determined the trends of the work of Klein and Lie and mutually complement each other.

These facts were mentioned in order to indicate how the theory of groups was becoming a general component and tool of the unification of mathematics in 1870. By this it acquired features which by their nature exceed the scope of the development of algebra which we are investigating. For the present purposes it remains to ask whether these generalizing features, when the algebraic structure was escaping its special form given by its origin, and beginning to acquire an abstract form, had appeared earlier. For this reason it is necessary to stop and consider the studies of Arthur Cayley.

This English mathematician stems from the special tradition of English mathematics. He undoubtedly knew the opinions of English algebraists of the thirties and forties, i.e., of Peacock, Morgan, Gregory and others (cf. Chap. 6, 2), well. He also studied continental authors in detail, including Gauss, Cauchy, Galois and Abel. As opposed to his English predecessors, in his studies he tried to penetrate into the pages of the leading continental journals. Although his publications included practically all branches of mathematics, perhaps in all can one observe his expressly algebraic way of thinking, trying to work his way to the most abstract expression of common features connecting various theories.[44] From the forties his main interest was the theory of invariants and the algebraic geometry of the time; in connection with these he developed the theories of determinants and matrices and in geometry pronounced a number of daring ideas, which included the study of n-dimensional geometry and the profound relations between projective and metric geometries, which led him to anticipate Klein's classification of geometry. His wide erudition was also displayed at the moment he approached the interpretation of the theory of groups, under the influence of Cauchy and with the knowledge of Galois' work. Following several preparatory

---

[43] Cf. Wussing [410].

[44] I am of the opinion that the general ideas on the nature of algebra, as expressed by Cayley's predecessors in England were markedly effective here. However, in contrast to them, Cayley embodied these ideas in the treatment of various mathematical problems. As regards this and Cayley's theory, cf. [316].

studies,[45] he published his interpretation of the group theory in 1854 in the treatise "*On the Theory of Groups, as Depending on the Symbolic Equation $\vartheta^n = 1$*" [96], which was continued in [97, 100, 101].

The main credit, widely acknowledged in literature[46], deriving from the first of Cayley's papers was the presentation of the definition of the abstract finite group[47] "A set of symbols 1, α, β, ... all of them different, and such that the product of any two of them (no matter in what order), or the product of one of them into itself, belongs to the set, is said to be a group" ([96] 124). He also required that the operation be associative and maintained that all elements of the group are obtained, albeit in a different order, if all elements of the group are multiplied by any one of them (he used circumlocution to describe left and right multiplication).

He used two means to define the group itself. They are the defining relation and the multiplication tables. For example, he gave the following defining relation for a group of order 6:

$$\alpha^3 = 1, \qquad \gamma^2 = 1, \qquad \alpha\gamma = \gamma\alpha \tag{7.2.9}$$

It was found that instead of the latter one may insert

$$\alpha\gamma = \gamma\alpha^2, \tag{7.2.10}$$

but then one gets a different group. He then illustrated these examples by means of the multiplication table, and indicated that both could be applied to the study of groups of higher orders.

If one compares this procedure of Cayley to that of Cauchy, it becomes clear that Cayley could have found a lot to stimulate him in Cauchy's work. Essentially Cauchy also used defining relations (cf. [96] 130) where he required for a group of order 6 that

$$U^3 = 1, \qquad V^2 = 1, \tag{7.2.11}$$

at the same time $UV = VU$. He also arrived at a table of the following type:

$$\begin{array}{ccc} 1 & U & U^2 \\ V & VU & VU^2 \end{array} \tag{7.2.12}$$

But there was a substantial difference in Cauchy's and Cayley's approach, apart from some obscurites in formulation. Cauchy considered the symbols U and V to be substitutions,[48] but Cayley only considered them to be symbols satisfying certain

---

[45] Apart from treatise [95] Cayley's immediate response to Hamilton's quaternions [91] and the publishing of Cayley's octaves [92] must also be taken into account. Cayley spoke directly of the theory of groups and with full knowledge he attributed the term "groupe" to Galois.

[46] Cf. Bourbaki [46] p. 76; Miller [286], p. 226, etc.

[47] For the discussion of this definition see [293].

[48] $U = (xzv)(yuw)$; $V = (xu)(yv)(zw)$. If U and V should represent other substitutions, and although the relations given in the text will still hold, Cauchy would have been considering a different group, and Cayley the same. As regards this, and the relation between Cauchy's and Cayley's theories of groups, cf. [317].

relations. Cauchy described groups of substitutions, Cayley the general structure.[49] This is connected with the new interpretation and the new tasks which Cayley imposed upon the theory of groups.

Cayley considered the defining relations or the multiplication tables as defining an abstract structure between symbols, for which one must first find an interpretation in the spirit of the ideas of the English school of symbolical algebra. Any interpretation which gives the symbols the meaning that the fundamental, defining relations hold is justified. In connection with this purposeful application of isomorphism, Cayley spoke of different interpretations; apart from the groups of substitutions he spoke of quaternions, of matrices, of transformations, of quadratic forms, and of combining quadratic forms. He was slightly obscure in mentioning groups represented by the theory of elliptical functions ([96] 126, 127, 129 – 130; [100] 89). These interpretations were not the subject of Cayley's study of the theory of groups. On the other hand, Cayley named the determination of various groups of an arbitrary given order as one of the important objectives of the theory of groups; he himself then went on to enumerate completely the groups of order 8. In this way he opened up the important topic of the further development of the theory of groups, the abstract topic, which did not deal with the interpretations of the groups found.

To conlude this brief list of the main features of the initial stage of development of the theory of groups, one should perhaps stress some of its more general features. First of all, the study of the theory of groups evolved mainly due to special data provided by the groups of substitutions. Since substitutions themselves have an exclusively unquantitative character, a theory was created by their study which differed substantially from the predominating nature of mathematics. The operation introduced for substitution was introduced independently of traditional operations and numerical fields. It is true that this is where the theory of groups may connect with combination calculus, and the multiplication of substitution was after all an established operation.[50] These features of the theory of groups of substitutions gradually found their place as the whole theory evolved. Several stages can be distinguished in the course of its development. From 1771 until the forties was the preparatory period. Only a few papers were published, which considered the theory of substitutions only in close contact with the requirement of the theory of equations. The concepts were frequently obscurely or fragmentarily defined, and it can be seen that they were only just being formed by the theory. In 1844 – 1846 Cauchy in fact returned to his earlier interpretation and his extensive studies initiatied the period of systematic research into the properties of finite groups which, however, was carried out on examples of the groups of substitutions. At this time the theory of groups was being

---

[49] That is why Cayley sometimes [101] spoke, quite in the spirit of the English school, only of the definition of a group "by means of the laws of combination of its symbols".

[50] This feature is being stressed because it disrupted traditional thinking considerably. In combining the classes of quadratic forms, and with operations in the theory of matrices alike, one encounters a similar introduction of operation.

presented as a relatively independent branch which was capable of at least co-determining the trends of its own research by means of the requirements of its subject. It was also found that an abstract procedure could be created within the theory of groups as regards the defining of finite groups, as well as establishing the methods and determining the objectives of this theory (Cayley 1854). Indications also appeared of exploiting this abstract structure for the study of other branches of mathematics which were clearly growing by the end of the sixties. In its abstract character the theory of groups got ahead of the study of other algebraic structures. Probably thanks to the fact that it had had an expressly untraditional character from the very beginning, and that its development had been stimulated by such a strong impulse as the study of the solution of algebraic equations, one could observe elements of purposeful abstract construction in the theory of groups quite long before 1870. In this way the theory of groups acquired a nature exceeding the scope of the stage of modern algebra which is being studied here, and presaged the further evolution of algebra.

7.3 Serret's and Salmon's textbooks of algebra

Long into the second half of the 19th century new editions of algebra textbooks from the turn of the century were being published only slightly altered. New books also appeared which did not differ from the latter in standard and contents. However, since the middle of the century some authors were trying to summarize the new results of some of the algebraic disciplines achieved. This gave rise to mono- graphs, or to textbooks with a different topical range. Among these, special mention is due to Serret's *Cours d'algèbre supérieur* [352], with which only *Modern Higher Algebra* [347] by G. Salmon can be compared as regards import. These books gradually took the place of the other textbooks in more involved university studies and became a compulsory part of the preparation of young algebraists at most of the universities in Europe. The dominating role of Serret's book was upset only towards the end of the century when it was replaced by Weber's *Lehrbuch der Algebra* [396].

Both books, Serret's and Salmon's, which, as we shall see, complement each other, were published rapidly in several subsequent editions and translation. Serret's *Cours* originated as a result of lectures given at the Sorbonne in 1848. After the first edition in the following year it was again published in 1854, slightly altered but mainly extended by Serret's new supplements. By the time of the 5th edition in 1885 it had reached the size of two volumes and nearly triple the original extent, two German editions having been published meanwhile. Salmon's book had a similar history; after the first edition in 1859 a second followed in 1866 and a third English edition in 1876, the French edition, extended by Hermite's comments, having been published in 1868. The translator, Wilhelm Fiedler, deserves credit for propagating Salmon's textbook (which also stemmed from several repeated lectures) in the world of the German mathematicians, since he published it in 1863.

As with all synthetic treatises, it is not important whether these books brought a new result at one time or another, or whether they forgot to mention a result achieved. Both the authors said that their books did not encompass the whole of algebra. However, it is important which sections of contemporary algebra the authors included in their books and what approach they chose for presenting them.

Both books differ in topic. Serret essentially dealt with the theory of equations, in other words, he explained how far the effort for an algebraic solution to equations had come. On the other hand, Salmon summarized the results of the trend in algebra represented by Cayley and Sylvester, to whom the book was inscribed; it encompasses the theory of determinants and invariants in particular.

In the introduction to his book Serret defined the concept of algebra as the analysis of equations[1] and listed the authors which he had expanded upon. In the first instance they were Lagrange and Gauss. Gauss'·ideas were extended by Abel, who determined the classes of algebraically solvable equations. However, as regards the equations which were solvable algebraically Serret said "cette question difficile a été résolues complétement... par Évariste Galois" ([352] 4) and he quoted Liouville's edition.[2] He also mentioned that the impossibility of a general solution of an equation of a higher order than the fourth was expressed by Ruffini, and became certain ·in the newer studies of Abel. The names Serret mentioned in the introduction to his book, describe sufficiently the trend his interpretation had. One can see that Serret was aware of the objectives towards which research was progressing; in the text he complemented the said authors by quoting many papers published in the latter decades.

The actual contents of the book, expressed briefly and slightly schematically, are roughly this: in the first lessons he discussed symmetric functions, and beginning with the third he concentrated on eliminations: here he mentioned the older methods of Bézout, as well as the newer ones (Abel, Liouville, etc.). Only incidentally he expressed a number of theorems here on polynomials, on fractional rational functions, on the development of the ratio of polynomials into a series. Not until the eleventh lesson (pp. 128 − 146) did he begin to deal with substitutions and functions of n letters, also repeating (apart from the results of Lagrange) some ideas of Cauchy, whom he quoted. In the next lesson he followed Galois' interpretation of the functions of roots and proved among other things the following theorem (152): Étant données tant d'irrationnelles algébriques qu'on voudra, on peut toujours les exprimer toutes en fonction rationnelle d'une même irrationnelle. However, he interrupted the interpretation of this topic and on the following pages he discussed the roots of binomial equations. Beginning with the fifteenth lesson he followed the pattern of Lagrange's work of 1770 and analysed the methods of solving cubic equations, treating biquadratic equations in

---

[1] L'Algèbre est, à proprement parler, l'Analyse des équations; les diverses théories partielles qu'elle comprend se rattachent toutes, plus ou moins, à cet objet principal ([352a] 1). This definition was preserved in the remaining editions as well.

[2] He characterized Galois with the words: "...ancien élève de l'École Normale, et l'un des géomètres les plus profonds que la France ait produit."

a similar way in the seventeenth lesson. He summarized these results (like Lagrange in his fourth section) in a general instruction on the solution of equations in the 18th lesson (pp. 229—247). He did not add anything to what Lagrange had already written. He saw the main problem in finding the resolvent of the least possible degree.[3] Therefore, in the next two lessons he went back to functions of n letters, where he recalled Ruffini's and Abatti's efforts and theorems, and Cauchy's proof, which he modernized slightly by means which were used later,[4] and the work of Abel, i.e., that part of the proof of insolvability of the general equation of the fifth (or higher) degree which belonged to the theory of substitutions. By then he had prepared everything for lessons 21 and 22, which rounded of this proof. In the former he drew on Abel's study of 1826 and defined (nearly in Abel's words) the concept of an "algebraic function",[5] and following Abel's example he distinguished among these functions according to order and degree. He also arrived at "the general form of algebraic functions" here. In the latter lesson mentioned he quoted Wantzel's proof of insolvability. This first climax of the book is followed by three lessons (297—343) which were devoted to the theorems of the theory of numbers,[6] which mostly represent preparatory means for the interpretation of the results stemming from Gauss, and which, apart from Abel, Kronecker was working on at the time. Of course, the conluding lessons of the book were also under the influence of Galois' interpretations,[7] but the interpretion of that which forms the nucleus of Galois' studies and which one would call Galois' theory, is lacking.

On the whole fifteen extensive "Notes", differeing considerably in topic, were added

---

[3] Toutes les méthodes connues que les géomètres ont essayé d'appliquer à la résolution algébrique des équations, et il en serait nécessairement de même des nouvelles qu'on pourrait imaginer, reviennent à faire dépendre la résolution de l'équation plus facile à résoudre, et dont les racines sont des fonctions de cleles de la proposée ([352a], 229).

[4] He also recalled Bertrand's studies [29] of 1845.

[5] Soient $x_1, x_2, x_3, \ldots, x_k$, k quantités quelconques indépendantes, et γ une fonction de ces quantités; γ sera une fonction algégrique, si l'on peut l'exprimer en x_1, x_2, x_3, etc., à l'aide des opérations suivantes, effectuées un nombre fini de fois; $1° 1'$addition ou la soustraction; $2°$ la multiplication; $3°$ la division; $4°$ l'extraction des racines d'indices premiers. Serret [352a], p. 276—277.

[6] Here he introduced the concept of congruence (he spoke of "des nombres congrus ou équivalents"), operations with congruences. He proved the minor Fermat theorem in order to be able to devote himself to the solution of binomial equations with respect to a prime number modulus. The twenty fifth lesson discussed the problem of the theory of numbers which had no connection with the algebraic contents of the book. That is probably why he replaced it in the second edition with a different topic; he explained the principal ideas of Galois' study *Sur la théorie des nombres* here.

[7] The note Serret attached to the beginning of his text ([352a], 344) is interesting: Galois a donné la condition nécessaire et suffisante pour qu'une équation irréductible de degré premier soit résoluble par radicaux.... Mon ami M. Liouville m'a annoncé l'intention où il était de publier un jour des développements relatifs à ce remarquable travail. Ce n'est que par ces développements, dont M. Liouville a bien voulu me communiquer une partie, que je suis parvenu à comprendre certains points du Mémoire de Galois, dont la lecture ne peut être abordée que par les géomètres qui se sont occupés d'une manière toute spéciale de la théorie des équations. On voit par quelle réserve je suis empêché de présenter ici la découverte de Galois.

to the second edition of the book. Some of them complement various points in the text, others are independent of it. However, all of them—albeit in an altered form—repeat that which had already been published elsewhere, whether Lagrange's, Waring's, or Serret's newer studies of the theory of substitutions, or Hesse's study of inflexion points of curves of the third degree were concerned. The word for word translation of Kronecker's study of 1853, expanding on Abel and Galois, and accompanied by Hermite's supplement (pp. 560 – 575), was very important.

In the interpretation of the contents of Serret's book some features were stressed, which we will now try to summarize. Serret wrote a book of a high standard, which was very modern for its time. He rejected nearly all of what we called the traditional approach and traditional interpretation of algebra, and he only mentioned what had been created in the theory of equations[8] since Lagrange's paper of 1770. His summary, however, differs from Lacroix's *Supplement*, since Serret did not burden the new results with the old concepts and contents. He tried to arrange the new results in a new way which is manifest especially in the division of the contents, so that the book has a clear-cut logical backbone which tends towards Galois' theory. However, Serret attached to this backbone seemingly raw tissue, taken from and prepared elsewhere: it is formed by the work of authors from whose studies he drew. It is true that he mostly interpreted them in his own words, but he did not reform the whole into a uniform logical unit. That is why he is terminologically and conceptually inconsistent. This can best be seen in his interpretation of numerical fields. Most of the book is based on concepts which he only introduced in connection with reproducing Abel's results. The greatest weakness was displayed in time by the insufficient exploitation of Galois' work which, as Serret's own words indicate, was perhaps connected with the dificulties in understanding it. Thus, although all them components which algebraists beginning with Lagrange had contributed to reconstructing algebra are included, nevertheless a new algebra was not created. The accumulation of new elements actually provided an impulse for a new and comprehensive interpretation of algebra. A key was given implicitly, which led to the reconsideration of the whole and to mastering new concepts which were indicated here in various forms and which, after all, can only be understood fully in the theory of equations by dealing with the Galois theory. This was not carried out here; the first to attempt this was Weber at the end of the century, but not even he was fully successful because his interpretation was not clear and abstract enough. All this, including the new logical finish, was to become the work of the next stage in the development of modern algebra.

Now let us go back to Salmon's book. We have said that it summarized the results of a different group of topics, a group which was close to the algebraic interests of Sylvester and Cayley. Serret and Salmon alike, summarized all important results of the English school, but Salmon also drew on the studies of Hermite, Clebsch,

---

[8] One need not even mention that Serret did not discuss the successes of the numerical solution of equations, or of the other parts of algebra.

Aronhold, and others. The topic was really modern, but the treatment fully bore the traces of the current state of the treated disciplines. We shall only mention two features. In the interpretation of determinants Salmon essentially fluctuated between two ways of understanding them. In some cases he attributed to them the meaning of a square or even rectangular arrangement of elements, i.e., the meaning of a matrix, and elsewhere only the meaning of the number attributed to this arrangement. That is why he defined operations, as we sould say, with matrices and their modification (all in the first few lessons), in order to put the sign of equality in the numerical case (2nd edition, p. 12),

$$\begin{vmatrix} 4 & -1 & -1 \\ -7 & -2 & +2 \\ 0 & 0 & 1 \end{vmatrix} = \begin{vmatrix} 4 & -1 \\ -7 & -2 \end{vmatrix}$$

and in this way to decrease considerably the "order" of the original determinant in the course of the subsequent calculation. But determinants in this concept are mere numbers.

Most of the book is devoted to the theory of forms and of their invariants and covariants. Although he explained the symbolics introduced by Aronhold here, he essentially kept to the symbolics and formulations introduced by Cayley. In looking for the invariants, the "calculation" aspect prevailed with Salmon, where the results were verified (and also sought) by means of extensive calculations.

Although Salmon summarized and classified the results, it did not lead him to a more comprehensive and general interpretation. It seems, therefore, that the fact that his textbook became widely read is more due to the new topical results with which the mathematicians in various branches of the science wanted to acquaint themselves. Thus, Serret's *Cours* was complemented by a book which presented an interpretation of other algebraical problems. In spite of its novelty and modernity, it only substantiated the different theoretical standard of treatment of those problems for which a more general approach and corresponding terminology had not yet been applied. On the contrary, although both books have "Higher Algebra" in their titles, there is no visible connection between them. The authors, in spite of sometimes drawing on the work of the same authors. talk as if in two different languages.

CHAPTER 8

CONCLUSION

We have investigated the content and meaning of changes which took place over an interval of approximately one hundred years. This was based on the hypothetical idea of the origin and existence of modern algebra. Clearly the use of the term "modern" without more specific annotation is too vague and uncertain, especially in historical research. It is the historical point of view which demonstrates the relativity and transience with which one may attribute the term "moderness" to any body of scientific knowledge; during the life of the generation which created it, modern knowledge is usually replaced by an even more modern knowledge.

In spite of the constantly changing state of science, which is due to the constant and infinite development of scientific knowledge, historical research is trying to discover the principal and fundamental stages of the development of science. The irremovable disadvantage of this endeavour is the fact, that not even the most serious research may dispense with the contemporary state of science, which necessarily tends to appear as the climax of the development in spite of its transiency, in defining those stages which delimit the periods in development.

All these difficulties are reflected in the study of the history of algebra which is one of the oldest spheres of interest in mathematics. However, the purpose here was not to determine the character of the main stages of the development of algebra. When the term "modern algebra" was used, it was understood to be the algebra which was being formed in the thirties of this century in accord with the influence of van der Waerden's book *Moderne Algebra*, which summarizes the previous tendences and results and which had the meaning given to it later by the work of Bourbaki. Rather, we have based our approach on the fact that one of the fundamental changes in the concept of the subject and methods of algebra took place between the second half of the 18th century and the time when algebra began to be understood consciously and purposefully to be the study of algebraic structures (as represented, e.g., by the work of Bourbaki). We tried to indicate that this change, which could be called a revolutionary reversal and which is a part of the parallel changes taking place in the overall approach to mathematics, occurred in two phases, of which only the first was considered in detail.

In this first phase the algebraic ideas of the earlier system were being disrupted and, new terms and elements of the new approach and new methods of algebra were being formed. It was not until the second phase that the new possibilities were elaborated, and, by building up a conceptually and logically comprehensive whole, that the individual results and theories were classified and assigned their appropriate place.

As we tried to point out, the meaning of the first phase became clear as the individual problems, some of which had already been formulated for a relatively long time, were solved. Fundamentally there were two kinds of problems: The first concerned the theory of equations itself, interest being concentrated on solving an algebraic equation of the 5th or higher degree. The second was concerned with conceptual foundations from which the theory of equations stemmed, which included the concept and definition of realms of numbers, or, more historically, of various types of numbers. A solution to both kinds of problems could not be given considering the ideas and manner of algebraic thinking prevalent in the 17th and 18th centuries. It is true that ideas had appeared in the course of these two centuries which were beyond the usual traditional frame. There was, for example, Wallis' idea of the geometrical interpretation of complex numbers, or some of Leibniz's attempts to understand – to use modern terms – the relations between the various fields of irrational numbers. However, these and similar "untraditional" ideas not only remained isolated, but the subsequent development did not tie up with them, finally leading up to them in a different way. Such "anomalies" are being born in scientific thinking constantly without disputing the prevalent concepts.

Although a historical process is never quite discontinuous, a more detailed analysis of the data might indicate the way in which sympathy for the necessity of an "untraditional" approach grew in the course of the 18th century, and the way the work of Lagrange, Vandermonde and Waring introduced new elements into algebraic deliberations. The question of their expansion and complementation then represents the key to further changes in algebra.

In their decisive papers the authors mentioned proceeded from an analysis of the task of providing an algebraic solution to equations of a degree even higher than the fourth. Their most important result was not scepticism about the solvability of equations of an higher degree than the fourth, but the indication that progress can be made in the problems of the theory of equations only, if other ideas and approaches are adopted than those used hitherto. Nevertheless, Lagrange's example is indicative of the fact that the anticipation of the nucleus of the problems remained very obscure. As a simile one might perhaps say that Lagrange and others opened up the way forward, but they were incapable of clearly determining its direction. It is characteristic that most mathematicians were still being educated in the misunderstanding of the possibilities the papers of the said mathematicians provided in the 1st half of the 19th century.

If one is able to distinguish various levels in the understanding of the work of Lagrange, Waring and Vandermonde (their papers are considered to be either an attempt at solving the well-known problems of the solution of equations, or to find the essence of the problem of the solvability of equations in which new ideas are also indicated), the difference in the levels at which Gauss' work is understood is even more marked.

Some of the solutions of partial problems were presented by Gauss in a convincing and comprehensive way, so that they were accepted and adopted by his contemporaries; this goes, e.g., for Gauss' solution of binomial equations which was explained, together with its consequences, in the 7th Section of his *Disquisitiones arithmeticae* (1801), although Vandermonde's results in this region had not been understood then. The second level contained in Gauss' papers and disclosed gradually during his lifetime by a larger number of mathematicians, was a wide range of special algebraic problems for which Gauss had also indicated possible solutions. However, besides these two levels on which the sense of Gauss' work should be understood, there existed yet a third, undisclosed: this was the endeavour, only discovered later, to study abstract algebraic structures stressing the isomorphism of their different representations. Gauss' work can be used as an example to show (and this also holds more or less for the works of other mathematicians) the delay with which the various intellectual levels of a work are understood and developed.

The thirties, and expecially the forties, play a very interesting role in the development of algebra. Allthough the overall concept and trend of algebra had not changed then, the topics studied and the intensity (absolutely and relatively) with which the individual fields were studied had gradually begun to change. The old unity of the arithmetico-algebraic sphere, which helped the mutual influence of the theory of numbers and of algebra, was to an advantage here. This could be observed markedly in Gauss' work and in that of his continuators. The study of the n-th roots of unity and of the theory of forms is an example of this. The problems of the theory of numbers, also studied in other fields than that of rational numbers, to use modern terms, considerably affected the elucidation of the different realms of numbers; the arithmetic of these realms with its paradoxes thus became one of the moments of the conceptual change of algebra. Several features were characteristic of this: Gauss still tried to formulate the arithmetic of algebraic numbers, which he used, e.g., in studying biquadratic residues, as much in accordance with the arithmetic of rational numbers as possible. In Dirichlet's work some concepts (e.g., the concept of unity) acquired a more general form; the transposition of the problems of divisibility, of Euclid's algorithm, or of the resolution into prime factors into other realms than that of rational numbers, disclosed a plentiful field of problems which provided abundant data for grasping the differences between the realms of numbers on the one hand, and, as further development showed, led to new general fields which made it possible to express in what way various numerical sets agreed and differed, on the other hand. This trend was stressed at the end of the period investigated in Dedekind's work, but the intellectual basis for it had already been provided in the forties.

As regards the other problems, data was also being accumulated in the course of the forties, and this made it possible to create a new conceptual system in algebra later on. New and gradually more general concepts were being worked out and points of view changed more or less by studying the special problems. The number of mathematicians who considered the problem of the solvability of equations of a higher

degree than the fourth solved, increased. This terminated one of the problems of traditional development. The problem as to which of the types of algebraic equations were
in fact solvable became the center of interest. The beginnings of the study of Galois'
results, the understanding of Abel's work, and the first papers of Kronecker also
belong to the forties.

However, the procedure in the various sets of topics was very different. For example,
the new concept of a group was not only adopted very quickly, but soon also acquired
a clear-cut abstract form. It is possible that a considerable role was played here by the
influence of Galois' personality and by the requirements resulting from the elaboration
of his theory. However, the fact that the concept had already been created outside the
tradition, which with most problems had a delaying effect due to its inertia, and that
the study of the properties of the group could thus develop on its own to a large extent,
which was proved by Cauchy's papers of the middle of the forties anyway, had a larger
effect. However, the abundance of comparative data had the largest effect on the rapid
generalization of the concept of a group, since in the various fields of mathematics
sets were being studied in which the same relations held as in groups of substitutions.
A careful reader of Gauss' *Disquisitiones arithmeticae* would substantiate this unhesita-
tingly today. But these sets were not only in the spheres of arithmetic and algebra.
Therefore Cayley (1854) in connection with defining an abstract finite group, could also
name several of their representations in various fields of mathematics. However, the
subsequent development showed that this abstract understanding of a group was possible, but that unification at this level was not necessary for the contemporary state
of mathematics thus Cayley's definition remained unnoticed.

The concept of a field was not subject to such rapid generalization or such accurate
definition. Several features could have played a part in establishing it, in particular
those in which it differed from the concept of a group. Besides, for a long time it was
possible to make do with the illustrative and traditionally used concept of a realm of
numbers; the differentiation between fields and even between adjunctions was very
illustrative, as was the case, e.g., in adjoining to a field of rational numbers of n-th
roots of unity when the simple case in adjunction $i = \sqrt{-1}$ had generally been accepted.
Neither had it a representation in fields of mathematics other than arithmetic and
algebra. However, as soon as this concept had been clearly defined, it was immediately
accepted. Moreover, Kronecker maintained with conviction that he had already used
it in the forties.

However, let us go back to the forties. Not only can one observe an intense development of new algebraic disciplines, inclusive of the theory of invariants, the theory
of substitutions, etc., which was an expression of the change in topic, but other quali-
tative changes also took place. Whereas earlier the properties of realms which were not
numerical in the traditional sense of the word were only studied rarely, this now became
current practice. The theories mentioned, to which one may add the theory of matrix
created a little later, are an example of this. The former numerical illustrativeness also
faded for other reasons. "The creation" of artificial systems, the best known of which

were the quaternions in the forties, not only added to the heterogeneity of the properties of the system (e.g. non-commutativity), but also showed the potential abundance of new trends in algebra. Simultaneously, but at a different level than before, the problem of proof and truth cropped up again: the requirement for the axiomatic principle was established in algebra as well.

These changes which were becoming felt under the influence of various stimuli from the forties onward, were not inherent in algebra alone. However, there is a question, which deserves special investigation, whether the other branches of mathematics also displayed an increase in the intensity of work in the forties, and whether this increase was accompanied by the development of new branches and changes in the concepts of disciplines and reasoning.

Without wanting to establish a separate time scale within the period being investigated, one must acknowledge the fact that the period beginning with the forties did differ in some features from the earlier period. The former period differs from the latter in the new discoveries made, in their acknowledgement, in the elaboration of new topics with their own terminology, in the study of untraditional ("non-numerical") systems with their special requirement for giving reasons for the truthfulness of results, and in the increased intensity of research. Whereas before one could observe these features with individual exceptional mathematicians, they were now the property of most who were taking a creative part in the development of agebra. But the overall concept of algebra and the fundamental terms and ideas were not changed. Even Serret's textbook presents algebra as a theory of solving equations although mathematicians then had reason for disagreeing with this concept. In accordance with this, Salmon's book then only presented topical complements. As regards the methods of the logical structure of algebra, the English School, besides the criticism of the contemporary state, only brought daring ideas which could be exploited in forming partial "artifical" systems, but which could not evoke a critical reconstruction.

To summarize, one might say that new data on algebraic results and ideas multiplied rapidly from the forties on, and step by step, drew algebra out of the captivity of traditional problems. Investigations in the individual branches of algebra were independent, frequently in contradiction to the system of ideas and concepts of the algebra of the 18th century.

The new approach, based on contemporary results and accompanied by a point of view which rejected the study of "numbers" as an algebraic problem, but requiring, perhaps very vaguely at the time, the investigation of the properties of sets with certain algebraic structures, became outstanding in the seventies. At the time, the foundations of a new terminology had also been laid for the following generations of mathematicians; the elaboration of Galois' theory inclusive of the topic it brought to life, became important.

This initiated the second phase of the origin of modern algebra. Its beginning cannot be determined uniquely. However, the beginning of the seventies is plausible enough, and not only for internal changes, which were taking place in algebra at the time and

which affected the thinking of the algebraists who were then beginning to assert themselves. The sixties and the seventies were also connected with certain conceptual
changes in the other branches of mathematics. As regards calculus the revision of its
foundations was being concluded, and further development was unthinkable without
mathematicians drawing on concepts and methods in the elaboration of which Riemann
and Weierstrass took part, and which differed from the previous in accuracy and content.
In geometry the changes in understanding were expressed especially in Klein's
Erlangen Programme, which classified and also unified the various spheres of geometry
from a more general point of view; if the non-Euclidean geometries had hitherto been
considered a kind of heterogeneous subject in geometrical research, they were now
being accepted as mathematically equal with Euclidean geometry. But Klein indicated
even more: he pointed out the possibilities of uniting the geometrical and algebraic
spheres. This union was being pushed through earlier in the work of some mathematicians (Cayley, Sylvester) in some of the disciplines (e.g., the theory of invariants), but
now it received an intellectual basis as well as the necessary program.

The subsequent development of algebra was then already based on the new
conceptual and topical basis, but the full awarness of this basis of the program and
contents of algebra was only realized within the subsequent time interval. Its main
stimuli were the discoveries and the development of new theories. However, the overall
development of mathematics, which was reconstructed on a more general basis,
undoubtedly also had its effect, together with the increasing attention devoted to the
study of the foundations of mathematics, its axiomatic method, and mathematical
logics. These factors were responsible for the more profound approach to algebra;
concepts, originally perhaps intuitive, gradually became more accurate and general.
These features of the development can well be seen in the concepts of the principal
algebraic structures and in the interpretations of the most important and widespread
textbooks. If we have said that generations of mathematicians have built their algebraic
work on the basis of new concepts from the seventies (this also includes groups and
fields), the first comprehensive interpretation of algebraic theories was only presented in
the nineties in the extensive textbook of algebra by H. Weber (*Lehrbuch der Algebra*,
1st edition, 1st volume 1895), which replaced the supplemented and altered editions
of Serret's textbook and the interpretations of the partial sections which were created
in the 2nd half of the 19th century. As the author himself said in the preface, its purpose
was to present "eine zusammenfassende Darstellung und Verknüpfung der verschiedenen
theoretischen Betrachtungen und mannigfachen Anwendungen", which the last decades
had brought in algebra. Although the author wanted to connect various theories, the
endeavour to interpret them to an extent filled the book with numerous details (including
numerical and approximative methods which did not belong). The study of algebraic
structures was still mostly hidden in details not sufficiently generalized. Even the fundamental concepts, including the concept of a field and a group, remained illustrative
and defined only, provided the theories which were being interpreted with their help
required it.

The axiomatic and abstract approach, which developed the unifying features of the various algebraic spheres, was more outstanding in monographs. An example of this is the algebraic field theory published by E. Steinitz in 1910, which represents a forecast of further development. However, only van der Waerden's book in 1930 presented a relatively comprehensive image of algebra, containing a classification of algebraic disciplines, their algebraic substance, and internal interconnection based on an abstract approach. Thus, it was not until the thirties that the transformation of algebra, the first phase of which we investigated, was concluded.

REFERENCES

[1] N. H. ABEL, Oeuvres complètes, Nouvelle édition, par MM. L. Sylow et S. Lie, Tome I, Christiania 1881, VIII + 621 pp. Tome II, Christiania 1881, 338 pp.

[2] —, Mémoire sur les équations algébriques, ou l'on démontre l'impossibilité de la résolution de l'équation général du cinquième degré. Christiania 1824; Oeuvres complètes I, pp. 28—33.

[3] —, a) Démonstration de l'impossibilité de la résolution algébrique des équations générales qui passent le quatrième degré. Journal für reine und angewandte Mathematik, Bd 1, Berlin 1826; Oeuvres I, pp. 66—87. b) Appendice — Analyse du mémoire précédent. Bulletin de Férussac, t. 6, p. 347 n., Paris 1826; Oeuvres I, pp. 87—94.

[4] —, Mémoire sur une propriété générale d'une classe très-étendue de fonctions transcendantes. Présenté à l'Académie des sciences à Paris le 30 octobre 1826. Oeuvres I, pp. 145—211.

[5] —, Recherches sur les fonctions elliptiques. Journal für die reine und angewandte Mathematik, Bd 2, 3, Berlin 1827, 1828; Oeuvres I, pp. 263—388.

[6] —, Mémoire sur une classe particulière d'équations résolubles algébriquement. Journal für die reine und angewandte Mathematik, Bd 4, Berlin 1829; Oeuvres I, pp. 478—507.

[7] —, Precis d'une théorie des fonctions elliptiques. Journal für die reine und angewandte Mathematik, Bd 4, Berlin 1829; Oeuvres I, 518—617.

[8] —, Sur la résolution algébrique des équations. Oeuvres II, pp. 217—243.

[9] NIELS HENRIK ABEL. Memorial publié à l'occasion du centenaire de sa naissance. Kristiania, Paris, Londres, Leipzig 1902, 119 + 135 + 61 + 59 pp.

[10] —, Abhandlung über cine besondere Klasse algebraisch auflösbarer Gleichungen. Ostwald's Klassiker, Nr. 111, Leipzig 1900, 50 pp.

[11] R. ARGAND, Réflexions sur la nouvelle théorie des imaginaires, suivies d'une application à la démonstration d'un théorème d'Analyse. Annales de mathématiques pures et appliquées, Tome V, 1814—1815, pp. 197—209.

[12] —, Essai sur une manière de représenter les quantités imaginaires dans les constructions géométriques, a) 1ère édition Paris 1806; b) 2e édition ... suivie d'un appendice, contenant des extraits des Annales de Gergonne, Paris 1877, XIX + 126 pp.

[13] —, Essai sur une manière de représenter les quantités imaginaires dans les constructions geométriques, Annales de mathématiques pures et appliquées, Tome IV, 1813—1814, pp. 133—147.

[14] A. AUBRY, Essai historique sur la théorie des équations, Journal de mathématiques spéciales, 4e série, Tomes III—V, 1894—1897.

[15] CH. BABBAGE, Observations on the Notation employed in the Calculus of Functions. Transactions of the Cambridge Philosophical Society, vol. 1, part 1, Cambridge 1821, pp. 63—76.

[16] —, On the Influence of Signs in Mathematical Reasoning. Transactions of the Cambridge Philosophical Society, Vol. II, Part II, Cambridge 1827, pp. 325—377; Read Dec. 16, 1821.

[17] —, An essay towards the calculus of functions. Part II. Philosophical Transactions of the Royal Society of London, For the Year 1816, London 1816, pp. 179—256; Read March 14, 1816.

[18] —, Observations on the Analogy which subsists between the Calculus of Functions and other

branches of Analysis. Philosophical Transactions of the Royal Society of London. For the Year 1817, London 1817, pp. 197—216; Read April 17, 1817.

[19] S. BACHELARD, La représentation géometrique des quantités imaginaires au début du XIXe siècle. Université de Paris, Palais de la Découverte, N° D 113, Paris 1966, 32 pp.

[20] I. BACHMACOVA, Le théorème fondamental de l'Algèbre et la construction des corps algébriques, Archives internationales d'histoire des sciences, Année XIIIe, N° 52—3, 1960, pp. 211—222.

[20a] —, О доказательстве основной теоремы алгебры. Историко-математические исследования, Вып. X, Москва 1957, pp. 257—304.

[21] —, Sur l'histoire de l'algèbre commutative, in: Revue de synthèse, IIIe S., N°s 49—52, Paris 1968, pp. 185—202.

[22] —, Об одном вопросе теории алгебраических уравнений в трудах И. Ньютона и Э. Варинга. Историко-математические исследования, Вып. XII, Москва 1959, pp. 431—456.

[23] , О нскоторых особенностях развития алгебры XVIII. в. Историко-математические исследования, Вып. XVII, Москва 1966, pp. 317—323.

[24] R. BALTZER, Theorie und Anwendung der Determinanten, Leipzig.

[25] E. T. BELL, Gauss and the Early Development of Algebraic Numbers, National Mathematics Magazine, Volume XVIII, October 1943, 10 May 1944, pp. 189—204, 219—233.

[26] —, The Development of Mathematics, Second Edition, New York—London 1945, XIII + 637 pp.

[27] —, Men of Mathematics, 1 ed. 1935; 3. ed. 1965.

[28] J. BERG, Bolzano's Logic, Stockholm 1962, 212 pp.

[29] J. BERTRAND, Mémoire sur le nombre de valeurs que peut prendre une fonction quand on y permute les lettres qu'elle renferme. Journal de l'École polytechnique, Tome XVIII, Paris 1845, pp. 123—140.

[30] E. BETTI, Les substitutions de 6 lettres, CR, 63, 1866.

[31] —, Sur les fonctions symétriques des racines des équations, Journal für reine und angewandte Mathematik, 1857.

[32] —, Sulla risoluzione dell'Equazioni algebriche, Annali di scienze matematiche e fisiche, T. III, Roma 1852, pp. 49—119.

[33] —, Sopra la risolubilita per radicali delle equazioni algebriche irriduttibile di grado primo, Annali di scienze matematiche e fisiche, T. II, Roma 1851, pp. 5—19.

[34] É. BÉZOUT, Théorie générale des équations algébriques, Paris 1779, 471 pp.

[35] —, Cours de mathématiques, à l'usage des Gardes du Pavillon et de la Marine. Troisième Partie, contenant l'Algèbre et l'application de cette science à l'Arithmétique et à la Géométrie, Paris, 1773, XII + 490 pp.

[36] —, Sur la résolution générale des équations de tous les degrès. Histoire de l'Acad. Roy. des Sciences, Année 1769, Paris, pp. 533—552.

[37] —, Sur plusieurs classes d'équations de tous les degrès qui admettent une solution algébrique. Histoire de l'Acad. Roy. des Sciences, Année 1762, Paris 1764, pp. 17—73.

[38] J. P. M. BINET, Mémoire sur un systè me de formules analytiques, et leur application à des considérations géométriques. Journal d e l'Ecole Polytechnique, Tome IX, 1812, pp. 280—302.

[38a] C. A. BJERKNES, N. H. ABEL. Tableau de sa vie et de son action scientifique. Paris 1885, III + 368 pp.

[39] B. BOLZANO, Rein analytischer Beweis, Prag 1816.

[40] —, Wissenschaftslehre. Sulzbach 1837.

[41] —, Functionenlehre. Praha 1930, 183 + 24 pp.

[42] —, Theorie der reellen Zahlen. Praha 1 962, 103 pp.

[43] G. BOOLE, Notes on linear transformation. Cambridge Mathematical Journal, vol. IV., 1844, pp. 167—171.

[44] —, The mathematical analysis of Logic. Being an essay towards a calculus of deductive reasoning, Cambridge 1847.

[45]   —, An investigation of the Laws of Thought. London-Cambridge 1854, 424 pp.

[46]   N. BOURBAKI, Éléments d'histoire des mathématiques. Paris 1960, 276 pp.

[47]   P. L. M. BOURDON, Éléments d'algèbre.
a) Paris 1817, XXII + 605 pp.
b) Ed. 7., Paris 1834, XVI + 719 pp.
c) Ed. 8., Bruxelles 1837, VII + 632 pp.
d) Ed. 19., Paris 1897, XII + 655 pp.

[48]   C. B. BOYER, Colin Maclaurin and Cramer's Rule. Scripta mathematica 27/1966, N° 4, pp. 377—9.

[48a] J. J. BRET, Sur les équations du quatrième degré, Correspondance sur l'école polytechnique, Tome III, pp. 217—219.

[49]   BUÉE, Mémoire sur les quantités imaginaires. Philosophical Transactions of the Royal Society of London, For the year 1806, 96 (1806), pp. 23—88.

[50]   W. BUCHHEIM, Wiliam Rowan Hamilton und das Fortwirken seiner Gedanken in der modernen Physik (Teil I.). Schriftenreihe für Geschichte der Naturwissenschaften, Technik und Medizin, Heft 12, Leipzig 1968, pp. 19—30.

[51]   H. BURKHARDT, Die Anfänge der Gruppentheorie und Paolo Ruffini. Abhandlungen zur Geschichte der Mathematik, Heft V, Leipzig 1892, pp. 119—159.

[52]   —, Endliche discrete Gruppen. Encyklopädie der mathematischen Wissenschaften, Bd I, Teil I, Leipzig 1899, pp. 208—226.

[53]   E. BURNS, The Foundation Period in the History of Group Theory. The American Mathematical Monthly, Volume XX (1913), pp. 141—148.

[54]   F. CAJORI, Historical Note on the Graphic Representation of Imaginaries Before the Time of Wessel, The American Mathematical Monthly, Vol. XIX, 1912, pp. 167—171.

[55]   M. CANTOR, Vorlesungen über Geschichte der Mathematik,
a) Dritter Band von 1668—1758, zweite Auflage Leipzig 1901, 923 pp.
b) Vierter Band von 1759 bis 1799, zweite Auflage Leipzig 1908, 1113 pp.

[56]   É. CARTAN, Un centenaire: Sophus Lie; in: Le Lionnais, Les grand courants de la pensée mathématique, 1948, pp. 253—257.

[57]   A. L. CAUCHY, Oeuvres complètes, 26 vol. (2 séries), Paris 1882—1958.

[58]   —, Mémoire sur les fonctions qui ne peuvent obtenir que deux valeurs égales et de signes contraires par suite des transpositions opérées entre les variables qu'elles renferment. Journal de l'École Polytechnique T. X, 1815, pp. 29—112.

[59]   —, Mémoire sur le nombre des valeurs qu'une fonction peut acquérir, lorsqu'on y permute de toutes les manières possibles les quantités qu'elle renferme. Journal de l'École Polytechnique, Dix-septième cahier, Tome X, Paris 1815, pp. 1—28.

[60]   —, Mémoire sur les arrangement que l'on peut former avec les lettres données et sur les permutations ou substitutions à l'aide desquelles on passe d'un arrangement à un autre. Exercices d'analyse et de physique mathématique Tome III, Paris 1844; Oeuvres S. II, T. XIII, pp. 171—282.

[61]   —, Sur le nombre des valeurs égales ou inégales que peut acquérir une fonction de variables indépendantes, quand on permute ces varibles entre elles d'une maniére quelconque. C. R., Tome XXI, (15. Septembre—6. Octobre 1845); Oeuvres S. I., Tome IX, pp. 277—341.

[62]   —, Mémoires sur diverses propriétés remarquables des substitutions régulières ou irrégulières, et des systèmes de substitutions conjuguées. C. R, Tome XXI, (13. octobre—10. novembre 1845); Oeuvres S. I., Tome IX, pp. 342—405.

[63]   —, Rapport sur un Mémoire présenté à l'Académie par M. Bertrand. CR, T. XXI, (10. novembre 1845); Oeuvres S. I., Tome IX, pp. 405—407.

[64]   —, Mémoire sur les premiers termes de la série des quantités qui sont propres à représenter le

nombre des valeurs distinctes d'une fonction des n variables indépendantes. CR, T. XXI, (17 november 1845); Oeuvres S. I., T. IX, pp. 408—417.

[65] —, Mémoire sur la résolution des équations linéaires symboliques. CR. T. XXI, (24 novembre 1845); Oeuvres S. I., T. IX, pp. 417—430.

[66] —, Mémoires sur les substitutions permutables entre elles. CR. T. XXI, (1 décembre 1845); Oeuvres S. I., T. IX, pp. 430—442.

[67] —, Note sur la réduction des fonctions transitives et sur quelques propriétés remarquables des substitutions qui n'altèrent pas la valeur d'une fonction transitive, CR. T. XXI, (1 décembre 1845); Oeuvres S. I., T. IX, pp. 442—444.

[68] —, Note sur les substitutions qui n'altèrent pas la valeur d'une fonction, et sur la forme régulière que prennent toujours celles d'entre elles qui renferment un moindre nombre de variables. CR. T. XXI, (8 décembre 1845); Oeuvres S. I., T. IX, pp. 444—448.

[69] —, Note sur les fonctions caractéristiques des substitutions. CR. T. XXI, (8 décembre 1845); Oeuvres S. I., T. IX, pp. 466—7.

[70] —, Mémoire sur diverses propriétés des systèmes de substitutions, et particulièrement de ceux qui sont permutables entre eux. CR. T. XXI, (8 décembre 1845); Oeuvres S. I. T. IX, pp. 449—465.

[71] —, Mémoire sur le nombre et la forme des substitutions qui n'altèrent pas la valeur d'une fonction de plusieurs variables indépendantes. CR. T. XXI, (15 décembre 1845); Oeuvres S. I. ,T. IX, pp. 467—482.

[72] —, Applications diverses des principes établis dans les précédents Mémoires. CR. T. XXI, (22 décembre 1845); Oeuvres S. I., T. IX, pp. 482—496.

[73] —, Mémoire sur les fonctions de cinq ou six variables, et spécialement sur celles qui sont doublement transitives. CR. T. XXI, (29 décembre 1845); T. XXI, (5 janvier 1846); Oeuvres S. I., T. IX, pp. 496—505; T. X, pp. 5—35.

[74] —, Mémoire sur un nouveau calcul qui permet de simplifier et d'étendre la théorie des permutations. CR, T. XXII, (12 janvier 1846); Oeuvres S. I. T. X, pp. 35—46.

[75] —, Applications diverses du nouveau calcul dont les principes ont été établis dans la séance précédente, CR. T. XXII, (18 janvier 1846); Oeuvres S. I., T. X, pp. 47—55.

[76] —, Recherches sur un système d'équations simultanées, dont les unes se déduisent des autres à l'aide d'une ou de plusieurs substitutions. CR. T. XXII, (26 janvier 1846); Oeuvres S. I., T. X, pp. 56—57.

[77] —, Note sur diverses propriétés de certaines fonctions algébrique. CR. T. XXII, (26 janvier 1846); Oeuvres S. I., T. X, p. 57.

[78] —, Sur la résolution directe d'un système d'équations simultanées, dont les unes se déduisent des autres à l'aide d'une ou de plusieurs substitutions. CR. T. XXI, (3 février 1846); Oeuvres S. I., T. X, pp. 57—61.

[79] —, Sur la résolution des équations symboliques non linéaires. CR. T. XXII, (9 février 1846); Oeuvres S. I., T. X, pp. 61—65.

[80] —, Note sur un théorème fondamental relatif à deux systèmes de substitutions conjuguées. CR. T. XXII, (11 avril 1846); Oeuvres S. I., T. X, pp. 65—68.

[81] —, Exercices d'analyse et de physique mathématique. Tome quatrième, Paris 1847, 404 pp.

[82] —, Mémoire sur la théorie des équivalences algébriques, substituée à la théorie des imaginaires. Exercices d'analyse et de physique mathématique, Tome 4, Paris 1847, pp. 87—110.

[83] —, Mémoire sur les quantités géometriques. Exercices d'analyse et de physique mathématique, Tome 4, Paris 1847, pp. 157—180.

[84] —, Sur les fonctions des quantités géométriques, Exercices d'analyse et de physique mathématique, Tome 4, Paris 1847, pp. 308—313.

[85] —, Mémoire sur les clefs algébriques. Exercices d'analyse et de physique mathématique, Tome 4, Paris 1847, pp. 356—400.

[86]   A. Cayley, The Collected Mathematical Papers. 14 vol., Cambridge 1889—1898.

[87]   —, On a theorem in the geometry of position, Cambridge Mathematical Journal, vol. II, 1841, pp. 267—271; Math. Papers, vol. 1, pp. 1—4.

[88]   —, Chapters in the analytical geometry of (n) dimensions, Cambridge Mathematical Journal, vol. IV, 1843, pp. 119—127; Math. Papers, vol. 1, pp. 55—62.

[89]   —, On the theory of determinants. Transactions of the Cambridge Philosophical Society, vol. VIII, 1843, pp. 1—16; Math. Papers, vol. 1, pp. 63—79.

[90]   —, On the theory of linear transformations, Cambridge Mathematical Journal, vol. IV, 1845, pp. 193—209; Math. Papers, vol. 1, pp. 80—94.

[91]   —, On Jacobi's elliptic functions, in reply to the Rev. B. Brouwin; and on quaternions. Philosophical Magazine, vol. XXVI (1845) pp. 211; Math. Papers, vol. 1, p. 128.

[92]   —, Note on a System of Imaginaries. Philosophical Magazine, vol. XXX (1847), pp. 257—258; Math. Papers, vol. 1, p. 301.

[93]   —, Note on the Theory of Permutations. Philosophical Magazine, vol. XXXIV (1848), pp. 527—529; Math. Papers, vol. 1, pp. 423—424.

[94]   —, On the triatic Arrangements of seven and fifteen things. Philosophical Magazine, vol. XXXVII (1850), pp. 50—53; Math. Papers, vol. 1, pp. 481—484.

[95]   —, On the Theory of Permutants. Cambridge and Dublin Mathematical Journal, vol. VII (1852), pp. 40—41; Math. Papers, vol. 2, pp. 16—26.

[96]   —, On the Theory of Groups, as depending on the Symbolic Equation $\vartheta^n = 1$. Philosophical Magazine, vol. VII, 1854, pp. 40—47; Mathematical Papers, vol. 2, pp. 123—130.

[97]   —, On the Theory of Groups, as depending on the Symbolic Equation $\vartheta^n = 1$. — Sécond Part. Philosophical Magazine, vol. VII, 1854, pp. 408—409; Mathematical Papers, vol. 2, pp. 131—132.

[98]   —, Remarques sur la notation des fonctions algébriques, Journal für die reine und angewandte Mathematik, Bd. L, 1855, pp. 282—285; Math. Papers, vol. 2, pp. 185—188.

[99]   —, A Memoir on the Theory of Matrices. Philosophical Transactions, vol. CXVIII, London 1858, pp. 17—37; Math. Papers, vol. 2, pp. 475—496.

[100] —, On the theory of Groups, as depending on the Symbolic equation $\vartheta^n = 1$. Third Part. Philosophical Magazine, vol. XVIII, 1859, pp. 34—37, 125—126; Mathematical Papers, Vol. 4, pp. 88—91.

[101] —, Recent terminology in Mathematics. English Cyclopaedia, vol. V, 1860, pp. 534—542; Mathematical Papers, Vol. 4, pp. 594—608.

[102] A. C. CLAIRAUT, Élèmens d'algèbre.
       a) Paris 1746,
       b) Quatrième édition, Paris 1768,
       c) Cinquième édition, Paris 1798,
       d) Dernière édition, Prais 1801, T. 1, 398 pp.
                             T. 2, 537 pp.

[103] E. B. DE CONDILLAC, La Langue des Calcules, Oeuvres de C,. Tome XXIII, Paris 1798, 487 pp.

[104] M. J. A. N. K. DE CONDORCET, Équations déterminées, Encyclopédie méthodique (Mathématiques), 1784, pp. 633 nn.

[105] —, Réflexions sur la forme des équations déterminées, la réduction et la solution de ces équations. Mélanges de Phil. et de Math. de la Soc. Roy. de Turin, pour les années 1770—1773, Classe Math., pp. 1—7.

[106] G. CRAMER, Introduction à l'analyse des lignes courbes algébriques. Génève 1750, XXIII + 680 pp.

[107] A. L. CRELLE, Encyklopädische und elementare Darstellung der Theorie der Zahlen, Journal für reine und angewandte Mathematik, Bd. 27, Berlin 1843; Bd. 28, Berlin 1844, pp. 111—178.

[108] M. J. CROWE, A History of Vector Analysis. The Evolution of the Idea of a Vectorial System. Notre Dame + London, 1967, XVII + 270 pp.

[109] A. Dalmas, Évariste Galois, révolutionaire et géomètre. Paris 1956, 175 pp.

[110] de la Caille,
a) Leçons élémentaires de mathématiques; ou éléméns d'algèbre et de géométrie. Nouvelle Édition, revue, corrigée et augmentée Paris 1764, 277 pp.
b) Lectiones elementares mathematicae, seu elementa algebrae, et geometriae, in latinum traductae et ad editionem parisinam anni MDCCLIX denno exactae. Viennae, Pragae, Tergesti 1762, 219 pp.

[111] Cl. F. M. Dechales, Cursus seu mundus mathematicus. Ludguni, 1674.

[112] R. Dedekind, Gesammelte mathematische Werke. vol. 1.—3., Braunschweig 1930—1932.

[113] —, Abriss einer Theorie der höheren Kongruenzen in bezug auf einen reellen Primzahl-Modulus, Journal für reine und angewandte Mathematik, Bd. 54, 1857, pp. 1—26; Werke I, pp. 40—67.

[114] —, Beweis für die Irreduktibilität der Kreisteilungs-Glcichung. Journal für reine und angewandte Mathematik, Bd. 54, 1854, pp. 27—30; Werke I, pp. 68—71.

[115] J. B. J. Delambre, Rapport historique sur les progrès des sciences mathématiques depuis 1789, et sur leur état actuel. Paris 1810, 272 pp.

[116] Б. Н. Делоне, Работы Гаусса по теории чисел. Карл Фридрих Гаусс, Сборник статей, Москва 1956, pp. 13—112.

[117] M. Despeyrous, Sur la théorie générale des permutations. Journal de mathématiques pures et appliquées, Série 2, Tome VI, 1861, pp. 417—439.

[118] L. E. Dickson, History of the Theory of Numbers, Vol. I—III, Washington 1919—1927.

[119] J. Dieudonné, Notes sur les travaux de C. Jordan relatifs à la théorie des groupes finis. In C. Jordan Oeuvres, T. I, pp. XVII—XLII.

[120] J. Dieudonné, L'oeuvre mathématique de C. F. Gauss. Conférence donnée au Palais de la Découverte, le 2 Décembre 1961, Université de Paris, Palais de la Découverte, N. D 79, Paris 1962, 18 pp.

[121] G. Lejeune-Dirichlet, Werke, vol. I.—II. Berlin 1889—1897.

[122] —, Vorlesungen über Zahlentheorie. Herausgegeben von R. Dedekind,
a) Braunschweig 1863;
b) Zweite umgearbeitete und vermehrte Auflage, Braunschweig 1871, XVIII + 497 pp.

[123] —, Mémoirc sur l'impossibilité de quelques équations indéterminées du cinquième degré, Werke I, pp. 1—20.

[124] —, Mémoire sur l'impossibilité de quelques équations indéterminées du cinquième degré. Journal für reine und angewandte Mathematik, Bd. 3, 1828, pp. 354—375; Werke I, pp. 21—46.

[125] —, Démonstration d'une propriété analogue à la loi de réciprocité qui existe entre deux nombres premiers quelconques. Journal für die reine und angewandte Mathematik, Bd. 9, pp. 379—389; Werke I, pp. 173—188.

[126] —, Untersuchungen über die Theorie der complexen Zahlen. Bericht über die Verhandlungen der Königl. Preuss. Akademie der Wissenschaften, Jahrg. 1841, pp. 190—194; Werke I, pp. 503—508.

[127] —, Untersuchungen über die Theorie der complexen Zahlen. Abhandlungen der Königlich Preussischen Akademie der Wissenschaften von 1841, pp. 141—161; Werke I, pp. 509—532.

[128] —, Recherches sur les formes quadratiques à coeficients et à indéterminées complexes. Journal für die reine und angewandte Mathematik, Bd. 24, 1842, pp. 291—371; Werke I, pp. 533—618.

[129] —, Sur la théorie des nombres (Extrait d'une lettre adressée à M. Liouville). Werke I, pp. 619—623.

[130] —, Zur Theorie der complexen Einheiten, Bericht über die Verhandlungen der Königl. Preuss. Akademie der Wissenschaften, Jahrg. 1846, pp. 103—107; Werke I, pp. 639—644.

[131] Dubourguet, Démonstration du principe qui sert de fondement à la théorie des équations. Annales de mathématiques pures et appliquées, T. II, 1811—1812, pp. 338—340.

[132] P, Dubreil, Aperçu historique sur le développement de l'algèbre. Bulletin de l'Association

des Professeurs de Mathématiques de l'Enseignement Public, 38e Année — N° 196, Janvier 1959, pp. 85—94.

[133] —, La naissance de deux jumelles: la logique mathématique et l'algèbre ordonnée, Revue de synthèse, IIIe S., Nos 49—52, Paris 1968, pp. 203—209.

[134] D. Encontre, Mémoire sur les principes fondamentaux de la théorie générale des équations. Annales de mathématiques pures et appliquées T. IV, 1813—1814, pp. 201—222.

[135] Encyclopédie des sciences mathématique pures et appliquées. Édition française, Paris—Leipzig.

[136] Encyklopädie der mathematischen Wissenschaften mit Einschluss ihrer Anwendungen, Bd. 1, Leipzig 1898—1904.

[137] L. Euler, De resolutione aequationum cuiusvis gradus. N. Comm. Petr. T. LX, pro annis 1762 et 1763, pp. 80—98.

[138] —, Introductio in analysis infinitorum. 1748.

[139] —, De fractionibus continuis. Commentarii Academiae Petropolitanae, ad annum 1737 (1744) T. IX, pp. 98—137.

[140] —, De formis radicum aequationum cujusque ordinis conjectatio. Commentarii Academiae Petropolitanae and annum 1732 et 1733, T. VI, pp. 216—231.

[141] —, Vollständige Anleitung zur Algebra. 1770.

[142] Ch. Fisher, The Death of a Mathematical Theory: a Study in the Sociology of Knowledge. Archive for History of Exact Sciences, Volume 3, Number 2, 1967, pp. 137—159.

[143] J. Folta, Bernard Bolzano and the foundations of geometry, Acta historiae rerum naturalium necnon technicarum, Special Issue 2, Prague 1966, pp. 75—104.

[144] —, The geometric and algebraic axiomatics and the generalization of the subject of geometry. Actes du XIIe Congrès international d'histoire des sciences, Paris 1968, Tome IV, Paris 1971, pp. 59—65.

[145] J. F. Français, Lettre au Rédacteur des Annales, sur la théorie des quantités imaginaires. Annales de mathématiques pures et appliquées, Tome IV, 1813—1814, pp. 222—228.

[146] —, Extrait d'une Lettre adressée au Rédacteur des Annales. Annales de mathématiques pures et appliquées, Tome IV, 1813—1814, pp. 364—367.

[147] —, Nouveaux principes de Géométrie de position, et interprétation géométrique des symboles imaginaires. Annales de mathématique pures et appliquées, Tome IV, 1813—1814, pp. 61—71.

[148] L. B. Francoeur, Cours complet de mathématiques pures. Ouvrage destiné aux élèves des écoles normales et polytechniques, et aux candidats qui préparent a y être admis.
a) Paris 1809
b) Second Édition, Revue et considérablement augmentée, Paris 1819, T. 1, XX + 479 pp.
T. 2, 515 pp.

c) Quatrième Édition, Revue et augmentée, Paris 1837, T. 1, 505 pp.
T. 2, 609 pp.

[149] H. Freudenthal, L'Algèbre topologique en particulier les groupes topologiques et de Lie, Revue de synthèse, IIIe S., Nos 49—52, Paris 1969, pp. 223—243.

[150] J. G. Garnier, Analyse algébrique faisant suite à la première section de l'algèbre. Deuxième édition, Paris 1814, 668 pp.

[151] H. G. Funkhouser, A short account of the history of symetric functions of roots of equations. The American Mathematical Monthly, Volume XXXVII, 1930, pp. 357—365.

[151a] П. И. Галченкова, Алгебра в неопубликованных работах Л. Эйлера. История и методология естественных наук, Выпуск V Математика, Москва 1966, pp. 45—61.

[152] É. Galois, Oeuvres mathématiques. Journal de mathématiques pures et appliquées, T. XI, 1846, pp. 381—444.

[153] —, Oeuvres mathématiques. Avec une introduction par M. Émile Picard. Paris 1897, X + 61 pp.

[154] —, Manuscrits. Publiés par Jules Tannery. Paris 1908, 67 pp.

[155] —·, Écrits et mémoires mathématiques. Édition critique intégrale de ses manuscrits et publications par R. Bourgne et J.—P. Azra. Préface de J. Dieudonné. Paris 1962, XXI + 541 pp.

[156] —, Sur la théorie des nombres. Bulletin des Sciences mathématiques de M. Férussac, t. XIII, 1830, pp. 428; Oeuvres, [153] pp. 15—23.

[157] —, Mémoire sur les conditions de résolubilité des équations par radicaux. Oeuvres, [153] pp. 33—50.

[158] —, Des équations primitives qui sont solubles par radicaux. Oeuvres, [153] pp. 51—61.

[159] G. F. Gauss, Werke. 12 vol., Göttingen 1870—1927.

[160] —, Dissertation. Helmstadt 1799.

[161] —, Disquisitiones arithmeticae. Leipzig 1801.

[162] —, Untersuchungen über höhere Arithmetik, Deutsch herausgegeben von H. Maser, Berlin 1889, XIII + 695 pp.

[163] —, Weitere Entwicklung der Untersuchungen über die reinen Gleichungen. Untersuchungen über höhere Arithmetik, pp. 630—652.

[164] —, Theoria residuorum biquadraticorum commentatio prima (Societati regiae tradita 1825, apr. 5). Commentationes societatis regiae scientiarum Gottingensis recentiores, Vol. VI, Gottingae 1828; Werke II, pp. 65—92.

[165] —, Theoria residuorum biquadraticorum commentatio secunda (Societati regiae tradita 1831, apr. 15). Commentationes societatis regiae scientiarum Gottingensis recentiores, Vol. VII, Gottingae 1832; Werke II, pp. 93—149.

[166] —, Theoria residuorum biquadraticorum, commentatio secunda (Anzeige). Göttingische gelehrte Anzeigen, 1831 April 23; Werke II, pp. 169—178.

[167] —, Demonstratio nova altera theorematis omnem functionem algebraicam rationalem integram unius variabilis in factores reales primi vel secundi gradus resolvi posse. Werke III, pp. 31—56.

[167a] —, Die vier Gauss'schen Beweise für die Zerlegung ganzer algebraischer Functionen in reele Factoren ersten oder zweiten Grades (1799—1849). Ostwald's Klassiker, Nr. 14, Leipzig 1890, 81 pp.

[168] J. D. Gergonne, De l'identité entre les produits qui résultent des mêmes facteurs différement multipliés entre eux, Annales de mathématiques pures et appliquées, Tome I, 1810—1811, pp. 52—59.

[169] —, De l'analyse et de la synthèse dans les sciences mathématiques. Annales de mathématiques pures et appliquées, Tome VII, pp. 345—372.

[170] —, Sur une réclamation de M. Hoëne—Wronski, contre quelques articles de ce recueil. Annales de mathématique pures et appliquées. Tome III, 1812—1813, pp. 206—209.

[171] —, Remarque sur la résolution des équations du quatrième degré par la méthode de M. Wronski. Annales de mathématiques pures et appliquées, Tome III, 1812—1813, pp. 137—139.

[172] —, Doutes et réflexions sur méthode proposé par M. Wronski pour la résolution générale des équations algébriques de tous degrès. Annales de mathématiques pures et appliquées, Tome III, 1812—1813, pp. 51—59.

[173] H. Grassmann, Gesammelte Werke. Bd. 1—3, Leipzig 1894—1911

[174] —, Die Wissenschaft der extensiven Grösse oder die Ausdehnungslehre, eine neue mathematische Disciplin, dargestellt und durch Anwendungen erläutert…, Erster Teil, die lineare Ausdehnungslehre enthaltend. Leipzig 1844; ed. 2, 1878. H. Grassmann gesammelte mathematische u. physikalische Werke, Ersten Bandes erster Theil, Leipzig 1894.

[175] R. Graves, Life of Sir William Rowan Hamilton… including Selections from his Poems, Correspondence, and Miscellaneous Writings.
Vol. 1 Dublin—London 1882, XVIII + 698 pp.
Vol. 2 Dublin—London 1885, XV + 719 pp.
Vol. 3 Dublin—London 1889, XXXV + 673 pp.

[176] D. F. Gregory, On the Nature of Symbolical Algebra. Trans. Roy. Soc. Edin., Vol. 14, 1840, pp. 208—216.

[177] —, On the elementary principles of the application of algebraical symbols to geometry. The Cambridge Mathematical Journal, Vol. II, No. VII, November 1839, pp. 1—9.

[178] —, On the solution of certain functional equations. The Cambridge Mathematical Journal, Vol. I, No III, 1843, pp. 239—246.

[179] —, On a difficulty in the theory of algebra. The Cambridge Mathematical Journal, Vol. I, No III, 1843, pp. 153—159.

[180] —, Note on a class of factorials. The Cambridge Mathematical Journal, Vol. I, No III, 1843, pp. 89—91.

[181] W. R. Hamilton, The Mathematical Papers, Volume III. Algebra. Cambridge 1967, XXIV + + 672 pp.

[182] —, Theory of Conjugate Functions, or Algebraic Couples; with a Preliminary and Elementary Essay on Algebra as the Science of Pure Time. Read November 4th, 1833 and June 1st, 1835; Transactions of the Royal Irish Academy, Vol. XVII, 1837, Part II, pp. 293—422.

[183] —, On conjugate functions, or algebraic couples, as tending to illustrate generally the doctrine of imaginary quantities..., Report of the British Association for the Advancement of Science for 1834, pp. 519—523; The Mathematical Papers, Volume III, pp. 97—100.

[184] —, On the Argument of Abel, respecting the Impossibility of expressing a Root of any General Equation above the Fourth Degree, by any finite Combination of Radicals and Rational Functions. Transactions of the Roy. Irish Academy, vol. XVIII, 1839, pp. 171—259; Math. Papers vol. III, pp. 517—569.

[185] —, Quaternions (1843), Math. Papers, Vol. III, pp. 103—105.

[186] —, On Quaternions; or on a new System of Imaginaries in Algebra (A letter to John T. Graves, dated 17 October 1843). Philosophical Magazine, vol. XXV, pp. 489—495; Math. Papers, vol. III, pp. 106—110.

[187] —, On a new species of Imaginary Quantities connected with the Theory of Quaternions. Proceedings of the Roy. Irish Academy, vol. II, 1844, pp. 424—434; Math. Papers, vol. III, pp. 111—116.

[189] —, Researches respecting Quaternions. First Series (1843). Transactions of the Roy. Irish Academy, vol. XXI (1848), pp. 199—296; Math. Papers, vol. III, pp. 159—226.

[189] —, On Quaternions; or a new System of Imaginaries in Algebra. Philosophical Magazine 1844—1850; Math. Papers, vol. III, pp. 227—297.

[190] —, On Equations of the Fifth Degree: and especially on a certain System of Expressions connected with these Equations, which Proffesor Badano has lately proposed. Transactions of the Roy. Irish Academy, vol. XIX, 1843, pp. 329—376; Math. Papers, vol. III, pp. 572—602.

[191] —, Lectures on quaternions. Dublin 1853.

[192] —, Preface to "Lecture on Quaternions". Mathematical Papers, vol. III, pp. 117—155.

[193] —, Account of the Icosian Calcus (Communicated 10 November 1856). Proceedings of the Royal Irish Academy, vol. VI, 1858, pp. 415—416; Math. Papers, vol. III, pp. 609.

[194] —, Letter to John T. Graves on the Icosian (17 Octobre 1856). Math. Papers, vol. III, pp. 612—625.

[195] H. Hancock, The Fundations of the Theory of Algebraic Numbers. Science, N. S., Vol. LXI, January—June 1925, pp. 5—10, 30—35.

[196] H. Hankel, Theorie des complexen Zahlensystems insbesondere der gemeinen imaginären Zahlen und der Hamilton'schen Quaternionen nebst ihrer geometrischen Darstellung. Vorlesungen über die complexen Zahlen und ihre Functionen in zwei Theilen. I. Teil. Theorie des complexen Zahlensystems. Leipzig 1867, 196 + XII pp.

[197] D. Harkin, The Development of Modern Algebra. Norsk matematisk tidsskrift, 33. Argang, Oslo 1951, pp. 17—26.

[198] C. HERMITE, Oeuvers, T. I.–IV., Paris 1905–1917.

[199] —, Considerations sur la résolution algébrique de l'équation du cinquième degré. Nouvelles Annales de Mathématiques, Tome I, (1842); Oeuvres, Tome I, pp. 3–9.

[200] —, Sur les fonctions de sept lettres. CR, Tome LVIII, 1863 (II), 750 nn; Oeuvres, Tome II, pp. 280–288.

[201] D. HILBERT, Die Theorie der algebraischen Zahlkörper. Jahresbericht der Deutschen Mathematiker-Vereinigung, Vierter Band, 1894–1895, Berlin 1897, pp. 175–539.

[202] M. HIRSCH, Sammlung von Aufgaben aus der Buchstabenrechnung und Algebra. Berlin 1804.

[203] Histoire de la science, Des origines au XXᵉ siècle, sous la direction de Maurice Daumas, Paris 1957, XLVIII + 1094 pp.

[204] Histoire générale des sciences, publié sous la direction de René Taton.
Tome II. La science moderne, Paris 1958, VII + 800 pp.
Tome III. La science contemporaine
    Volume I. Le XIXᵉ siècle, Paris 1961, 755 + VIII.

[205] A. E. L. HULBE, Analytische Entdeckungen in der Verwandlungs- und Auflösungs-Kunst der höheren Gleichungen. Berlin u. Stralsund 1794, 136 pp.

[206] J. ITARD, La théorie des nombres et les origines de l'algèbre moderne. Revue de synthèse, IIIᵉ S., Nos. 49–52, Paris 1968, pp. 165–184.

[207] J. IVORY, Equations, in: Supplement to the… Encyclopaedia Britannica, Vol. IV, Edinburgh 1824, pp. 669–708. Addendum to Volume Fourth.

[208] G. C. J. JACOBI, Gesammelte Werke, vol. I–VII, Berlin 1881–91.

[209] —, De formatione et proprietatibus Determinantium. Journal für die reine und angewandte Mathematik, Bd. 22, 1841, pp. 285–318; Werke III, pp. 355–392.

[210] —, De determinantibus functionalibus. Journal für die reine und angewandte Mathematik, Bd. 22, 1841, pp. 319–352; Werke III, pp. 393–438.

[211] —, De functionibus alternatibus carumquc divisionc per productum de differentiis elementorum conflatum. Journal für die reine und angewandte Mathematik, Bd. 22, 1841, pp. 360–371; Werke III, pp. 439–452.

[212] —, Sur les nombres premiers complexes que l'on doit considérer dans la théorie des résidus de cinquième, huitième et douzième puissance. Extrait de Journal de M. Crelle, Tome XI; lu à l'Académie de Berlin, le 16 mai 1839. Journal de mathématiques pures et appliquées Tome 8, Paris 1843, pp. 268–272.

[213] —, Über die Kreistheilung und ihre Anwendung auf die Zahlentheorie. Journal für die reine und angewandte Mathematik, Bd. 30, 1846, pp. 166–182.

[214] C. JORDAN, Oeuvres de C. J. Tome I, Paris 1961, XLII + 498 pp., Tome II, Paris 1961, 576 pp.

[215] —, Mémoire sur le nombre des valeurs des fonctions. Journal de l'École polytechnique, Tome XXII, Paris 1861, pp. 113–194.

[216] —, Mémoires sur les groupes de mouvements. Annales mathématiques (2), II, (1868–9), pp. 167–215, 322–345.

[217] —, Commentaire sur Galois. Mathematische Annalen. Erster Band, Leipzig 1869, pp. 141–160.

[218] —, Sur les équations de la division des fonctions abéliennes, Mathematische Annalen, Erster Band, Leipzig 1869, pp. 583–591.

[219] —, Traité des Substitutions et des équations algébriques. Paris 1870, 667 pp.

[220] А. П. ЮШКЕВИЧ, История математики в средние века, Москва 1961, 448 pp.

[221] A. G. KÄSTNER, Anfangsgründe der Mathematik.
Erste Auflage, Göttingen 1758 nn.
Sechste Auflage, Göttingen 1800.

[222] —, Theoria radicum in aequationibus. Lipsiae 1739, 31 pp.

[223] W. J. G. KARSTEN, Lehrbegriff der gesammten Mathematik, Bd. 1–8, Greifswald 1767–1777.

[224] —, Mathematische Abhandlungen, Halle 1786, 432 pp.

[225] —, Von verneinten und unmöglichen Wurzelgrössen, in: Karsten, Mathematische Abhandlungen, pp. 203—282.

[226] —, Von den Logarithmen der verneinten und unmöglichen Grössen, in: Karsten, Mathematische Abhandlungen, pp. 283—390.

[227] H. KINKELIN, Neuer Beweis des Vorhandenseins complexer Wurzeln in einer algebraischen Gleichung. Mathematische Annalen 1869, Bd. 1, pp. 502—506.

[228] T. P. KIRKMAN, On the theory of groups and many-valued functions. Memoirs of the Literary and Philosophical Society of Manchester, Third Series, First Volume 1862, London 1862, pp. 274—397.

[229] —, On Non-Modular Groups. Proceedings of Literary and Philosophical Society of Manchester, vol. II, 1860—62, Manchester 1862, pp. 254—253.

[230] —, Theorems on Groups. Proceedings of the Literary and Philosophical Society of Manchester, vol. II, 1860—62, Manchester 1862, pp. 73—97.

[231] —, On Maximum Groups. Proceedings of the Literary and Philosophical Society of Manchester, vol. III, 1862—64, Manchester 1864, pp. 59—65.

[232] —, The Complete Theory of Groups, being the Solution of the Mathematical Prize Question of the French Academy of 1860, Proceedings of the Literary and Philosophical Society of Manchester, Manchester 1864, pp. 133—152, 161—162.

[233] —, Theory of Groups. Corrigenda and Addenda. Proceedings of the Literary and Philosophical Society of Manchester, vol. IV, 1884—5, Manchester 1865, pp. 171—172.

[234] F. KLEIN, Vorlesungen über die Entwicklung der Mathematik im 19. Jahrhundert, Teil I, Berlin 1926, XIII + 385 pp.

[235] —, Vergleichende Betrachtungen über neuere geometrische Forschnugen, Erlangen 1872; Mathematische Annalen 43 (1893).

[236] G. S. KLÜGEL, Über die Lehre von den entgegengesetzten Grössen. Archiv der reinen und angewandten Mathematik, Bd. 1, Leipzig 1795, pp. 309—319; 470—481.

[237] —, Mathematisches Wörterbuch.

[238] R. KOCHENDÖRFFER, Gauss' algebraische Arbeiten, in: C. F. Gauss Gedenkband anlässlich des 100. Todestages am 23. Februar 1955, Berlin 1957, pp. 80—91.

[239] А. Н. КОЛМОГОРОВ, Математика, ин: Большая советская энциклопедия, 2. ed., Москва 1954.

[240] L. KÖNIGSBERGER, Berichtigung eines Satzes von Abel, die Darstellung der algebraischen Function betreffend, Mathematische Annalen, Band I. Leipzig 1869, pp. 168—9.

[241] Ф. Д. КРАМАР, Векторные иссчисления конца XVIII и начала XIX вв. Историко-математические исследования, Вып. XV, Москва 1963, pp. 225—290.

[242] —, От универзальной аритметики Ньютона к алгебре кватернионов Гамильтона. Историко-математические исследования, Вып. XVII, Москва 1966, pp. 309—316.

[243] L. KRONECKER, Werke, Bd. 1—5, Leipzig 1895—1930.

[244] —, Grundzüge einer arithmetischen Theorie der algebraischen Grössen, Berlin 1882, VIII + 122 pp.

[245] —, De unitatibus complexis dissertatio inauguralis arithmetica, §§ 1—15, Berlin 1845; Werke I, (§§ 1—20), pp. 5—73.

[246] —, Mémoire sur les facteurs irréductibles de l'expression $x^n - 1$. Journal de mathématiques pures et appliquées, Tome 19, 1854, pp. 177—192; Werke I, pp. 75—97.

[247] —, Über die algebraisch auflösbaren Gleichungen I. Abhandlung. Monatsberichte der Kön. Akademie der Wissenschaften zu Berlin vom Jahre 1853, pp. 365—374; Werke 4, pp. 1—11.

[248] —, Über die algebraisch auflösbaren Gleichungen II. Abhandlung. Monatsberichte der Kön· Akademie der Wissenschaften zu Berlin vom Jahre 1856, pp. 203—215; Werke IV, pp. 25—37.

[249] —, Note sur les fonctions semblables des racines d'une équation. Journal de mathématiques pures et appliquées, T. 19, 1854, pp. 279—280; Werke IV, pp. 15—16.

[250]   —, Sur la théorie des substitutions. Annali di matematica pura ed applicata, T. II, 1859, pp. 131; Werke IV, pp. 51—52.

[251]   —, Über complexe Einheiten. Journal für die reine und angewandte Mathematik, Bd. 53, 1857, pp. 176—181; Werke I, pp. 109—118.

[252]   —, Über die Klassenzahl der aus Wurzeln der Einheit gebildeten complexen Zahlen. Monatsberichte der Kön. Akad. der Wissenschaften zu Berlin vom Jahre 1863, pp. 340—345; Werke I, pp. 123—132.

[253]   —, Über bilineare Formen. Monatsberichte der Königl. Akademie der Wissenschaften zu Berlin vom Jahre 1866, pp. 597—612; Werke I, pp. 143—162.

[254]   —, Bemerkungen zur Determinanten-Theorie. Journal für die reine und angewandte Mathematik, Bd. 42, 1869, pp. 152—175; Werke I, pp. 235—270.

[255]   —, Auseinandersetzung einiger Eigenschaften der Klassenzahl idealer complexer Zahlen. Monatsberichte der Königl. Akademie der Wissenschaften zu Berlin vom Jahre 1870, pp. 881 bis 889; Werke I, pp. 271—282.

[256]   H. KÜHN, Meditationes de quantitatibus imaginariis construendis et radicibus imaginariis exhibendis. Novi Commentarii Academiae Petropolitanae ad annum 1750 et 1751, T. III, pp. 170—223.

[257]   E. E. KUMMER, De residuis cubicis disquisitiones nonnullae analyticae. Journal für reine und angewandte Mathematik, Bd. 32, Berlin 1846, pp. 341—359.

[258]   —, Zur Theorie der complexen Zahlen. Journal für reine und angewandte Mathematik, Bd. 35, Berlin 1847, pp. 319—326.

[259]   —, Über die Zerlegung der aus Wurzeln der Einheit gebildeten complexen Zahlen in ihre Primfactoren. Journal für reine und angewandte Mathematik, Bd. 35, Berlin 1847, pp, 327—367.

[260]   —, Extrait d'une lettre de M. Kummer à M. Liouville. Journal de mathématiques pures et appliquées, Tome 12, Paris 1847, p. 136.

[261]   —, De nummeris complexis, qui radicibus unitatis et numeris integris realibus constant. Gratulationsschrift der Breslauer Universität zum dreihundertjährigen Jubiläum der Universität Königberg, Breslau 1844; Journal de mathématiques pures et appliquées, Tome 12, Paris 1847, pp. 185—212.

[262]   S.-F. Lacroix, Éléments d'Algèbre à l'usage de l'école centrale des quatre nations,
        a) Paris (AN VII) 1799, 298 pp.
        b) Seizième édition, Revue et corrigée, Paris 1836, 374 pp.
        c) Dix-huitième édition, Revue et corrigée, Paris 1847, 375 pp.

[263]   —, Complément des Éléments d'Algèbre
        a) Éd. 1. Paris 1798, 250 pp.
        b) Éd. 2. Paris 1801, 280 pp.
        c) Éd. 3. Paris 1804, 315 pp.
        d) Éd. 5. Paris 1825, 332 pp.
        e) Éd. 6. Paris 1835, 374 pp.
        f) Éd. 7. Paris 1863, 342 pp,

[264]   J. L. Lagrange, Oeuvres. 14 vol., Paris 1867-1892.

[265]   —, Théorie des fonctions analytiques, Paris 1797.

[266]   —, Sur la solution des Problèmes indéterminés du second degré. Mémoires de l'Académie royale des Sciences et Belles-Lettres de Berlin, T. XXIII, 1769; Oeuvres, Tome 2, pp. 377-535.

[267]   —, Réflexions sur la résolution algébrique des équations. Nouveaux Mémoires de l'Académie royale des Sciences et Belles-Lettres de Berlin, années 1770 et 1771; Oeuvres de Langrange, Tome 3, pp. 205—421.

[268]   —, Recherches d'arithmétique. Nouveaux Mémoires de l'Académie royale des Sciences et Belles-Lettres de Berlin, années 1773 et 1775; Oeuvres T. 3, pp. 693—795.

[269]   —, Traité de la résolution des équations numériques de tous les degrès avec des notes sur plusieurs points de la théorie des équations algébriques.
a) Deuxième Édition, Paris 1798.
b) Troisième Édition, conforme à celle de 1808, et procédée d'une Analyse de l'Ouvrage, par M. Poinsot.
c) Paris 1826, XX + 314 pp.
d) Oeuvres, Tome 8, pp. 1—367.

[270]   V. A. LEBESGUE, Théorèmes nouveaux sur l'équations indéterminée $x^5 + y^5 = az^5$. Journal de mathématiques pures et appliquées, Tome VIII, Paris 1843, pp. 49—70.

[271]   L. LEFÉBURE de FOURCY, Leçons d'algèbre
a) 1st ed. Paris 1833, XIV + 584 pp.
b) 2nd ed. Paris 1835, VIII + 680 pp.
c) 9th ed. Paris 1880, VI + 535 pp.
d) 10st ed. Paris 1893

[272]   —, Théories du plus grand commun diviseur algébrique et de l'élimination entre deux équations à deux inconnues, Paris 1827, IV + 47 pp.

[273]   A. M. LEGENDRE, Essai sur la théorie des nombres
a) Paris 1798
b) Ed. 2nd 1808
c) Théorie des nombres, 1830

[274]   LE LIONNAIS, Les grands courants de la pensée mathématique, Paris 1948.

[275]   G. LIBRI, Journal für reine u. angewandte Mathematik, Bd. 10, 1822, pp. 167nn.

[276]   —, Notice biographique sur N. H. Abel. Extrait de la Biographie universelle, Tome LVI, 1833, 7 pp.

[277]   A. LOEWY, Inwieweit kann Vandermonde als Vorgänger von Gauss bezüglich der algebraischen Auflösung der Kreisteilungsgleichungen $x^n = 1$ angesehen werden? Jahresbericht der Deutschen Mathematiker-Vereinigung, Bd. 27, 1918, pp. 189—195.

[278]   C. MACLAURIN, A treatise of algebra in three parts, 1748, 366 pp.

[279]   G. MALFATTI, De aequationibus quadrato—cubicis disquisitio analytica. Atti dell'Accademia delle Scienze di Siena detta dei Fisio-critici. L'anno 1771, T. IV., pp. 129—184.

[280]   É. MATHIEU, Sur le nombre de valeurs que peut acquérir une fonction quand on y permute ses variables de toutes les manières possibles. Journal de mathématiques pures et appliquées, Séries 2, Tome V, 1860, pp. 9—42.

[281]   —, Sur l'étude des fonctions de plusieurs quantités, sur la manière de les former et sur les substitutions qui les laissent invariables. Journal de mathématiques pures et appliquées, Série 2, Tome VI, 1861, pp. 241—323.

[282]   L. MATTHIESSEN, Grundzüge der antiken und modernen Algebra der literalen Gleichungen, Leipzig 1878, XVI + 1001 pp.

[283]   K. O. MAY, Growth and Quality of the Mathematical Literature, Isis, vol. 59, No. 199, 1968, pp. 363—371.

[283a] F. X. MAYER, Eduard Warings „Meditationes algebraicae". Dissertation, Universität Zürich; 1923; 61 pp.

[284]   F. W. MEYER, Bericht über den gegenwärtigen Stand der Invariantentheorie. Jahresbericht der Deutschen Mathematiker-Vereinigung, 1, 1892, pp. 72—292.

[285]   G. A. MILLER, The Collected Works.
Volume I, University of Illinois, Urbana 1935, 475 pp.
Volume II, University of Illinois, Urbana 1938, XI + 535 pp.

[286]   —, Definitions of abstract groups. Annals of Mathematics, S. II, Vol. 29, 1927—1928, pp. 223—228.

[287] —, Early definitions of the mathematical term abstract group, Science, New Series, Vol. LXXII, July—December 1930, NY, pp. 168—9.

[288] —, Historical note on the determination of abstract groups of given orders. The Collected Works, Vol. I, pp. 91—98.

[289] —, Historical note on the Determination of all the permutation groups of low degrees. The Collected Works, Vol. I, pp. 1—9.

[290] —, Historical Note of the Determination of all the Permutation Groups of Low Degrees. The Tôhoku Mathematical Journal, Vol. 39, June 1934, pp. 60—65.

[291] —, History of the theory of groups to 1900. The Collected Works, Vol. I, pp. 427—467.

[292] —, Note on the history of groupe theory during the period covered by this volume. The collected Works, Vol. II, pp. 1—18.

[293] —, The Founder of Group Theory. The American Mathematical Monthly, Vol. XVII (1910), pp. 162—165.

[294] —, Errors in the Literature on Groups of finite Order, in: The American Mathematical Monthly, Vol. XX (1913), pp. 14—20.

[295] В. Н. Молодший,
  a) Основы учения о числе в XVIII веке. Москва 1953, 179 pp.
  b) Основы учения о числе в XVIII и начале XIX века. Москва 1963, 261 pp.

[296] A. de Morgan, The Elements of Algebra preliminary to the differential calculus, and fit for the higher classes of schools in which the principles of arithmetic are taught, London 1835, XI + 248 pp.

[297] —, Treatise on the Calculus of Functions. Extracted from the Encyclopedia Metropolitana London 1836, 88 pp.

[298] —, On the Foundation of Algebra. Transactions of the Cambridge Philosophical Society, Vol. VII, Part II, pp. 173—188; 1841, Read Dec. 9, 1839.

[299] —, On the Foundation of Algebra. No. II. Transactions of the Cambridge Philosophical Society, Vol. VII, Part III, pp. 287—300; Read November 29, 1841.

[300] —, On the Foundation of Algebra, No. III. Transactions of the Cambridge Philosophical Society, Vol. VIII, Part II, 1844, pp. 139—142; Read Nov. 27, 1843.

[301] —, On the Foundation of Algebra, No. IV, on Triple Algebra. Transaction of the Cambridge Philosophical Society, Vol. VIII, Part III, 1847, pp. 241—254; Read October 28, 1844.

[302] —, Trigonometry and double Algebra, London 1849, XI + 167 pp.

[303] —, A Budget of Paradoxes. Reprinted, with the author's additions, from the „Athenaeum", London 1872, 511 pp.

[304] —, Memoir of Augustus de Morgan, By his Wife Sofia Elizabeth de Morgan. With Selections from his Letters. London 1882, 415 pp.

[305] F. X. Moth, Theorie der Differential-Rechnung ... Prag 1827, 260 pp.

[306] C. V. Mourey, La vrai théorie des quantités négatives et des quantités prétendues imaginaires, Paris 1828.

[307] T. Muir, The Theory of Determinants in the Historical Order of Development. Vol. I—IV, London 1906—1923.

[308] —, Contributions to the History of Determinants 1900—1920, London 1930.

[309] A. Natucci, Il Concetto di Numero e le sue estensioni. Torino 1923, 474 pp.

[310] I. Newton, Arithmetica universalis. Cantabrigiae 1707.

[311] N. Nielsen, Géomètres français sous la révolution. Copenhague 1929, VIII + 250 pp.

[312] M. Noether, Arthur Cayley. Mathematische Annalen, Bd. 46, 1895, pp. 462—480.

[313] L. Nový, Matematika v Čechách v 2. polovině 18. století. Část 1., Sborník pro dějiny přírodních věd a techniky, N° 5, Praha 1960, pp. 9—113.

[314] —, К вопросу о возникновении систематических исследований по теории групп, Историко-математические исследования Вып. XVII, Москва 1966, pp. 31—56.

[315] —, Ke vzniku systematického zkoumání v teorii grup. Sborník pro dějiny přírodních věd a techniky, No. 11, Praha 1966, pp. 135—168.

[316] —, Arthur Cayley et sa définition des groupes abstraits finis. Acta historiae rerum naturalium necnon technicarum, Special Issue 2, Prague 1966, pp. 105—151.

[317] —, Cauchy a Cayleyho definice konečné grupy. Dějiny věd a techniky, Vol. 1, No. 1, Praha 1968, pp. 51—54.

[318] —, Anglická algebraická škola. Dějiny věd a techniky, Vol. 1, No. 2, Praha 1968, pp. 88—105.

[319] —, L'école algébrique anglaise. Revue de Synthèse IIIᵉ S., Nᵒˢ 49—52, Janvier—Décembre 1968, pp. 211—222.

[320] M. Онм,
a) Lehrbuch der gesamten höheren Mathematik in zwei Bänden. Zum Gebrauche für die oberen Klassen der Gymnasien und anderen höheren Lehr-Anstalten, Leipzig 1839, Bd. 1, XVI + 476 pp., Bd. 2, XII + 489 pp.
b) Die reine Elementar-Mathematik, zum Gebrauche an höheren technischen Lehr-Anstalten, Berlin 1825;
Ed. 2nd Berlin 1834, Band 1, XV + 476 pp.
        Berlin 1835, Band 2, XII + 436 pp.

[321] —, Der Geist der mathematischen Analysis und ihr Verhältniss zur Schule, Berlin 1842, XVI + 159 pp.

[322] —, Kritische Beleuchtungen der Mathematik überhaupt und der Euklidischen Geometrie insbesondere. Berlin 1819, XII + 84 pp.

[323] —, Versuch eines vollkommen consequenten Systems der Mathematik.
a) Erster Theil, Arithmetik u. Algebra enthaltend. Zweyte umgearbeitete, durch viele neue erläuternde Beyspiele verdeutlichte Ausgabe, Berlin 1828, 418 + XXXIV pp.
b) Zweyter Theil, Algebra und Analysis des Endlichen enthaltend. Zweite umgearbeitete, durch viele neue erläuternde Beyspiele verdeutlichte und mit einer Figurentafel versehene Ausgabe. Berlin 1829, XXX + 455 pp.

[324] O. Ore, Niels Henrik Abel, Mathematician Extraordinary, Univ. of. Minnesota Press, Minneapolis 1957, IV + 273 pp.

[325] Ozanam, Nouveaux éléments d'algèbre ou principes generaux pour résoudre toutes sortes de problèms de mathematique. Amsterdam 1702, 668 pp.

[326] А. В. ПАПЛАУСКАС, Тригонометрические ряды от Эйлера до Лебега, Москва 1966, 276 pp.

[327] G. Peacock, A Treatise on Algebra,
a) Cambridge 1830, XXXVIII + 685 pp.
b) Vol. I. Arithmetical Algebra. Reprinted from the 1842 Edition, New York 1940, XVI + + 399 pp.
c) Vol. II. On symbolical Algebra and its Applications to the Geometry of Position. Reprinted from the 1846 Edition. X + 355 pp.

[328] —, Report on the recent progress and present state of certain branches of analysis. From the Report of the British Association for the Advancement of Science for 1833, London 1834, pp. 185—352.

[329] B. Peirce, Linear Associative Algebra. 1870; American Journal of Mathematics, Vol. 4, 1881, pp. 97—229.

[330] Ch. L. de Peslouän, N. H. Abel. Sa vie et son oeuvre. Paris 1906, 168 pp.

[331] E. Picard, Préface, in: Oeuvres de Ch. Hermite, Paris 1905, Tome I, pp. VII—XL.

[332] J. Pierpont, Zur Geschichte der Gleichung des V. Grades (bis 1858). Monatshefte für Mathematik u. Physik, Jahrgang VI, 1895, pp. 15—68.

[333] Pilatte, Méthode nouvelle et fort simple pour la résolution de l'équation générale du quatrième degré. Annales de mathématiques pures et appliquées, Tome II, 1811—1812, pp. 152—154.

[334] G. Prasad, Some great mathematicians of the nineteenth century, Benares 1933—4, Vol. 1 + 2.

[335] D. J. Price, Quantitative Masures of the Development of Science. Archives Internationales d'Histoire des Sciences, Vol. 4, 1951, pp. 85—93.

[336] A. A. L. Reynaud, Théorie du plus grand commun diviseur et de l'élimination: précédée de Ia règle des signes de Descartes. Paris 1833, 92 pp.

[337] —, Élémens d'algèbre et introduction au calcul différentiel. Paris 1810, 416 pp.

[338] —, Traité d'algèbre à l'usage des élèves qui se destinent à l'école royale polytechnique et des élèves de l'école spéciale militaire. Cinquième édition. Paris 1821, XVI + 336 pp.

[339] —, Élémens d'algèbre à l'usage des élèves, qui se destinent à l'école polytechnique, à la marine, à l'école militaire de St. Cyr et à l'école forestière. Septième édition, Paris 1826, XVI + 592 pp.

[340] —, Notes sur l'Algèbre à l'usage des élèves, qui se destinent à l'école polytechnique, à la marine, à l'école militaire de Saint-Cyr et à l'école forestière. Septième Édition, Paris 1834, 244 pp.

[342] P. Ruffini, Opere matematiche.
a) Tomo Primo, Palermo 1915 (reprint Roma 1953), 421 + XIII pp.
b) Tomo Secondo, Roma 1953, XIII + 507 pp.
c) Tomo Terzo, Carteggio matematico, Roma 1954, XVII + 254 pp.

[343] —, Teoria generale delle equazioni, in cui si dimostra impossibile la soluzione algebraica delle equazoni generali di grado superiore al quarto. Bologna 1799, 2 vol., 516 pp; Opere matematiche, Tomo primo, pp. 1—324.

[344] —, Della soluzione delle equazioni algebraiche determinate particolari di grado superiore al quarto, Memorie di Matematica e di Fisica della Società Italiana delle Scienze, Tomo IX, Modena 1802, pp. 444—526 (Riccvuta il 21 Ottobre 1801); Opere matematica, Tomo primo, pp. 343—406.

[345] —, Beweis der Unmöglichkeit vollständige alg. Gleichung mit einer unbekannten Grösse, deren Grad den vierten übersteigt, durch eine geschlossene algebraische Formel aufzulösen, Zeitschrift für Physik und Mathematik, Bd. I, Wien 1826, pp. 253—262.

[345a] —, Riposta . . . ai dubj propostigli dal socio G. — Fr. Malfatti sopra la insolubilità algebraica delle Equazioni di grado superiore al quarto.
Memorie di Matematica e di Fisica della Società Italiana delle Scienze, Tomo XII, Modena 1805, pp. 213—267.

[346] K. Rychlík, Preisaufgabe der königlichen böhmischen Gesellschaft der Wissenschaften zu Prag über das Jahr 1834, Časopis pro pěstování matematiky, 86 Vol., Praha 1961, pp. 76—89.

[347] G. Salmon, Modern Higher Algebra,
a) Dublin 1859.
b) Lessons Introductory to the Modern Higher Algebra, Dublin 1866, 296 pp.
c) Leçons d'algèbre supérieure, augm. de notes par M. Hermite, Paris 1868.
d) Vorlesungen zur Einführung in die Algebra der linearen Transformationen, Bearbeitet von W. Fiedler, Leipzig 1863.

[348] E. Schering, Die Fundamental-Classen der zusammensetzbaren arithmetischen Formen. Abhandlungen der Königlichen Gesellschaft der Wissenschaften zu Göttingen, Bd. 14, Göttingen 1869, pp. 3—13.

[349] T. Schönemann, Über die Beziehungen, welche zwischen den Wurzeln irreductibler Gleichungen stattfinden, insbesondere wenn der Grad derselben eine Primzahl ist. Denkschriften der kaiserlichen Akademie der Wissenschaften. Mathematiknaturwissenschaftliche Classe, Fünfter Band, Wien 1853, Abth. II, pp. 143—156.

[350] K. Schott, Cursus mathematicus sive absoluta omnium mathematicarum disciplinarum encyclopaedia. Herbipoli 1661.

[351] Séances des écoles normales recueillies par des sténographes et revues par les professeurs. Premières partie: Leçons. Tome I—VIII, Paris.

[352] J. A. Serret, Cours d'Algèbre supérieur.

a) Paris 1849, 400 pp.
b) Deuxième ed. Paris 1854, 600 pp.
c) Cinquième ed. Paris 1885;
d) Handbuch der höheren Algebra, Leipzig 1868;
e) Zweite Auflage, 2 Bde, 1878—9, 528 + 574 pp.

[353] —, Mémoire sur les fonctions de quatre, cinq et six lettres. Journal de mathématiques pures et appliquées, XV, Année 1850, pp. 45—70.

[354] —, Remarque sur un mémoire de M. Bertrand. Journal de mathématiques pures et appliquées, Tome XIV, Année 1849, pp. 135—136.

[355] —, Mémoire sur le nombre de valeurs que peut prendre une fonction quand on y permute les lettres qu'elle renferme. Comptes rendus, Tome XXIX, Paris 1849, pp. 10—11.

[356] —, Sur le nombre de valeurs que peut prendre une fonction quand on y permute les lettres qu'elle renferme. Journal de mathématiques pures et appliquées, Tome XV, Année 1850, pp. 1—44.

[357] —, Développements sur une classe d'équations, Journal de mathématiques pures et appliquées, Tome XV, Année 1850, pp. 152—168.

[358] F. J. SERVOIS, Lettre au Rédacteur des Annales, sur la théorie des quantités imaginaires. Annales de mathématiques pures et appliquées, Tome IV. 1813—1814, pp. 228—235.

[359] —, Essai sur un nouveau mode d'exposition des principes du calcul différentiel, in: Annales de mathématiques pures et appliquées, Tome V, 1814—15, pp. 93—170.

[360] Л. А. СОРОКИНА, Работы Абеля об алгебраической разрешимости уравнений, Историко-математические исследования, XII, Москва 1959, pp. 457—480.

[361] D. J. STRUIK, A Concise History of Mathematics, 2 nd Ed., London 1956.

[362] F. J. STUDNIČKA, A. L. Cauchy als formaler Begründer der Determinantentheorie, Praha 1876.

[363] L. STURM, Tractatus de natura et constitutiones Matheseos, Francofurct 1702.

[364] A. SUREMAIN-MISSERY, Théorie purement algébrique des quantités imaginaires et des fonctions qui en résultent, Paris 1801, 299 pp.

[364a] L. SYLOW, Anmerkungen zu der hinterlassenen Abhandlung von Abel.
Abel N. H. und Galois E., Abhandlungen über algebraische Auflösung der Gleichungen, Berlin 1889, pp. 57—81.

[365] J. J. SYLVESTER, The Collected Mathematical Papers, 4 vol., Cambridge 1804—1911.

[366] —, On the intersections contacts, and other correlations of two conics expressed by indeterminate coordinate. Cambridge and Dublin Mathematical Journal, 1850, pp. 262—282; Math. Papers, vol. I, pp. 119—137.

[367] —, Additions to the articles „On a new class of theorems", and "On Pascal's theorems". Philosophical Magazine, vol. XXXVII, 1850, pp. 363—370; Math. Papers, vol. I, pp. 145—151.

[368] —, On the relation between the minor determinants of linearly equivalent quadratic functions, Philosophical Magazine, 1850, pp. 295—305; Math. Papers, vol. I, pp. 241—250.

[369] —, On a theorem concerning the combination of determinants. Cambridge and Dublin Mathematical Journal, vol. VIII, 1853, pp. 60—62; Math. Papers, vol. I, pp. 399—401.

[370] —, Memoir on the dialytic method of elimination. Part 1. Philosophical Magazine, XXII, 1842, pp. 534—539; Math. Papers, vol. 1, pp. 86—90.

[371] —, Elementary researches in the analysis of combinatorial aggregation. Philosophical Magazine, XXIV (1844), pp. 285—296; Math. Papers, vol. 1, pp. 91—102.

[372] —, Généralisation d'un théorème de M. Cauchy et Addition. CR LIII (1861), pp. 644—645; 1722—1725; Math. Papers, vol. II, pp. 245—246, 247—249.

[373] —, Note on the historical origin of the unsymmetrical six-valued function of six lettres. Philosophical Magazine, XXI 1861, pp. 369—377; Math. Papers, vol. II, pp. 264—271.

[374] —, On a Problem in Tactic which serves to disclose the existence of a four-valued function of

three sets of three letters each. Philosophical Magazine, XXI, 1861, pp. 515—520; Math. Papers, vol. II, pp. 272—276.

[375] —, Concluding Paper on Tactic. Philosophical Magazine XXII, 1861, pp. 45—54; Math. Papers, vol. II, pp. 277—285.

[376] —, Remark on the Tactic of nine Elements. Philosophical Magazine XXII, 1861, pp. 144—147; Math. Papers, vol. II, pp. 286—289.

[377] —, On a Generalization of a Theorem of Cauchy on arrangements. Philosophical Magazine XXII, 1861, pp. 378—382; Math. Papers, vol. II, pp. 290—293.

[378] Theorie der Gleichungen aus den Schriften der Herren Euler u. de la Grange von J. A. C. Michelsen in das Deutsch übersetzt, Berlin 1791.

[379] J. Tropfke, Geschichte der Elementarmathematik. Bd. I—VII, Zweite Auflage, Leipzig 1921—1924.

[380] J. Tvrdá, Vznik teorie matic, Dějiny věd a techniky, Vol. 3, Praha 1970, pp. 11—23.

[381] H. Umpfenbach, Lehrbuch der Algebra. Giessen 1825, VI + 503 pp.

[382] C. A. Valson, La vie et les travaux du baron Cauchy. Paris 1868, Tome I + II.

[383] A. T. Vandermonde, Mémoire sur la résolution des équations. Histoire de l'acad. roy. des sciences, année 1771, Paris 1774, pp. 365—416.

[384] —, Mémoire sur l'élimination. Histoire de l'acad. roy. des sciences, année 1772, II. Partie, Paris 1776, pp. 516—532.

[385] —, Abhandlungen aus der reinen Mathematik. In deutscher Sprache herausgegeben von Carl Itzigsohn, Berlin 1888, 104 pp.
a) Abhandlung über die Auflösung der Gleichungen.
b) Abhandlung über die irrationalen Grössen verschiedener Ordnung nebst einer Anwendung auf den Kreis.
c) Abhandlung über die Elimination.

[386] H. Vogt, Sur les groupes finis discontinus (d'après l'article allemand de H. Burkhardt), in: Encyclopédie des sciences mathématiques, Tome I, Vol. 1, Paris—Leipzig, 1909, 532 nn.

[387] J. Vuillemin, La philosophie de l'algèbre. Tome Premier: Recherches sur quelques concepts et méthodes de l'Algèbre moderne, Paris 1962, 582 pp.

[388] —, La philosophie de l'algèbre de Lagrange. (Réflexions sur la mémoire de 1770—1771). Les Conférences du Palais de la Découverte, Série D, N° 71, Paris 1960, 24 pp.

[389] P. L. Wantzel, De l'impossibilité de résoudre toutes les équations algébriques avec des radicaux, Nouv. Annales de mathématiques pures et appliquées, T. 4, 1845, pp. 57—65.

[390] E. Waring, Meditationes algebraicae
a) Editio secunda, Cantabrigiae 1770,
b) Editio tertia recensita et aucta, Cantabrigiae 1782, XLIV + 389 pp.

[391] —, Miscellanea analytica de aequationibus algebraicis et curvarum proprietatibus. Cantabrigiae 1762, 162 pp.

[392] J. Warren, A Treatise on the Geometrical Representation of the Square Roots of Negative Quantities, Cambridge 1828.

[393] —, Considerations of the Objections Raised Against the Geometrical Representation of the Square of Negative Quantities. Philosophical Transactions of the Royal Society of London, 119 (1829), pp. 241—254.

[394] —, On the Geometrical Representation of the Power of Quantities Whose Indices Involve the Square Roots of Negative Quantities, Philosophical Transactions of the Royal Society of London, 119 (1829), pp. 339—359.

[395] H. Weber, Leopold Kronecker, Mathematische Annalen, Bd. 43, Leipzig 1893, pp. 1—25.

[396] —, Lehrbuch der Algebra, Braunschweig 1895—6; 2. Aufl. 1898—1899.

[397] C. Wessel, Om Directionens analytiske Betegning. Nye Samling af det Kongelige Danske Videnskabernes Selskabs Skrifter, Vol. V, 1799.

[398] —, Essai sur la représentation analytique de la direction, avec des applications, en particulier
       à la détermination des polygones planes et des polygones sphériques. Trad. par. H. Valentiner
       et T. N. Thiele. Copenhaque 1897, XIV + 60 pp.
[399] H. WEYL, Algebraic Theory of Numbers, Princeton 1940, VIII + 233 pp.
[400] —, The Classical Groups, their Invariants and Representations. Princeton 1946, XIII + 320 pp.
[401] H. WIELEITNER, Geschichte der Mathematik. Von 1700 bis zur Mitte des 19. Jahrhunderts.
       Berlin und Leipzig 1923.
[402] —, Geschichte der Mathematik. II. Teil. Von Cartesius bis zur Wende des 18. Jahrhunderts.
       Leipzig 1911—1921.
[403] C. WOLFF, Anfangsgründe aller mathematischen Wissenschaften. Halle 1710.
[404] —, Mathematisches Lexicon, Leipzig 1716.
[405] J. WOOD, The Elements of Algebra designed for the Use of Students in the University. The
       second edition. Cambridge 1798, 307 pp.
[406] M. HOËNE—WRONSKI, Résolution générale des équations. Paris 1811.
[407] H. WUSSING, Über den Einfluss der Zahlentheorie auf die Herausbildung der abstrakten
       Gruppentheorie. Beiheft der Schriftenreihe für Geschichte der Naturwissenschaften, Technik
       und Medizin, Leipzig 1964, pp. 71—88.
[408] —, Zur Entstehungsgeschichte der abstrakten Gruppentheorie. Zeitschrift für Geschichte
       der Naturwissenschaften, Technik und Medizin, Heft 5, Leipzig 1965, pp. 1—16.
[409] —, О генезисе обстрактного понятия труппы, Историко-математические исследования,
       Выпуск XVII, Москва 1966, pp. 11—30.
[410] —, Die Genesis des abstrakten Gruppenbegriffes. Ein Beitrag zur Enstehungsgeschichte der
       abstrakten Gruppentheorie. Berlin 1969, 258 pp.
[411] Jaroslava ŽÁČKOVÁ, Problém kvadratury kruhu a Lambertův důkaz iracionality čísla π. Po-
       kroky matematiky, fyziky a astronomie, Praha 1966, Vol. 11, pp. 240—250.

INDEX OF NAMES